Contributions
to Current Research
in Geophysics (CCRG)
1

Earthquake Prediction and Rock Mechanics

Editor: Max Wyss
Cooperative Institute for Research in Environmental Sciences,
University of Colorado, Boulder, Colorado, USA

1975
BIRKHÄUSER VERLAG BASEL
UND STUTTGART

Reprinted from Pure and Applied Geophysics (PAGEOPH)
Volume 113 (1975)

ISBN 3-7643-0809-5

PURE AND APPLIED GEOPHYSICS
Special Issue: Earthquake Prediction and Rock Mechanics
Editor: Max Wyss
Vol. 113 Nos. 1/2 (1975) Pages 1–330

CONTENTS

PURE AND APPLIED GEOPHYSICS

Pageoph,
Vol. 113 (1975) Birkhäuser Verlag, Basel

V_p/V_s Anomalies in Dilatant Rock Samples

By KATE HADLEY[1])

Summary – In a series of triaxial experiments we have measured V_p, V_s and volumetric strain simultaneously in dilating dry and saturated rocks. For the first time these data permit quantitative comparison of seismic velocities or their ratio and dilatant volumetric strain. In air-dry samples V_p/V_s decreases by a few per cent at strains of 10^{-3}; in saturated materials with high pore pressure, V_p/V_s increases by a comparable amount. Decreases in seismic velocity ratio are difficult to generate in initially saturated rocks even with low pore pressures and at strain rates of 10^{-4}/sec. A liquid–vapor transition will not produce a significant drop in V_p/V_s. If dilatancy and fluid flow are responsible for seismic travel time anomalies prior to earthquakes, our results suggest that such anomalies will occur only in regions where pore fluid source to sink dimensions are of the order of 10 km or more, or in regions where the rocks are not saturated to begin with.

1. *Introduction*

Ever since Russian seismologists first correlated large changes in the travel time ratio of transverse to longitudinal seismic waves with the time of occurrence of earthquakes [1, 2] considerable controversy has existed about their cause. New field data from California and New York State [3, 4] imply that the drop in t_s/t_p is due to a large drop in compressional wave velocity coupled with a somewhat smaller drop in shear wave velocity.

Although it is widely assumed in this country that dilatancy is responsible for the field observations, various combinations of dry, partially saturated or vapor-filled cracks have been invoked to account for the t_s/t_p decrease [5, 6, 3]. Existing laboratory data are insufficient to eliminate any one of these mechanisms or even to answer the more fundamental questions: Starting with a saturated rock, can dry or partially saturated cracks be created such that t_s/t_p will drop? Or, starting with a water-saturated rock above 100°C, can enough of that water be converted to vapor in the dilating material such that t_s/t_p will decrease? To date, with one exception [7], the only laboratory measurements of dilatancy-related velocity changes have been conducted on *dry* materials [8–10]. Only in two investigations was the shear wave velocity measured [7, 11], and only in Gupta's work were V_p and V_s (of unknown polarization) measured simultaneously. Since strains were not recorded simultaneously in any of these experiments, one cannot quantitatively relate velocity changes to dilatant strains.

[1]) Department of Earth and Planetary Sciences, Massachusetts Institute of Technology, Cambridge, Massachusetts 02139.

In this paper, results are presented from a set of laboratory experiments in which V_p and V_s and volumetric strains have been simultaneously recorded in dilating dry and saturated rocks. For the first time this permits quantitative comparison of seismic velocities or their ratio and dilatant strains. Also for the first time, the crux of the dilatancy-fluid flow model has been tested.

2. *Procedure*

The sample configuration is shown in Fig. 1. Flats were ground into the surface of a right circular cylinder of Westerly granite or San Marcos gabbro 6.3 cm long and 3.2 cm in diameter. A thin (0.05 mm) copper jacket surrounded the sample, excluding the confining pressure medium. Two HDT-31 ceramic transducers, one shear and one compressional, were mounted on the copper jacket in each flat region with conducting epoxy. The frequency of each was approximately 1 MHz. Only one transducer was driven at a time; velocities were determined at discrete points in the loading sequence using a mercury delay-line technique [12]. Although the P arrival was clear, the P coda prevented exact determination of the first arrival of S, especially at high stress

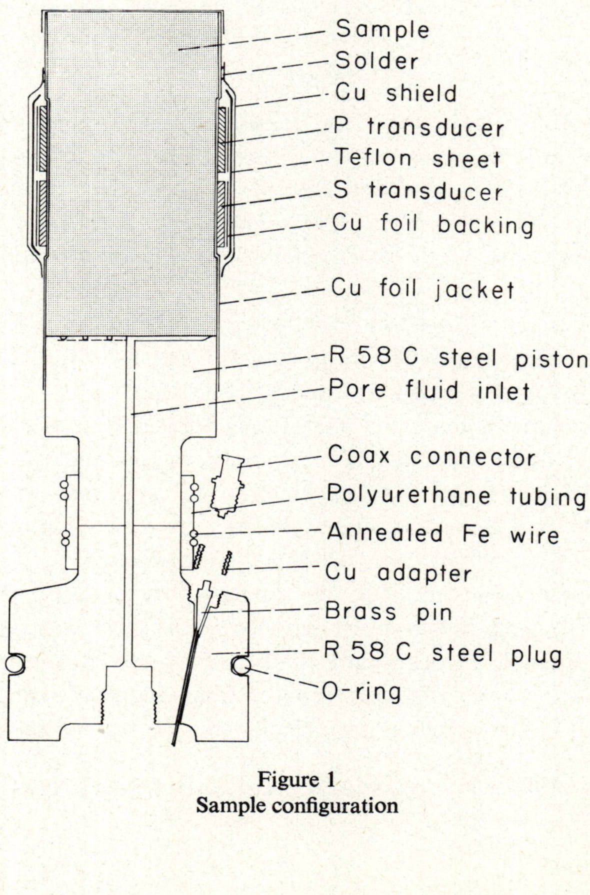

Figure 1
Sample configuration

when the signals became attenuated. For this reason, the peak of the first S pulse was chosen as the S arrival. This results in 10–12% lower values of V_s than are correct. Thus, the per cent change in V_s will be somewhat smaller than reported here.

The effect of folding the P coda into the S arrival usually works in the opposite sense. The shear transducers generate a low amplitude compressional wave signal which precedes the larger shear pulse. Depending upon the changing delay in P arrival relative to S, this P signal may increase the apparent S velocity by about 1% as loading progresses. Thus the true decrease in V_s may be greater than the decrease reported here by as much as 10% of the reported value.

Velocities were measured in a direction transverse to the greatest principal stress. Large changes in velocity are observed in this direction compared with the direction parallel to the greatest principal stress [9, 11]. The polarization of the shear waves was also chosen to be in the plane perpendicular to this stress, as this polarization was anticipated to give the greatest possible change in shear wave velocity in response to the opening of axial cracks under load. That it does indeed do so is shown by the results in Fig. 2, and confirmed by BONNER [7].

Confining pressure was 390 bars. This low value was chosen so that large dilatant strains could be achieved easily and large variations in seismic velocity could be

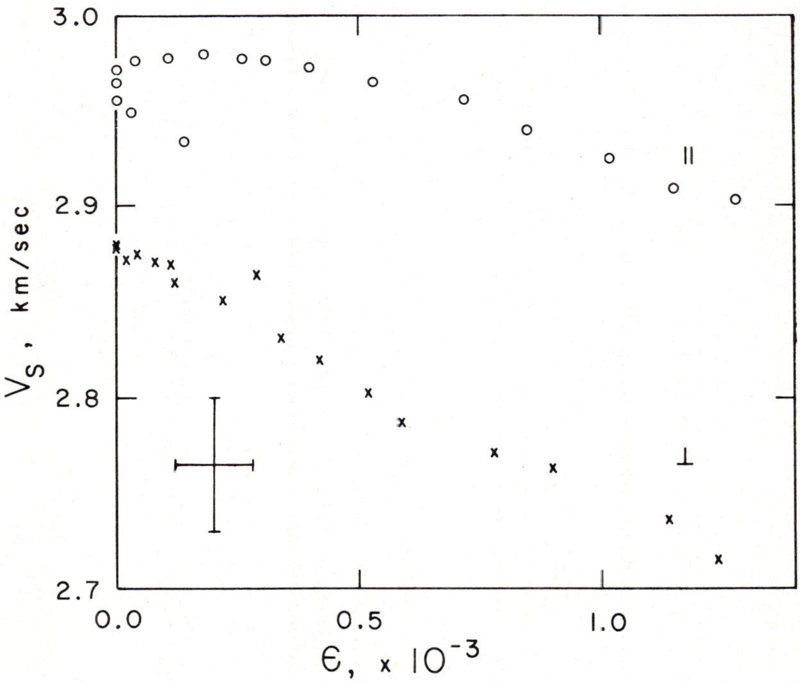

Figure 2

Change in shear wave velocity with ϵ for two polarizations of S. Circles, polarization parallel to σ_{max}; crosses, polarization perpendicular to σ_{max}

observed. Pore water pressure was initially chosen to be 350 bars. Later experiments were carried out with a pore pressure of 1 bar when it became apparent that partial saturation could not be developed at available strain rates with the higher pore pressure.

Because it was not possible to heat the sample assembly to temperatures in excess of 100°C, CO_2 was used instead of water to determine the effect of a liquid vapor pore fluid transition on seismic velocities. Sixty-two bars is the liquid–vapor transition pressure of CO_2 under ambient laboratory temperatures. Initially, CO_2 pore pressures used were in excess of this value, 112 and 68 bars, but when no decrease in V_p/V_s developed an experiment was run at the transition pressure.

All of the experiments run with CO_2 as the pore fluid were run with a constant mass of fluid available to the sample. When water was used as the pore fluid the reservoir fluid pressure was usually maintained at a constant value by addition or subtraction of water from the system. Pore pressure values stated in this report are those measured at the reservoir: local fluid pressures within the rock departed from those values during the experiments.

For comparison with wet results some tests were performed on air-dry samples. These rocks were soaked in acetone for several days after machining to dissolve any oil in them and then evacuated for 24 hours at room temperature. They were then exposed to the atmosphere before jacketing. Relative humidity in the laboratory is about 60–80%, but these rocks will be referred to as dry in this paper.

Strain rate is not so important in the case of dry rocks because no kinetics of pore fluid motions are involved but it becomes crucial when pore fluids are present as seismic velocities may be expected to behave differently when there is insufficient time for fluids to communicate between cracks. The fastest controllable loading rate available for these experiments was 1.5×10^{-4}/sec; the slowest rate practical was 1.5×10^{-7}/sec. Unloading had to be performed manually with pauses at each stress where velocities were measured.

Using a permeability of 10^{-6} darcies ($=10^{-14}$cm/sec) for the granite at 40 bars effective pressure and a fluid viscosity of 1.4×10^{-8} bar-sec for water at 350 bars pressure [13], equilibrium of pore pressure to within 10% throughout the rock specimen requires a time, t, given by:

$$t > l^2 \mu\beta/k$$

where l is the sample length, μ the fluid viscosity, β the aggregate compressibility and k the permeability [14]. Since this time is approximately 2 min it is unlikely that the wet rock could be driven to undersaturation even at the fastest strain rates. Therefore, the wet granite was run at 1.5×10^{-5} and 1.5×10^{-6}/sec, the most convenient strain rates to use with the expectation that saturated behavior would be observed as the rock dilated. Experimentally, this was the observed case.

Based on electrical resistivity measurements [15–16] and using the empirical formula of Brace *et al.* [13] relating resistivity and permeability, the initial permeability of the gabbro is probably 20 times less than that of the granite. The compressibility is probably half as large, so the resulting equilibration time for the gabbro under comparable conditions is about 80 min. Therefore it seemed likely that the gabbro would be undersaturated at strain rates of 10^{-4}/sec, when peak load was reached in a few minutes but saturated at 10^{-6}/sec, when 5 or 6 hours were required to reach peak load.

Rock specimens were not loaded to failure but rather cycled in stress. Apart from the convenience of being able to use a single sample for many different tests, repeatedly stressed specimens more accurately reflect the properties of *in situ* rocks near active fault zones than does the virgin material. During each stress cycle, therefore, the rock was loaded to within 80% of its initial fracture strength. There is no need to investigate higher stresses, as in an active shallow to mid-crustal fault zone, where pre-existing planes of weakness exist, stresses could reach no more than 60% to 70% of the intact fracture strength before sliding on fault surfaces occurred [17].

Axial and circumferential strains were monitored continuously with conventional foil gauges glued to the sample jacket. These were summed electronically to produce a volumetric strain: $\varepsilon_v = 2\varepsilon_\theta + \varepsilon_z$. Because this study sought to clarify the effect of dilatancy on seismic velocities, it was necessary, for each experiment, to determine an elastic reference line [18] from which all dilatant volumetric strains, ϵ, could be measured

$$\left(\epsilon = \frac{\Delta V_{\text{elastic}} - \Delta V_{\text{observed}}}{V}\right).$$

This reference line presumably represents an intrinsic property of the material and should not change from cycle to cycle or from sample to sample of the same homogeneous rock type. However, several kilobars of confining pressure are required to close the crack space in most rocks, pressures clearly not achieved in these experiments. Due to the presence of open cracks small differences in the elastic reference line were noted from sample to sample regardless of rock type. Larger differences showed up in individual samples from cycle to cycle because effective confining pressure and therefore the number of open cracks varied. Moreover, the population of cracks in the sample may have changed slightly during each loading. However, consecutive cycles run at the same effective pressure and strain rate showed nearly identical behavior in strain and velocities (Fig. 3, open circles and crosses). It was thus decided to take one elastic reference line for each specimen, specifically the one determined from the stress–volumetric strain curve for the second stress cycle. For any single specimen, therefore, all dilatant strains are referred back to the elastic reference line for the second stress cycle, and that cycle was conducted dry, so that the effective confining pressure was equal to the total confining pressure, 390 bars. When referenced that way, dilatant volumetric strain and velocities match closely for two different samples having similar stress histories (open and filled circles, Fig. 3).

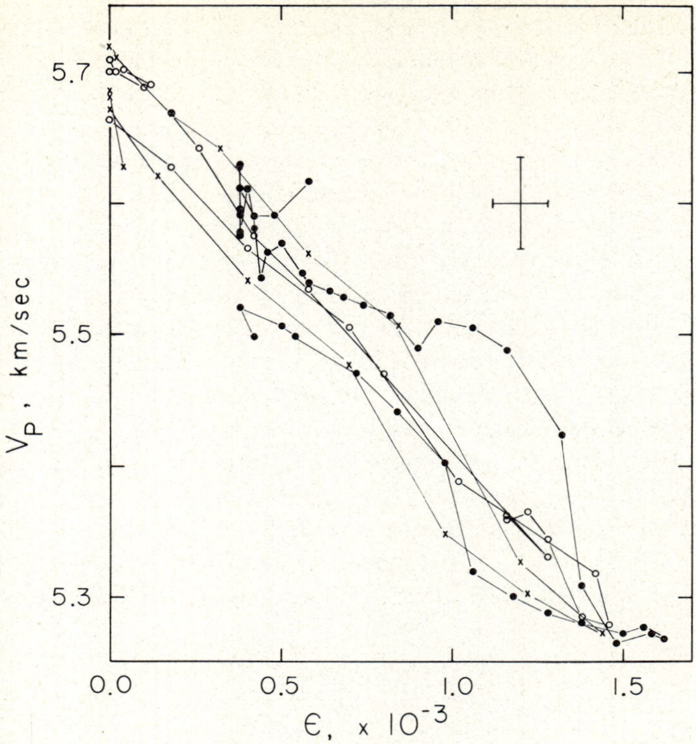

Figure 3
Change in compressional wave velocity with ϵ in dry granite. Crosses, sample W1, cycle 3; open circles, sample W1, cycle 2; filled circles, sample W2, cycle 2

3. Results

Results are presented in Table 1. In all samples studied both shear and compressional wave velocities decreased with increasing dilatant volumetric strain regardless of degree of saturation. For the same degree of saturation velocities appear to be nearly monotonic functions of dilatant volumetric strain for a given rock type. This is somewhat surprising as one might have expected minor variations in crack geometry from sample to sample to introduce scatter in the measurements. At low pressures it is difficult to separate out the effect of different mineralogies from the effect of different crack and pore porosities in order to compare velocities between the gabbro and the granite, but the sensitivity of the individual velocities to ϵ seems to be similar for the two rocks. Moreover, in the range studied, the velocities are insensitive to effective confining pressure except as it alters ϵ. The run designated 'variable' (Fig. 4, crosses) had a pore pressure which was allowed to deviate from the initial 350 bar value in response to opening and closing of cracks within the rock. In fact it rose to 360 bars during

Table 1

Sample # W1, Cycle # 12, Dry, $\dot{\varepsilon} = 10^{-6}$/sec

Stress kb	V_p, km/sec	V_s, km/sec	V_p/V_s	$V_p/(V_p)_0$	$V_s/(V_s)_0$	$(V_p/V_s)/(V_p/V_s)_0$	$\epsilon \times 10^{-3}$
0.0	5.347	2.372	2.254	1.0	1.0	1.0	1.40
0.0	5.322	2.339	2.375	0.995	0.986	1.009	1.40
0.07	5.278	2.294	2.301	0.987	0.967	1.018	1.44
0.14	5.262	2.315	2.273	0.984	0.976	1.008	1.44
0.21	5.295	2.325	2.277	0.990	0.980	1.010	1.44
0.28	5.295	2.326	2.276	0.990	0.981	1.010	1.44
0.37	5.284	2.355	2.286	0.988	0.993	1.014	1.52
0.47	5.236	2.322	2.255	0.979	0.979	1.000	1.64
0.57	5.183	2.291	2.262	0.969	0.966	1.004	1.88
0.69	5.150	2.270	2.269	0.963	0.957	1.007	2.14
0.75	5.112	2.263	2.259	0.956	0.954	1.002	2.20
0.85	5.091	2.263	2.250	0.952	0.954	0.998	2.36
0.93	5.060	2.253	2.246	0.946	0.950	0.996	2.56
1.06	5.049	2.239	2.255	0.944	0.944	1.000	2.86
1.13	5.004	2.257	2.217	0.936	0.952	0.984	3.04
1.25	4.975	2.214	2.247	0.930	0.933	0.997	3.29
1.32	4.948	2.208	2.241	0.925	0.931	0.994	3.41
1.43	4.975	2.244	2.217	0.930	0.946	0.984	3.47
1.51	4.928	2.230	2.210	0.922	0.940	0.980	3.57
1.58	4.917	2.225	2.210	0.920	0.938	0.980	3.71
1.63	4.921	2.225	2.212	0.920	0.938	0.981	3.81
1.70	4.904	2.224	2.205	0.917	0.938	0.978	3.93
1.77	4.869	2.216	2.197	0.911	0.934	0.975	4.07
1.84	4.780	2.200	2.173	0.894	0.927	0.964	4.23
1.91	4.624	2.175	2.126	0.865	0.917	0.943	4.51
1.98	4.574	2.138	2.139	0.855	0.901	0.949	4.73
1.98	4.489	2.129	2.109	0.840	0.898	0.937	4.83
1.84	4.485	2.127	2.109	0.839	0.897	0.937	4.71
1.58	4.574	2.124	2.153	0.855	0.895	0.955	4.53
1.58	4.569	2.128	2.147	0.854	0.897	0.953	4.57
1.27	4.692	2.141	2.191	0.878	0.903	0.972	4.19
1.27	4.720	2.147	2.198	0.883	0.905	0.975	4.13
0.99	4.877	2.165	2.253	0.912	0.913	1.000	3.71
0.99	4.889	2.185	2.238	0.914	0.921	0.993	3.57
0.71	4.887	2.164	2.258	0.914	0.912	1.002	3.35
0.71	4.898	2.155	2.273	0.916	0.910	1.008	3.39
0.57	4.885	2.145	2.277	0.914	0.904	1.010	3.29
0.57	4.921	2.179	2.258	0.920	0.919	1.002	2.94
0.42	4.922	2.157	2.282	0.921	0.909	1.012	2.90
0.42	4.942	2.183	2.264	0.924	0.920	1.004	2.72
0.28	5.007	2.170	2.307	0.936	0.915	1.024	2.64
0.28	5.023	2.200	2.283	0.939	0.927	1.013	2.40
0.14	5.008	2.183	2.294	0.937	0.920	1.018	2.42
0.14	4.996	2.174	2.298	0.934	0.917	1.020	2.38
0.0	5.002	2.146	2.331	0.935	0.905	1.034	2.28
0.0	4.912	2.105	2.333	0.919	0.887	1.035	2.38

Table 1—*continued*

Sample # W2, Cycle # 3, Dry, $\dot{\varepsilon} = 10^{-6}$/sec

Stress, kb	V_p, km/sec	V_s, km/sec	V_p/V_s	$V_p/(V_p)_0$	$V_s/(V_s)_0$	$(V_p/V_s)/(V_p/V_s)_0$	$\epsilon \times 10^{-3}$
0.0	5.617	3.047	1.843	0.998	0.992	1.007	0.58
0.09	5.590	3.055	1.830	0.993	0.994	0.999	0.48
0.19	5.590	3.055	1.830	0.993	0.994	0.999	0.42
0.43	5.627	3.073	1.831	1.0	1.0	1.0	0.38
0.60	5.629	3.046	1.848	1.000	0.991	1.009	0.38
0.72	5.591	3.026	1.848	0.994	0.985	1.009	0.38
0.90	5.596	3.017	1.855	0.994	0.982	1.013	0.38
1.04	5.577	3.015	1.850	0.991	0.981	1.010	0.38
1.18	5.575	3.008	1.853	0.991	0.979	1.012	0.38
1.30	5.611	3.007	1.866	0.997	0.979	1.019	0.40
1.47	5.580	3.003	1.858	0.992	0.977	1.015	0.42
1.59	5.542	2.996	1.850	0.985	0.975	1.010	0.44
1.73	5.563	2.989	1.861	0.989	0.973	1.016	0.46
1.88	5.569	2.983	1.867	0.990	0.971	1.020	0.50
2.02	5.546	2.977	1.863	0.986	0.969	1.017	0.56
2.17	5.539	2.971	1.864	0.984	0.967	1.018	0.58
2.31	5.533	2.964	1.867	0.983	0.965	1.020	0.64
2.46	5.528	2.956	1.870	0.982	0.962	1.021	0.68
2.60	5.522	2.948	1.873	0.981	0.959	1.023	0.74
2.75	5.515	2.940	1.876	0.980	0.957	1.025	0.82
2.89	5.490	2.930	1.874	0.976	0.953	1.024	0.90
3.03	5.510	2.918	1.888	0.979	0.950	1.031	0.96
3.18	5.505	2.907	1.894	0.978	0.946	1.034	1.06
3.32	5.489	2.891	1.897	0.975	0.941	1.036	1.16
3.52	5.424	2.868	1.891	0.963	0.933	1.033	1.32
3.61	5.309	2.854	1.866	0.943	0.929	1.019	1.38
3.68	5.264	2.845	1.863	0.935	0.926	1.017	1.48
3.76	5.277	2.831	1.864	0.938	0.921	1.018	1.56
3.78	5.264	2.826	1.863	0.935	0.920	1.014	1.62
3.61	5.267	2.831	1.860	0.936	0.921	1.016	1.58
3.32	5.272	2.833	1.861	0.937	0.922	1.016	1.50
3.03	5.280	2.840	1.859	0.938	0.924	1.015	1.38
2.75	5.287	2.845	1.858	0.940	0.926	1.015	1.28
2.46	5.301	2.851	1.859	0.942	0.928	1.015	1.18
2.17	5.320	2.860	1.860	0.945	0.931	1.016	1.06
1.18	5.403	2.868	1.884	0.960	0.933	1.029	0.98
1.59	5.441	2.878	1.891	0.967	0.937	1.033	0.84
1.30	5.471	2.887	1.895	0.972	0.939	1.035	0.72
1.01	5.498	2.893	1.900	0.977	0.941	1.038	0.54
0.72	5.507	2.902	1.898	0.979	0.944	1.037	0.50
0.43	5.520	2.915	1.894	0.981	0.949	1.034	0.38
0.14	5.505	2.920	1.885	0.978	0.950	1.029	0.38
0.0	5.498	2.925	1.880	0.977	0.952	1.027	0.42

Table 1—*continued*

Sample #W3, Cycle #6, p_{CO_2} = 62 bars (transition) and lower at highest stress, $\dot{\varepsilon} = 10^{-5}$/sec

Stress kb	V_p, km/sec	V_s, km/sec	V_p/V_s	$V_p/(V_p)_0$	$V_s/(V_s)_0$	$(V_p/V_s)/(V_p/V_s)_0$	$\epsilon \times 10^{-3}$
0.0	5.566	2.952	1.855	0.997	1.0	1.0	0.80
0.08	5.581	—	—	1.0	—	—	0.42
0.21	—	2.951	1.891	—	1.000	1.003	0.32
0.29	5.579	—	—	1.000	—	—	0.32
0.41	—	2.950	1.891	—	0.999	1.003	0.29
0.52	5.579	—	—	1.000	—	—	0.29
0.60	—	2.943	1.897	—	0.997	1.006	0.26
0.73	5.588	—	—	1.001	—	—	0.28
0.82	—	2.936	1.904	—	0.995	1.010	0.29
0.96	5.576	—	—	0.999	—	—	0.33
1.06	—	2.923	1.908	—	0.990	1.012	0.35
1.21	5.481	—	—	0.982	—	—	0.40
1.32	—	2.901	1.890	—	0.983	1.002	0.44
1.46	5.435	—	—	0.974	—	—	0.54
1.58	—	2.868	1.895	—	0.972	1.005	0.64
1.70	5.367	—	—	0.962	—	—	0.72
1.79	—	2.839	1.891	—	0.962	1.003	0.77
1.98	5.266	—	—	0.945	—	—	0.94
2.06	—	2.800	1.881	—	0.949	0.997	1.04
2.17	5.221	—	—	0.935	—	—	1.12
2.25	—	2.768	1.886	—	0.938	1.000	1.20
2.34	5.160	—	—	0.924	—	—	1.31
2.45	—	2.748	1.878	—	0.931	0.996	1.46
2.50	5.087	2.730	1.863	0.911	0.925	0.988	1.50
2.54	5.083	—	—	0.911	—	—	1.58
2.67	—	2.710	1.876	—	0.918	0.995	1.70
2.76	5.028	—	—	0.901	—	—	1.84
2.85	—	2.684	1.874	—	0.909	0.994	1.94
2.97	4.949	—	—	0.887	—	—	2.12
3.08	—	2.659	1.861	—	0.901	0.987	2.28
3.22	4.870	—	—	0.873	—	—	2.52
3.33	—	2.620	1.859	—	0.886	0.986	2.74
3.33	4.835	2.606	1.855	0.886	0.883	0.984	2.76
3.33	4.835	—	—	0.886	—	—	2.78
3.44	—	2.601	1.860	—	0.881	0.986	3.00
3.51	4.790	—	—	0.858	—	—	2.15
3.60	—	2.573	1.862	—	0.871	0.988	3.48
3.60	4.749	2.548	1.864	0.851	0.863	0.989	3.72
3.60	4.706	2.541	1.852	0.843	0.861	0.982	3.82
3.59	4.690	2.537	1.849	0.840	0.859	0.981	3.90
3.58	4.636	2.533	1.830	0.831	0.858	0.971	4.00
3.56	4.634	2.526	1.834	0.830	0.856	0.973	4.08
3.55	4.624	2.521	1.834	0.829	0.854	0.973	4.16
commence p_{CO_2} drop							
3.55	4.628	2.526	1.832	0.829	0.856	0.972	4.24
3.55	4.624	2.525	1.831	0.829	0.855	0.971	4.24
3.55	4.619	2.524	1.830	0.828	0.855	0.970	4.24

Table 1—*continued*

Stress kb	V_p, km/sec	V_s, km/sec	V_p/V_s	$V_p/(V_p)_0$	$V_s/(V_s)_0$	$(V_p/V_s)/(V_p/V_s)_0$	$\epsilon \times 10^{-3}$
3.53	4.619	2.523	1.830	0.828	0.855	0.971	4.28
3.53	4.630	2.522	1.836	0.830	0.854	0.974	4.28
3.53	4.626	2.519	1.836	0.829	0.853	0.974	4.28
3.53	4.615	2.511	1.838	0.827	0.851	0.975	4.28
3.53	4.588	2.496	1.838	0.822	0.846	0.975	4.28
3.53	4.519	2.482	1.820	0.810	0.841	0.965	4.28
3.53	4.483	2.481	1.806	0.803	0.840	0.958	4.26
3.53	4.483	2.483	1.805	0.803	0.841	0.957	4.26
3.53	4.502	2.485	1.811	0.807	0.842	0.961	4.26
3.53	4.508	2.489	1.811	0.808	0.843	0.960	4.24
3.53	4.523	2.491	1.815	0.810	0.844	0.963	4.24
3.53	4.523	2.492	1.815	0.810	0.844	0.962	4.24
3.53	4.533	2.494	1.817	0.812	0.845	0.964	4.20
3.53	4.544	2.496	1.820	0.814	0.846	0.965	4.20
3.53	4.568	2.499	1.828	0.818	0.847	0.969	4.16
3.53	4.560	2.499	1.824	0.817	0.847	0.968	4.16
$p_{CO_2} = 1$ bar							
3.37	4.571	2.500	1.829	0.814	0.847	0.970	4.10
3.09	4.578	2.503	1.829	0.822	0.848	0.970	3.94
2.82	4.594	2.511	1.830	0.825	0.851	0.970	3.72
2.54	4.626	2.522	1.834	0.831	0.854	0.973	3.44
2.27	4.643	2.542	1.827	0.834	0.861	0.969	3.16
1.99	4.705	2.560	1.838	0.845	0.867	0.975	2.84
1.72	4.793	2.595	1.847	0.861	0.879	0.979	2.30
1.44	4.880	2.624	1.860	0.877	0.889	0.986	1.96
1.27	5.009	2.665	1.880	0.900	0.903	0.997	1.58
0.89	5.089	2.699	1.886	0.914	0.914	1.000	1.20
0.62	5.216	2.750	1.897	0.937	0.931	1.006	0.88
0.34	5.341	2.778	1.923	0.960	0.941	1.020	0.60
0.07	5.422	2.813	1.927	0.974	0.953	1.022	0.50
0.00	5.463	2.816	1.940	0.981	0.954	1.029	1.05
Sample #SMG1, Cycle #9, $p_w = 350$ bars, $\dot{\varepsilon} = 10^{-6}$/sec							
0.0	6.325	3.146	2.009	0.993	1.000	0.992	0.68
0.0	6.368	3.146	2.024	1.0	1.0	1.0	0.68
0.13	6.332	3.138	2.018	0.993	0.997	0.997	0.52
0.27	6.329	3.130	2.022	0.994	0.995	0.999	0.52
0.42	6.325	3.118	2.029	0.993	0.991	1.002	0.44
0.63	6.314	3.108	2.032	0.992	0.988	1.004	0.38
0.84	6.297	3.102	2.030	0.989	0.986	1.003	0.37
1.07	6.293	3.098	2.031	0.988	0.985	1.003	0.42
1.25	6.293	3.087	2.039	0.988	0.981	1.007	0.43
1.46	6.280	3.086	2.035	0.986	0.981	1.005	0.48
1.67	6.274	3.082	2.036	0.985	0.980	1.006	0.49
1.92	6.266	3.077	2.036	0.984	0.978	1.006	0.56
2.09	6.232	3.070	2.030	0.979	0.976	1.003	0.59
2.30	6.232	3.064	2.034	0.979	0.974	1.005	0.64

Table 1—*continued*

Stress, kb	V_p, km/sec	V_s, km/sec	V_p/V_s	$V_p/(V_p)_0$	$V_s/(V_s)_0$	$(V_p/V_s)/(V_p/V_s)_0$	$\epsilon \times 10^{-3}$
2.51	6.193	3.053	2.028	0.973	0.970	1.002	0.70
2.72	6.174	3.042	2.030	0.970	0.967	1.003	0.74
2.84	6.149	3.034	2.027	0.966	0.964	1.001	0.78
2.99	6.150	3.026	2.032	0.966	0.962	1.004	0.83
3.09	6.145	3.013	2.039	0.965	0.958	1.007	0.89
3.22	6.102	3.004	2.031	0.958	0.955	1.003	0.92
3.13	6.128	2.991	2.049	0.962	0.951	1.012	0.99
3.42	6.106	2.968	2.057	0.959	0.943	1.016	1.11
3.22	6.007	2.941	2.043	0.943	0.935	1.009	1.29
3.21	6.025	2.916	2.066	0.946	0.927	1.021	1.40
3.20	6.008	2.895	2.075	0.943	0.920	1.025	1.45
3.19	5.946	2.885	2.061	0.934	0.917	1.018	1.65
3.19	5.946	2.857	2.081	0.934	0.908	1.028	1.81
3.17	5.922	2.842	2.071	0.930	0.903	1.023	2.01
3.16	5.916	2.814	2.102	0.929	0.894	1.039	2.17
3.14	5.856	2.764	2.119	0.920	0.879	1.047	2.53
3.13	5.829	2.727	2.138	0.915	0.867	1.056	2.75
3.11	5.810	2.691	2.159	0.912	0.855	1.067	3.03
3.10	5.801	2.671	2.172	0.911	0.849	1.073	3.17
2.72	5.767	2.656	2.247	0.906	0.844	1.110	3.13
2.30	5.750	2.635	2.182	0.903	0.838	1.078	3.13
1.88	5.767	2.618	2.149	0.906	0.832	1.062	3.13
1.46	5.728	2.600	2.203	0.899	0.826	1.088	3.07
1.05	5.748	2.577	2.231	0.903	0.819	1.102	3.05
0.63	5.769	2.542	2.269	0.906	0.808	1.121	3.01
0.21	5.780	2.535	2.280	0.908	0.806	1.126	2.93
0.0	5.785	2.558	2.262	0.908	0.813	1.118	2.75
0.0	5.918	2.721	2.175	0.929	0.865	1.074	2.09

initial loading and dropped to 307 bars at the peak axial load. Yet the behavior of velocities vs. ϵ during this loading cycle is nearly identical to that for the previous loading cycle (open circles, Fig. 4) during which the pore pressure was held constant at 350 bars. Confining pressure in both cases was 390 bars. The major difference between the two cycles is that less dilatant volumetric strain and less change in velocities were observed as a result of applying the *same peak load* when the pore pressure was allowed to vary; i.e. dilatancy hardening [19] occurred. Similarly, the results of two dry runs, one conducted at 40 bars and one at 390 bars confining pressure, may be taken to show the independence of velocities on effective stress (Fig. 5).

Hysteresis in velocity-ϵ curves. From Table 1, systematic differences may be noted in the velocity vs. ϵ relationship between the loading and unloading cycles. Such discrepancies occurred for both wet and dry stress cycles, with velocity invariably being less for a given value of ϵ during unloading. While some of the difference can be explained by time-dependent pore fluid migrations in wet rocks or time-dependent

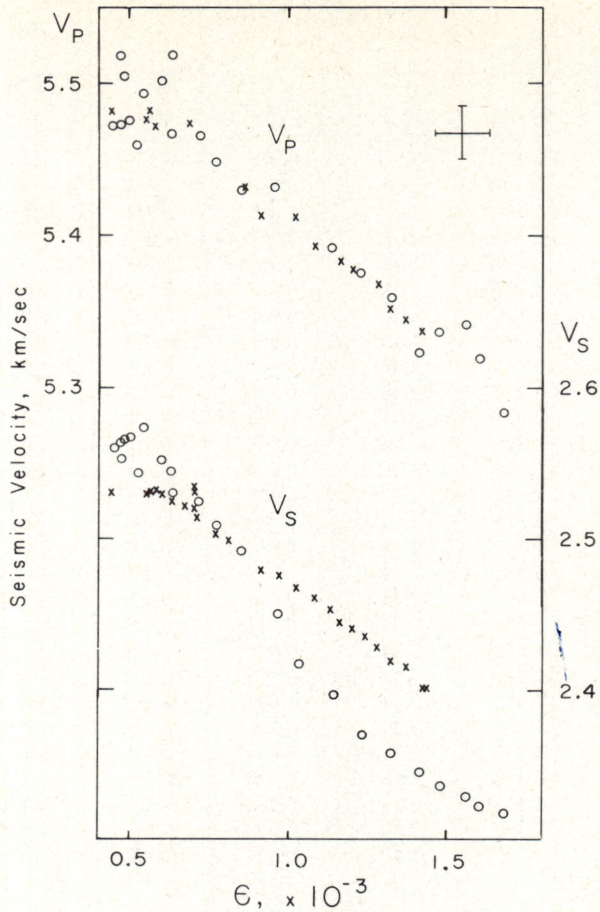

Figure 4
Change in seismic velocities with ϵ in saturated granite. Crosses, variable pore pressure; circles, 350 bars pore pressure

crack closure in wet and dry rocks all of the hysteresis does not disappear with time. This is shown for a wet sample by the arrows in Fig. 6.

During the first of the stress cycles illustrated (open circles) unloading was interrupted to follow the changes of ϵ, V_p and V_s with time. Similarly, during the second of the stress cycles illustrated (closed circles) the loading at 1.5×10^{-5}/sec was twice arrested and the migrations of ε followed for a two-hour interval under constant axial stress. There is no indication from the directions of migration that the observed hysteresis would disappear if loading and unloading were performed infinitely slowly.

A second possibility, loosening of the transducer bond, can be ruled out as the zero stress velocities are more or less recoverable from cycle to cycle provided the dilatancy can be recovered.

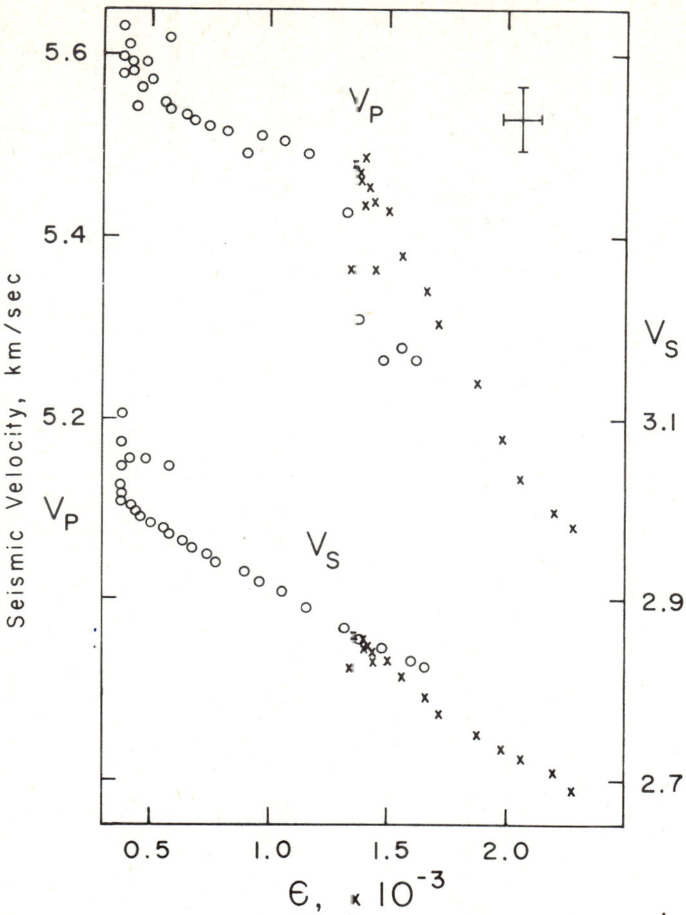

Figure 5

Change in seismic velocities with ϵ in dry granite. Crosses, 40 bars confining pressure; circles, 390 bars confining pressure

The most likely explanation for the hysteresis lies in uncertainty as to the elastic line for unloading. Probably, frictional resistance must be overcome during unloading before dilatant cracks can close [20]. This means that the rock will begin to unload 'elastically', without undergoing significant changes in crack distribution or shape. The true volumetric strain should, therefore, not change much, contrary to what is found if one refers the stres-volumetric strain curve back to the elastic line for loading in computing ϵ. As long as the cracks remain open, seismic velocities should remain low. This appears to be what is happening at points A–C in Fig. 6. Once the applied stress has dropped enough that the frictional resistance can be overcome, the velocities will begin to recover (somewhere between C and D in Fig. 6). As unloading proceeds from D, cracks begin to close and the velocities should recover along the same path

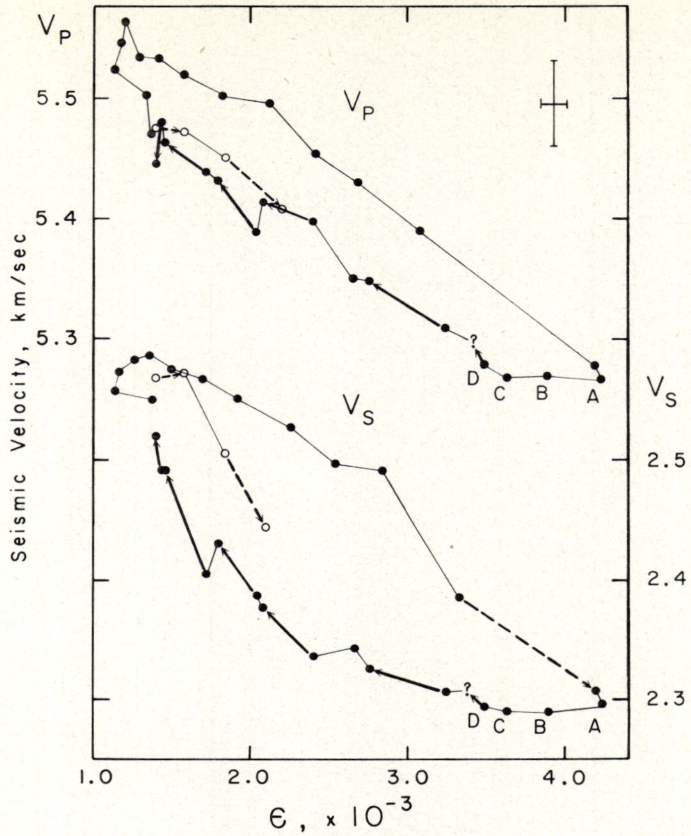

Figure 6

Hysteresis in seismic velocities vs. ϵ in saturated granite. See text for explanation

followed during loading. Regardless of rate the V_p or V_s vs. ϵ curves should look identical during stress application and release as long as the degree of saturation of the rock remains unchanged. That they fail to do so, as in Fig. 6, is an artefact of using the linear elastic line for loading to determine the dilatant volumetric strain during unloading.

Except where noted, the rest of this paper will treat only the results of measurements made on loading paths because of the uncertainty in the elastic line for unloading.

Velocity ratios. Although differences in degree of saturation of the rock samples are not striking when one looks at the behavior of the individual velocities as a function of dilatant volumetric strain, a sharp distinction may be drawn between saturated, partially saturated and dry rocks based upon the behavior of the ratio V_p/V_s. If, for each stress cycle, we define the parameter $(V_p/V_s)_0$ to be the value of the velocity ratio when the shear wave velocity is a maximum, and if we take ϵ_0 to be the dilatant volumetric strain corresponding to that value of V_s (ϵ_0 may be non-zero owing to crack orientation effects), then we may easily compare the variation of V_p/V_s and ϵ during different stress cycles by normalizing them with respect to these parameters. The percent change in the velocity ratio relative to the value it has at the shear wave velocity maximum is simply $(V_p/V_s)/(V_p/V_s)_0 - 1$. Results for dry rocks have been plotted in Fig. 7; V_p/V_s drops at large dilatant strains. For rocks with high pore pressure (Fig. 8) the ratio rises. These rocks presumably remain saturated. Again, there is evidence of the independence of velocity–dilatant strain behavior on rock type because the gabbro (closed circles, Fig. 9) run at 1.5×10^{-6}/sec behaved comparably to the granites run at 1.5×10^{-5} and 1.5×10^{-6}/sec.

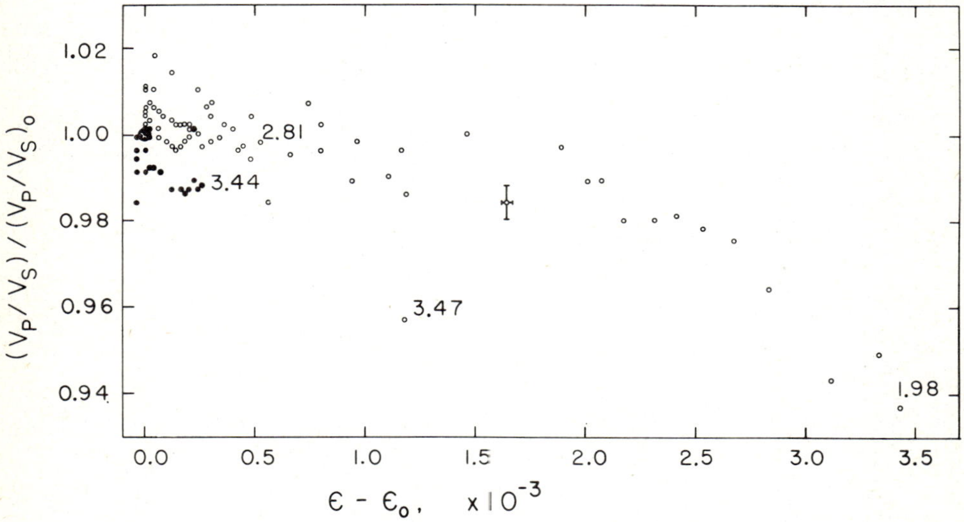

Figure 7
Change in seismic velocity ratio with ϵ in dry rock. Strain rates 1.5×10^{-5} and 1.5×10^{-6}/sec. Open circles, granite, filled circles, gabbro. Numbers shown are maximum stress attained during loading cycle, in kb

A further word about the gabbro is in order as its behavior may contradict some models of precursory phenomena of earthquakes [21]. Points A and beyond (Fig. 8) were obtained in the post-peak stress region of the loading cycle. The loading was arrested at a maximum stress of 3.21 kb and the rock allowed to relax for 2 hours due to accelerating creep. During this time, the stress dropped 140 bars, V_p decreased by 3.2%, but V_s decreased by 6.8% causing a net increase in V_p/V_s. It seems that a rather significant drop in stress would be required before cracks stopped propagating, let alone before frictional resistance could be overcome enough to permit crack closure and a V_s rise.

Results for stress cycles with 1 bar water pressure are rate-dependent (Fig. 9), a phenomenon not observed in the dry and high pore pressure cases. At slow rates when the experiment was conducted with the pore fluid reservoir connected to the sample so that additional water was available to fill opening cracks V_p/V_s rose. The rock properties followed the trend of the previous results for samples with 350 bars water pressure. At fast rates the results are more complex. When excess water was available, the V_p/V_s ratio tended to remain constant, but when the pore fluid reservoir was not connected to the rock the ratio dropped with increasing dilatant strain. However, the drop was not as pronounced as in the dry case. Under these fast rate conditions at the low pore pressure, the dilatant rock may be partially saturated, the amount of partial saturation depending upon the availability of additional pore fluid. The velocity ratio reflects the degree of partial saturation. This is in agreement with the theoretical results of O'CONNELL and BUDIANSKY [22].

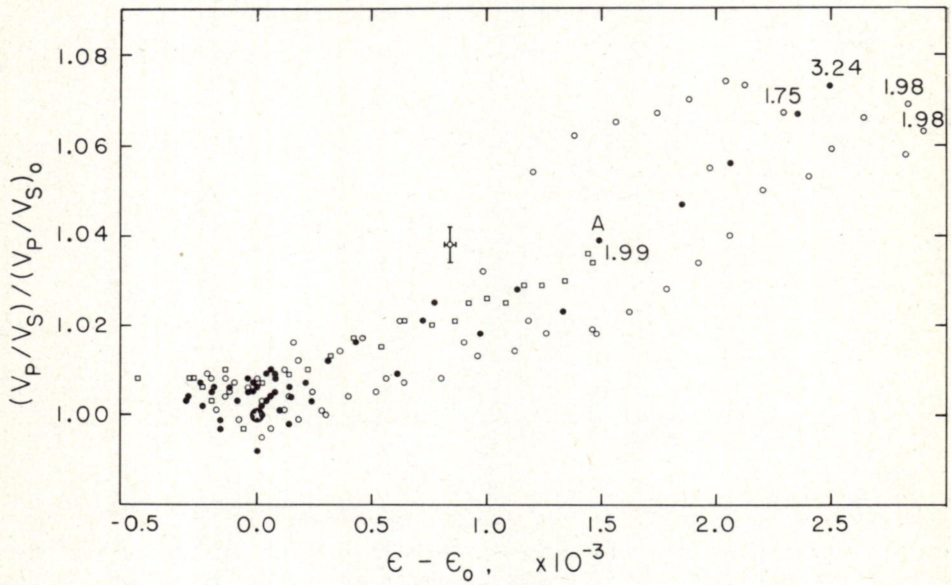

Figure 8
Change in seismic velocity ratio with ϵ in rocks with high pore pressure. Strain rates 1.5×10^{-5} and 1.5×10^{-6}/sec. Open circles, granite, 350 bars pore pressure; open squares, granite, variable pore pressure; filled circles, gabbro, 350 bars pore pressure. Numbers shown in figure are maximum stress attained during loading cycle, in kb

The liquid–vapor transition. The water–steam transition offers another possible mechanism for decreasing seismic velocity ratio. It is attractive because it provides an alternative to the opening of dry cracks in rock. Presumably, wet cracks would affect seismic velocities like dry cracks if the water in them could be converted to vapor through dilatancy-induced volume changes. Temperatures and hydrostatic pressures at mid-crustal depths are thought to be near the liquid vapor phase boundary for H_2O [23], so at any given temperature only a negligible drop in pressure might suffice to vaporize the pore fluid. The resulting change in the compressibility of the pore fluid, and consequently, of the compressional wave velocity of the aggregate could be large.

Because our apparatus is unable to reach the liquid–vapor transition temperature of H_2O, CO_2 was used to model the water–steam transition. Although the room temperature viscosity of liquid CO_2 near the liquid–vapor transition is somewhat lower than that of liquid water at, say, 200°C [24], the decrease in bulk modulus going through the transition is comparable [25].

Some differences between the behavior of CO_2-saturated and water-saturated rocks are apparent from Fig. 10. At $p_{CO_2} = 112$ bars, well above the transition pressure, the

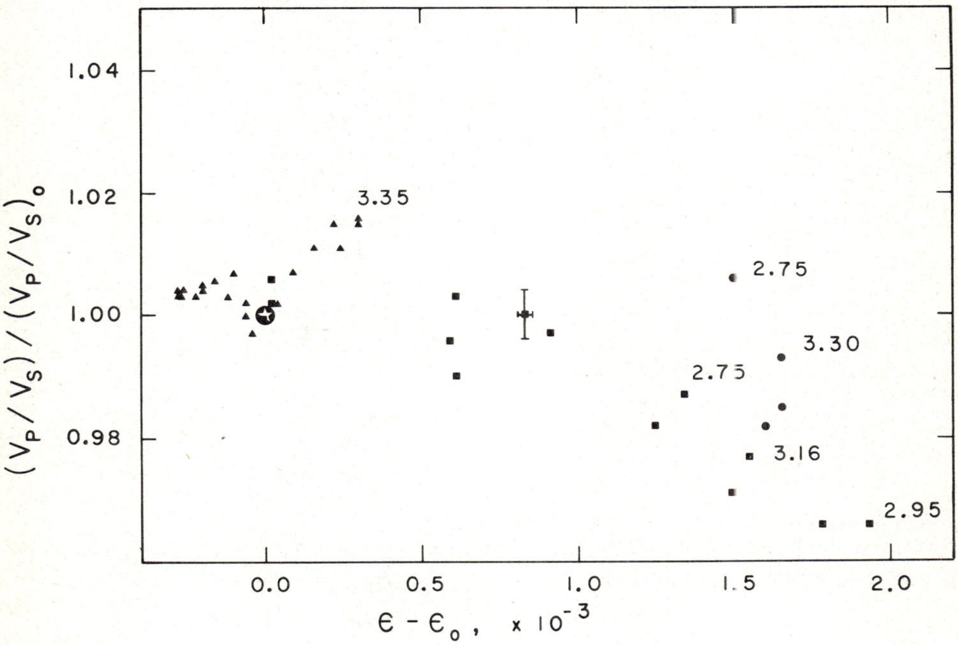

Figure 9

Change in seismic velocity ratio with ϵ in gabbro with 1 bar pore pressure. Squares, strain rate 1.5×10^{-4}/sec., pore fluid reservoir disconnected; circles, strain rate 1.5×10^{-4}/sec., reservoir connected; triangles, strain rate 1.5×10^{-6}/sec., reservoir connected. Numbers shown in figure are maximum stress attained during loading cycle, in kb

initial rise in V_p/V_s with dilatant volumetric strain appears to level off or reverse itself at higher strains. The cause of this is in doubt. Since no drop in V_p/V_s was observed for comparable dilatant volumetric strains when the starting CO_2 pore pressure was only slightly above the transition pressure, it is unsatisfying to try to explain the higher pore pressure behavior by appeal to the liquid-vapor transition. In fact, even running with CO_2 pore fluid at the transition pressure, the large drop expected in V_p/V_s failed to materialize. Although the drop in V_p was comparable to that for a dry rock with the same dilatant volumetric strain, the drop in V_s was somewhat larger so that the ratio of the velocities, although it decreased, decreased less than it would in a dry rock (compare Table 1, entry W1, cycle 12, with entry W3, cycle 6).

To remove any doubt about whether or not the liquid–vapor transition had been achieved in the dilatant rock with initial CO_2 pore pressure equal to 62 bars the pore pressure system was reconnected to the sample at the peak of the loading cycle by opening a valve and the reservoir pressure was monitored. After allowing time for any equilibration the pore pressure was bled off deliberately. The velocity ratio–dilatant strain behavior followed the path shown by the arrows in Fig. 10. The first star represents the point where the reservoir was reconnected to the sample. No change in

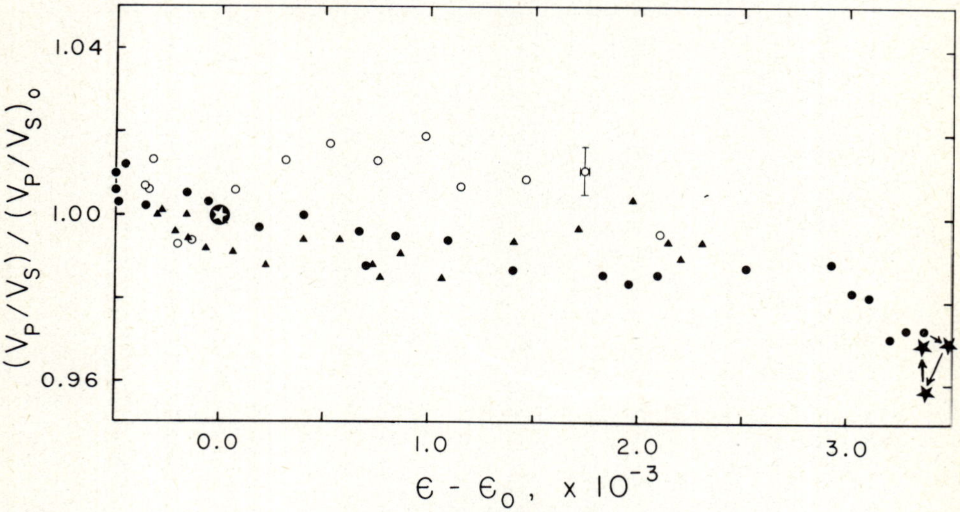

Figure 10
Change in seismic velocity ratio with ε in granite with CO_2 pore fluid. Open circles, 112 bars CO_2 pressure; traingles, 68 bars CO_2 pressure; solid circles, 62 bars (liquid-vapor transition) CO_2 pressure. See text for explanation of stars

velocities was observed within the 10-min interval thereafter and the pore pressure held steady at 62 bars. The CO_2 was then bled off slowly. The second star in Fig. 10 shows the last velocity ratio determination made while the pore pressure maintained its original 62 bar value. The third star represents the lowest value of V_p/V_s measured; pore pressure had by then dropped to 57 bars clearly below the transition pressure. Thereafter, V_p/V_s began to rise, finally reaching the value indicated by the last star at 1 bar pore pressure. The rise in V_p/V_s was due to an increase in both V_p and V_s caused by crack closure in response to the increasing effective confining pressure on the rock sample.

4. Discussion

Extrapolation of the laboratory results to the earth

Although the relative magnitudes and the signs of the experimentally observed changes in V_p, V_s and their ratio may apply to the field the absolute magnitudes may not.

Near-surface, *in situ* velocities can be two to three orders of magnitude more sensitive to changes in ambient stress than velocities measured in the laboratory [26, 27]. If stress–strain behavior remains about the same in the laboratory and the field, then the conclusion is that rocks in the field may be two to three orders of magnitude more sensitive to ε than the results reported here suggest.

The most likely cause of this enhanced sensitivity is the difference in aspect ratio, width length, of the cracks or joints *in situ* and those occurring in the laboratory rock samples. WALSH [28–30] has shown that the effective elastic moduli vary as the crack length cubed; therefore long, thin cracks will have a greater effect on seismic velocities than short, fat ones for the same crack axis orientations and total crack volume. In the dry laboratory samples, for $\epsilon \simeq 10^{-3}$ only a 10% change in V_p is observed. Applying this dependence directly to the field case requires that 10 m of uplift be associated with a 20% drop in V_p assuming a dilatant volume with a radius of 10 km. This, clearly, is impossible. If, however, the average aspect ratio of cracks in the field were two or three orders of magnitude smaller than that in our granite and gabbro cylinders, then only 1 to 10 cm of uplift would suffice. Two orders of magnitude decrease in average aspect ratios from laboratory to field appears plausible not merely because of the greater length scales possible *in situ*, but because of the difference in lateral stiffness conditions between our experiments (constant confining pressure) and the earth (somewhere between constant confining pressure and uniaxial strain).

When looking for precursory changes in travel time ratios in the field, care must be taken to separate out differences due to changing path and/or polarization from those associated with changing crack density or degree of saturation of the material. The former need not signify a change in stress state *in situ*, while the latter probably do. The laboratory experiments, of course, are conducted under constant path and polarization conditions. Thus changes in V_p/V_s are related only to changing rock properties under loading. If, however, the polarization of the S wave had been changed by 90° in these experiments, the decrease in shear wave velocity would have been about 60% of the observed value (Fig. 3 and [7]). This means that decreases (increases) in the ratio V_p/V_s would have appeared to be greater (less) than were in fact measured (corresponding to a clockwise rotation of the points in Figs. 7–10).

The effect of measurement path on the magnitude of V_p/V_s can be considered most easily under the assumption of axial symmetry. If we make that simplification then from these experiments and those of BONNER [7], MATSUSHIMA [9], and GUPTA [11], V_p/V_s surfaces for the rock may be constructed (Fig. 11b). These will be ellipsoids of rotation with the magnitude of the radius vector equal to the value of V_p/V_s measured in that direction. Because the magnitude of V_s is polarization as well as direction dependent a range of values of V_p/V_s are possible in any one direction. If axial cracks (semi-major axis in the 3 direction) are opening in the dilatant material with random orientations of semi-minor axes (Fig. 11a), then $V_{s_{13}}$[2]) will be equal to $V_{s_{31}}(=V_{s_{32}})$, while $V_{s_{12}}$ will

[2]) The first numerical subscript gives the propagation direction; the second, the polarization direction.

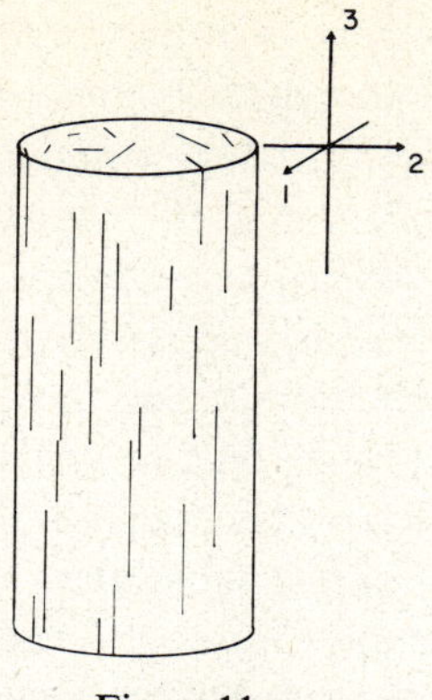

Figure 11a
Idealized representation of cracks in dilatant crack cylinder

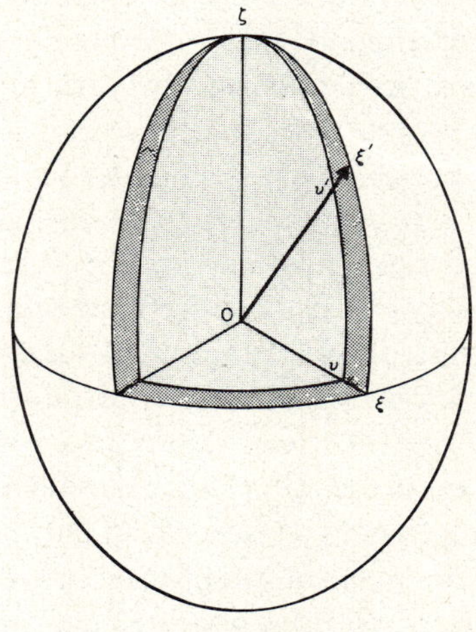

Figure 11b
Normalized velocity ratio $[(V_p/V_s)/(V_p V_s)_0]$ ellipse for dry granite with ε about 2×10^{-3} strain.

$$\zeta = (V_{p3}/V_{s31})/(V_{p3}/V_{s31})_0 = 1.01;$$
$$\xi = (V_{p1}/V_{s12})/(V_{p1}/V_{s12})_0 = 0.97;$$
$$\nu = (V_{p1}/V_{s13})/(V_{p1}/V_{s13}) = 0.95$$

Range of permissible normalized velocity ratios, ν' to ξ', in any given direction is indicated by the darkly shaded region

be least of all. Similarly, V_{p_1} will be less than V_{p_3}; in fact, V_{p_3} will be nearly indistinguishable from the velocity in the unstressed material. For dry granite at about 400 bars p_c, Fig. 11b illustrates the likely range of normalized V_p/V_s values.

In the field the condition of axial symmetry may not be met. In a tectonic region, two of the three principal stresses need not generally be equal. Moreover, even a slight foliation such as is found in Westerly granite can produce more than 10% departures from axial symmetry of cracks if that foliation does not coincide with the unique stress direction. This means that the V_p/V_s surfaces will in general be more complex forms and that the principal axes of these surfaces need not coincide with the principal stress directions. However, the surfaces and their orientation can be determined for any rock with existing laboratory techniques. Although existing laboratory data are insufficient to permit the complex constructions at present, the means exist to separate out effects due to changing path from effects due to changing material properties.

5. *Conclusions*

At low confining pressures, changes in seismic velocity depend upon both dilatant volumetric strain, ε, and degree of saturation of the material. Both V_p and V_s decrease with increasing ε. V_p decreases more rapidly in dry or nearly dry rocks, while V_s decreases more rapidly in saturated or nearly saturated rocks. Intrinsic velocity changes are completely over-shadowed by the effect of cracks; thus seismic velocities are insensitive to changes in effective confining pressure except as it affects ϵ.

Seismic velocity ratios show a comparable dependence on ϵ in both the granite and the gabbro under similar conditions of saturation. Although a liquid–vapor transition will cause a decrease in V_p/V_s, the seismic velocity ratio drops over twice as much in dry rock for the same dilatant volumetric strain.

In initially-saturated material, decreases in seismic velocity ratio occur under rather limited conditions. If strain rates of 10^{-4}/sec are required to lower V_p/V_s in dilatant laboratory specimens where the distance to the reservoir was 6 cm, then to produce this effect in the earth, where strain rates are likely to be at least 6 to 8 orders of magnitude lower, either the distances over which pore fluid must equilibrate are on the order of 10 to 100 km, or the average permeability of the rock *in situ* is 6 to 8 orders of magnitude less than that of repeatedly stressed laboratory samples of San Marcos gabbro. The first alternative appears more likely. Fluid source to fluid sink distances of 6 km are not grossly out of line with anomalous zones of about 10 km linear dimension, such as are suggested by certain field data. However, if the earth's surface is taken to be a fluid source, then significant dry crack space cannot occur in initially saturated material at depths shallower than about 6 km. Dilatant zones forming in saturated regions of the crust at depths less than 6 km will not cause decreases in seismic velocity ratio.

Since dilatant porosity increases probably do not exceed 10–100% *in situ* [31], a liquid–vapor transition is not an attractive means of generating V_p/V_s decreases, as only a small fraction of the pore fluid could be vaporized by the enlargement of liquid-filled cracks. Thus, even if *in situ* velocities are much more sensitive to changes in pore fluid state than those measured in the laboratory, the effect of a liquid–vapor transition is likely to be small. However, it would be in the right direction to produce the observed premonitory decreases in travel time ratios, and large changes in the individual seismic velocities can be involved.

Because the opening of dry cracks produces a much larger drop in V_p/V_s than any other mechanism, the introduction of dry cracks *in situ* must be regarded as the most effective means of generating anomalously low travel time ratios. Of all the possible combinations of dilatancy and pore fluids studied in our experiments, dry cracks can most easily reproduce the relative drops in V_p and V_s reported in the field. In certain localities such as Blue Mountain Lake where hypocentral temperatures are less than 100°C, or in places where uplift data sharply constrain the amount of new crack space that can be accommodated in the *in situ* material, opening of dry cracks may be the only one of the currently understood mechanisms capable of explaining the seismic observations.

Acknowledgements

Y. Aggarwal and C. H. Scholz participated in the early stages of these experiments. Discussion with R. J. O'Connell and B. Bonner was helpful. C. Goetze suggested the use of CO_2 to model the liquid–vapor transition. W. F. Brace and M. Wyss critically read the manuscript. This work was supported by the US Geological Survey under Contracts No. 14-08-0001-13229 and No. 14-08-0001-14551 while the author was an NSF Graduate Fellow.

References

[1] A. N. Semenov, *Variations in the travel time of transverse and longitudinal waves before violent earthquakes*, Izv. Phys. Solid Earth *3* (1969), 245.

[2] I. L. Nersesov, A. N. Semenov and I. G. Simbireva, *Space–time distribution of the travel time ratios of transverse and longitudinal waves in the Garm area*, in *The Physical Basis of Foreshocks* (Nauka Publ., Moscow 1969).

[3] D. L. Anderson and J. H. Whitcomb, *The dilatancy–diffusion model of earthquake prediction*, in Proc. of Conf. on Tectonic Problems of the San Andreas Fault System, edited by R. L. Kovach and A. Nur (Stanford University Press, California 1973), p. 417.

[4] Y. P. Aggarwal, L. Sykes, D. W. Simpson and P. G. Richards, *Spatial and temporal variations in t_s/t_p and in P-wave residuals at Blue Mountain Lake, New York: Application to earthquake prediction* (Preprint, 1974).

[5] A. Nur, *Dilatancy, pore fluids, and premonitory variations of t_s/t_p travel times*, Bull. Seismol. Soc. Amer. *62* (1972), 1217.

[6] C. H. Scholz, L. R. Sykes and Y. P. Aggarwal, *Earthquake prediction: A physical basis*, Science *181* (1973), 803.

[7] B. Bonner, *Shear wave birefringence in dilating granite*, Geophys. Res. Lett. *1* (1974), 217 and pers. comms.

[8] D. Tocher, *Anisotropy in rocks under simple compression*, AGU Trans. *38* (1957), 89.

[9] S. Matsushima, *Variation of the elastic wave velocities of rocks in the process of deformation and fracture under high pressure*, Dis. Prev. Res. Inst. Bull *32* (1960), 2.

[10] R. E. Thill, *Acoustic methods for monitoring failure in rock*, in Proc. of the 14th Sympos. on Rock Mech. edited by H. R. Hardy, Jr. and R. Stefanko (Am. Soc. Civil. Eng., N.Y., 1973).

[11] I. N. Gupta, *Seismic velocities in rock subjected to axial loading up to shear fracture*, J. Geophys. Res. *78* (1973), 6936.

[12] F. Birch, *The velocity of compressional waves in rocks to 10 kilobars*, J. Geophys. Res. *66* (1961), 2199.

[13] W. F. Brace, J. B. Walsh and W. T. Frangos, *Permeability of granite under high pressure*, J. Geophys. Res. *73* (1968), 2225.

[14] H. S. CARSLAW and J. C. JAEGER, *Conduction of Heat in Solids*, 2nd ed. (Oxford Univ. Press 1959), 510 pp.

[15] W. F. BRACE, A. S. ORANGE and T. R. MADDEN, *The effect of pressure on the electrical resistivity of water-saturated crystalline rocks*, J. Geophys. Res. *70* (1965), 5669.

[16] W. F. BRACE and A. S. ORANGE, *Further studies of the effects of pressure on electrical resistivity of rocks*, J. Geophys. Res. *73* (1968), 5407.

[17] K. HADLEY, *Laboratory investigation of dilatancy and motion on fault surfaces at low confining pressures*, in Proc. of Conf. on Tectonic Problems of the San Andreas Fault System, edited by R. L. KOVACH and A. NUR (Stanford University Press, California 1973), p. 427.

[18] W. F. BRACE, B. W. PAULDING, JR., and C. SCHOLZ, Dilatancy in the fracture of crystalline rocks, J. Geophys. Res. *71* (1966), 3939.

[19] W. F. BRACE and R. J. MARTIN III, *A test of the law of effective stress for crystalline rocks of low porosity*, Int. J. Rock Mech. Min. Sci. *5* (1968), 415.

[20] C. H. SCHOLZ and R. KRANZ, *Notes on dilatancy recovery*, J. Geophys. Res. *79* (1974), 2132.

[21] B. T. BRADY, *Theory of earthquakes, Part I: A scale independent theory of rock failure*, Pure and Appl. Geophys., in press.

[22] R. J. O'CONNELL and B. BUDIANSKY, *Seismic velocities in dry and saturated cracked solids*, Preprint, 1974.

[23] D. L. ANDERSON and J. H. WHITCOMB, *Time dependent seismology*, Preprint, 1974.

[24] E. W. WASHBURN (editor), *International Critical Tables: Vol. V* (McGraw-Hill, N.Y. 1929, for the National Research Council).

[25] G. C. KENNEDY and W. T. HOLSTER, *Pressure–volume–temperature and phase relations of water and carbon dioxide*, in *Handbook of Physical Constants*, Geol. Soc. Amer. Mem. *97* (1966).

[26] T. L. DE FAZIO, K. AKI and J. ALBA, *Solid earth tide and observed change in the* in situ *seismic velocity*, J. Geophys. Res. *78* (1973), 1319.

[27] P. REASENBERG and K. AKI, *A precise, continuous measurement of seismic velocity for monitoring* in situ *stress*, J. Geophys. Res. *79* (1974), 399

[28] J. B. WALSH, *The effect of cracks on the compressibility of rock*, J. Geophys. Res. *70* (1965), 381.

[29] J. B. WALSH, *The effect of cracks on the uniaxial elastic compression of rocks*, J. Geophys. Res. *70* (1965), 399.

[30] J. B. WALSH, *The effect of cracks on Poisson's ratio*, J. Geophys. Res. *70* (1965), 5249.

[31] T. C. HANKS, *Constraints on the dilatancy–diffusion model of the earthquake mechanism*, J. Geophys Res. *79* (1974), 3023.

(Received 28th October 1974)

Pageoph,
Vol. 113 (1975) Birkhäuser Verlag, Basel

V_p/V_s in Saturated Granodiorite Loaded to Failure

By B. P. BONNER[1])

Summary – Ultrasonic compressional and shear travel times have been measured simultaneously along the direction of maximum compression for granodiorite loaded in uniaxial stress. Excess pore fluid (kerosene) is available during the deformations which are performed under light confinement (2.5 MPa) at an average strain rate of about 5×10^{-6} s^{-1}. V_p/V_s increases monotonically with shear stress (~13%) until failure. These data are consistent with either: 1) a high rate of fluid flow of the same order as the increase in dilatant volume with time; or 2) an increase in total porosity with a decrease in mean aspect ratio of the pores (MPa = 10^6 N/m^2 – 10 bars).

1. Introduction

Seismic methods (NERSESOV *et al.* [1], WYSS and HOLCOMB [2]) have demonstrated potential as earthquake prediction tools in certain situations. Dilatancy – an inelastic volume increase which appears as shear stress increases (BRACE *et al.* [3]) – and its effect on fluid transport and properties, offers a possible explanation for seismic anomalies (NUR [4], WHITCOMB *et al.* [5]) and may be consistent with a broad range of precursory phenomena (SCHOLZ *et al.* [6]). Laboratory measurements of elastic velocities during deformation (TOCHER [7], MATSUSHIMA [8], GUPTA [9] have added to an understanding of the physical basis for earthquake prediction. The present study includes the effect of pore fluid on observed velocities.

2. Experimental details

Ultrasonic measurements were performed on a granodiorite from northern Nevada (JEROME [10]) with an interconnected porosity of ~1% as estimated by evacuation and water immersion. Jacketed specimens were loaded to failure under low effective pressure. Changes in compressional velocity produced by fluid saturation are greatest at low effective pressure (see the data of NUR and SIMMONS [11]), and the effects of dilatant strain should also be maximixed. Pore fluid was introduced by exposing the specimen directly to the pressure medium, kerosene, at 50 MPa for several hours. Observed increases in compressional velocity (15–20%) suggest that the kerosene filled a significant fraction of the porosity. Kerosene was used as the pore fluid for several reasons: 1) viscosity under standard conditions is comparable, although slightly higher than water; 2) chemical reactivity is low; and 3) the effects of kerosene on ultrasonic velocities,

[1]) University of California, Lawrence Livermore Laboratory, Livermore, California 94550.

as compared to water, are similar and well documented (King [12]). Excess pore fluid is introduced into the specimen assembly; these experiments are 'undrained' tests.

The pressure system is conventional except for the size of the cylindrical specimens (10 cm diameter by 16 cm long) (Bonner *et al.* [13]). Uniaxial stress is applied by hydraulically driven hollow steel pistons (Rc 58) which contain the piezoelectric ceramic ($Pb(ZrTi)O_3$) elements which generate and receive ultrasonic pulses. The transmitting transducer (dominant frequency 250 kHz) is polarized for transverse shear, but produces significant compressional energy. The receiving transducer contains three active elements; a combination of mechanical and electronic techniques enhance either the shear or compressional arrival (Bonner [14]). Delays are timed with and without a sample placed between the loading pistons. Accuracy is estimated at ~4%, but the precision during a single experiment is ~2%.

3. *Results*

Measured travel times for shear (t_s) and compressional (t_p) waves are plotted in Fig. 1 along with approximate velocities. After an initial decrease, t_s shows a significant increase with differential stress to failure (~12%) beginning at about 60% of the fracture stress. In contrast t_p increases only slightly (~3%) just before failure. The high compressional velocity is characteristic of saturated low-porosity rock. The magnitude of the

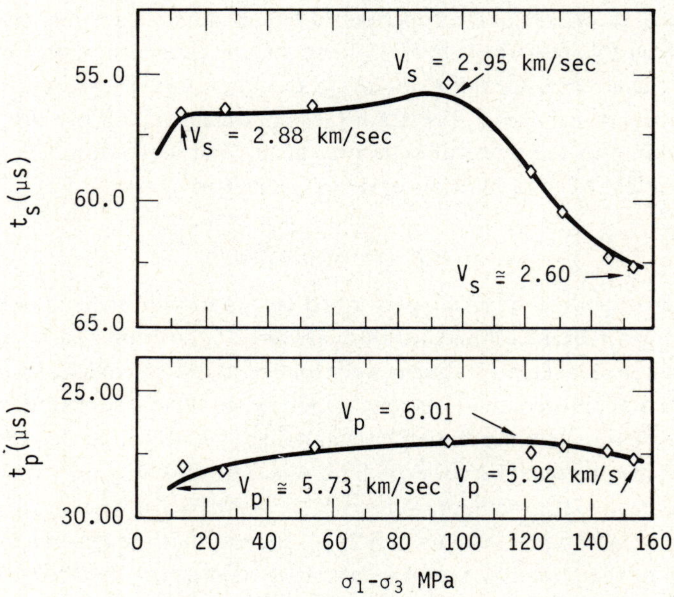

Figure 1

Compressional and shear travel-times for kerosene saturated granodiorite loaded to failure in uniaxial stress. Initial velocities are $V_p \cong 5.32$ km/s, $V_s \cong 2.59$ km/s; the correction for length change, increasing to ~2% at failure, is not included

t_p increase is consistent with velocity decreases observed prior to the failure of low-porosity rock (THILL [15]). The ratio V_p/V_s, numerically equal to t_s/t_p, is plotted in Fig. 2 as a function of shear stress normalized to fracture stress (τ/τ_F). High values of V_p/V_s at low shear stress are consistent with laboratory observations for granites (NUR and SIMMONS [11]), although low estimates of V_s could lead to systematically high ratios at all shear stresses. V_p/V_s increases monotonically from about 0.5 τ_F, the approximate stress at which dilatancy is expected to occur (BRACE *et al.* [3]).

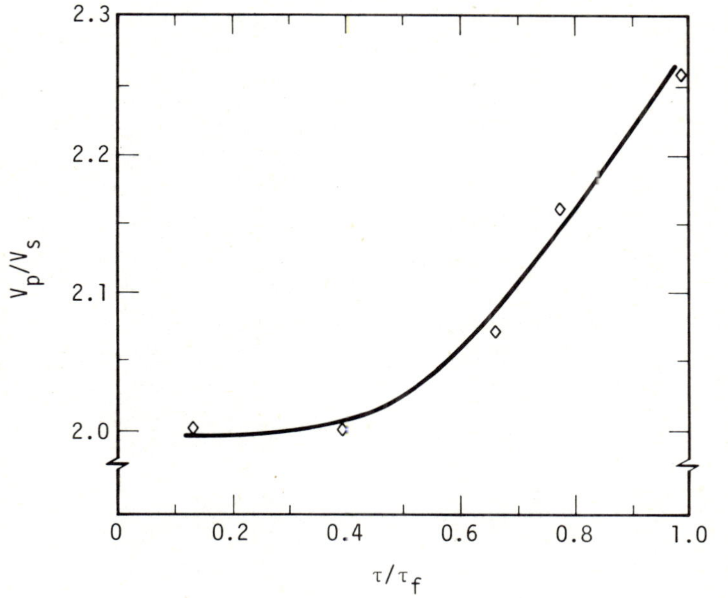

Figure 2
Velocity ratio V_p/V_s as a function of shear stress normalized to fracture stress (τ/τ_F), computed from Fig. 1 data

At least two processes are consistent with these observations. Fluid saturation increases V_p; the increase correlates with microcrack porosity (NUR and SIMMONS [11]). If the crack porosity which causes dilatant strain fills rapidly with pore fluid, V_p should remain relatively high. Since V_s decreases with dilatant strain (BONNER *et al.* [16]) and is insensitive to the bulk modulus of the pore fluid, the velocity ratio should increase at high shear stress. The implied assumption is that the flow rate of pore fluid is of the same order as the increase of dilatant volume with time. Dramatic increases in permeability with dilatancy in granite (ZOBACK and BYERLEE [17]) suggest that this condition may be met.

The second hypothesis arises from elastic anisotropy caused by dilatant strain and is supported by calculations extending the Eshelby–Walsh theory to solids containing oriented cracks (ANDERSON *et al.* [18]). Dilatancy occurs when tensile cracks form parallel to σ_1. Porosity increases, and it is reasonable to infer that the mean aspect ratio (minimum/maximum dimension of a flaw) decreases with shear stress. Increasing porosity and decreasing aspect ratio can lead to significant increases in V_p/V_s (nearly independent of the pore fluid bulk modulus) for waves propagating along the long axis of the cracks (Table 2, ANDERSON *et al.* [18]).

The preliminary results presented here, based on six experiments, do not uniquely determine if either of these processes dominates. These results, together with an estimate of V_p/V_s for another dry granodiorite loaded to failure (BONNER *et al.* [16]), suggest that both may be significant.

4. Conclusions

Increases in the ratio V_p/V_s have been observed along the direction of maximum principal stress for kerosene-saturated granodiorite loaded to fracture in uniaxial stress. These results may be due to either: 1) a high rate of pore fluid flow which nearly equals the increase of dilatant volume with time; or 2) an increase in porosity accompanied by a decrease in mean aspect ratio. Increases in the velocity ratio, at least along the direction of maximum compression, are compatible with dilatancy in this laboratory experiment, and may be possible in the field case. Such increases would be most likely in areas where rapid flow accompanies dilatancy, an hypothesis advanced to explain observations during the Matsushiro earthquake swarm (NUR [19]).

Acknowledgements

A. ABEY, A. DUBA, H. HEARD, A. PIWINSKII, R. SCHOCK and H. WEED critically reviewed the manuscript, which was prepared by Mrs. D. JACKSON. This work was performed under the auspices of the US Atomic Energy Commission.

REFERENCES

[1] I. L. NERSESOV, A. M. SEMONOVA and I. G. SIMBIREVA, in *Physical Basis of Foreshocks* (Nauka, Moscow 1969).

[2] M. WYSS and D. J. HOLCOMB, *Earthquake prediction based on station residuals*, Nature *249* (1973), 193–140.

[3] W. F. BRACE, B. W. PAULDING, JR. and C. H. SCHOLZ, *Dilatancy in the fracture of crystalline rocks*, J. Geophys. Res. *71* (1966), 3939–3953.

[4] A. NUR, *Dilatancy, pore fluids, and premonitory variations of t_s/t_p travel times*, Bull. Seism. Soc. Am. *62* (1972), 1217–1222.

[5] J. H. WHITCOMB, J. D. GARMANY and D. L. ANDERSON, *Earthquake prediction: variation of seismic velocities before the San Fernando earthquake*, Science, *180* (1973), 632–635.

[6] C. H. SCHOLZ, L. R. SYKES and Y. P. AGGARWAL, *Earthquake prediction: a physical basis*, Science *181* (1973), 803–810.

[7] D. TOCHER, *Anisotropy in rocks under simple compression*, Trans. Amer. Geophys. U. *38* (1957), 89–94.

[8] S. MATSUSHIMA, *Variation of the elastic wave velocities of rocks in the process of deformation and fracture under high pressure*, Disaster Prevention Res. Inst., Kyoto Univ. Bull. *32* (1960), 1–8.

[9] I. N. GUPTA, *Seismic velocities in rock subjected to axial loading up to shear fracture*, J. Geophys. Res. *78* (1973), 6936–6943.

[10] S. E. JEROME, *Geological geophysical, chemical, and hydrological investigations of the Sand River Range, Fairview Valley, and Fourmile Flat, Churchill County, Nevada*, University of Nevada Report. VUF-1001, Reno, Nevada (1965), 367 pp.

[11] A. NUR and G. SIMMONS, *The effect of saturation on velocity in low porosity rocks*, Earth Planet. Sci. Lett. *7* (1970), 183–193.

[12] M. S. KING, *Wave velocities in rock as a function of changes in overburden pressure and fluid saturants*, Geophys. *31* (1966), 50–73.

[13] B. P. BONNER, A. E. ABEY, H. C. HEARD and R. N. SCHOCK, *High pressure mechanical properties of merlin alluvium*, Lawrence Livermore Laboratory, Report UCRL-51252, Livermore, CA (1972), 23 pp.

[14] B. P. BONNER (1974) in preparation.

[15] R. E. THILL, *Acoustic methods for monitoring failure in rock*, in *Proceedings of the Fourteenth Symposium on Rock Mechanics*, edited by H. R. HARDY, Jr. and R. STEFANKO (Am. Soc. Civil Eng., New York 1973), pp. 649–687.

[16] B. P. BONNER, W. M. BENZING and R. N. SCHOCK, *Ultrasonic travel times in rock under uniaxial compression* (abs.), Trans. Amer. Geophys. U. *54* (1972), 467.

[17] M. D. ZOBACK and J. D. BYERLEE, *The effect of stress on the permeability of westerly granite* (abs.), Trans. Amer. Geophys. U. *55* (1974), 431.

[18] D. L. ANDERSON, B. MINSTER and D. COLE, *The effect of oriented cracks on seismic velocities*, J. Geophys. Res. *79* (1974), 4011–4016.

[19] A. NUR, *Matsushiro, Japan, earthquake swarm: confirmation of the dilatancy–fluid diffusion model*, Geology *2* (1972), 217–221

(Received 11th November 1974)

Pageoph,
Vol. 113 (1975)
Birkhäuser Verlag, Basel

Acoustic Emission During High-Temperature Frictional Sliding

By ROBERT M. STESKY[1,2])

Summary – Acoustic emissions during high temperature frictional sliding in faulted granite were detected with a piezoelectric transducer placed outside the furnace and pressure vessel. Suitable choice of materials allowed the transmission of signals sufficiently large for measurement and analysis. Acoustic emissions were also monitored during the fracture runs at room temperature and could be compared with earlier studies. As observed previously, the emission rate increased dramatically prior to failure and the Ishimoto–Iida relationship between emission amplitude, A, and number of emissions, $N(A)$, of amplitude, A, i.e., $\log N(A) = K' - (b + 1)\log A$, was followed. The constant, b, was typically about 1.4, higher than the value of 1 or less indicated by previous experimental and theoretical studies.

Acoustic emissions were detected during frictional sliding at all temperatures from 25° to 700°C. The activity began at low stresses and accelerated as the stress increased. During sliding the emission rate was nearly constant (with large counts during stick-slip events) at a level which decreased with increasing temperature. This apparent decrease may be explained by a real decrease in the number of emissions or by an increase in attenuation at higher temperatures.

The power-law relationship between amplitude and emission rate seemed to hold for sliding as for fracture. High b-values, between 1.2 and 2.1, are in agreement with earlier studies and are little affected by temperature within the scatter of the measurements. Where the number of emissions allowed its measurement, the b-value of emissions during loading prior to slip was consistently, although only slightly higher than that during steady sliding.

1. *Introduction*

Acoustic emission during rock deformation has been the subject of much research. It is based on the idea that microfracturing and crack propagation produce elastic wave pulses which are somewhat characteristic of the deformation process. Study of these pulses had led to a further understanding of tensile and shear fracturing, dilatancy, and creep of brittle rocks [1]. However, application of the technique to rock friction is sparse and to friction at high temperature non-existent.

SCHOLZ [2, 3] examined acoustic emissions occurring prior to stick-slip sliding in granite, but limited his study to sawcut samples at room temperature. He found that the level of activity was not reproducible between stick-slip cycles, although it began relatively abruptly during loading and stayed nearly constant as load increased until the sudden stress drop. There was no apparent increase in emission rate prior to sudden

[1]) Department of Earth and Planetary Sciences, Massachusetts Institute of Technology, Cambridge, Massachusetts, USA.

[2]) Present address: Erindale College, University of Toronto, Mississauga, Ontario, Canada L5L 1C6.

slip. The b-values were determined from the relationship of amplitude to emission rate [3], $N(A) = KA^{-(b+1)}$, where $N(A)$ is the number of emissions having peak amplitude A, and K and b are constants, b is typically greater than 1.0 for frictional sliding, rather than the values of less than 1.0 characteristic of brittle fracture. This result implies that the proportion of small to large events is greater during frictional sliding than during fracture. Byerlee and Peselnick [4] recorded acoustic emissions during sliding of cut glass surfaces at room temperature. They were unable to discern the details of the sliding process, but found that the cumulative results of six cycles of loading and unloading gave a b-value of about 0.9.

These studies represent pioneering work in acoustic emission during frictional sliding, but their results are difficult to apply to natural faults for two main reasons. The acoustic emission response from materials is apparently very structure sensitive [5]; the emissions from clean sawcut surfaces may not have the same characteristics as those from gouge-filled faults, perhaps the more common type in the earth. Secondly, the temperatures at depth on natural faults may be considerably higher than in these room-temperature experiments. Do acoustic emissions occur at high temperature and do their characteristics change? In particular, are there differences in emission activity between low-temperature stick-slip and high-temperature stable sliding? To examine these and other questions I applied Scholz's techniques to high-temperate friction experiments. Because of design constraints for high temperature, the sensitivity of the emission-detection apparatus was less than that of Scholz.

2. *Experimental technique*

The high-temperature friction apparatus is described elsewhere [6] and illustrated in Fig. 1. Briefly, a rock cylinder, 3.5 cm long and 1.6 cm diameter, was placed in a thin graphite sleeve and inserted in an annealed copper foil jacket. Titanium carbide and polycrystalline alumina (Lucalox) were placed above and below the rock in the jacket. A furnace surrounded the jacket and the entire assembly was set inside a gas medium pressure vessel. The axial load was applied through a steel piston and measured with a strain-gauge load cell. Because the machine motor produced acoustic noise, the piston was advanced by hand at a nearly constant rate of about 10^{-4} cm/sec, and its displacement was monitored with a time-chart recorder. The temperature was measured with a thermocouple placed at the base of the sample through a 1.6 mm diameter hole in the lower piston and spacers.

For acoustic emission work the problem lay in the location of the detecting transducer. The lack of room, the high temperatures, and the presence of the graphite sleeve prevented bonding the transducer to the side of the sample, as Scholz [2] had done. A simple, yet effective, method of detecting the emissions is shown in Fig. 1. A PZT-4 transducer (compressional mode, 1 MHz) was bonded to the steel piston within a cavity in the upper moving platen about 20 cm from the rock sample. Copper foil provided the electrodes and a piece of thick rubber and a brass backing minimized back reflections. Throughout the experiment the transducer remained unloaded and at room temperature and pressure. The materials between the rock and the transducer were selected to maximize acoustic–impedence matching with the additional constraints of high strength and suitable thermal properties [6]. Despite the mis-matching that remained and despite the separation of rock and transducer, the acoustic emission signals were sufficiently strong for detection and analysis.

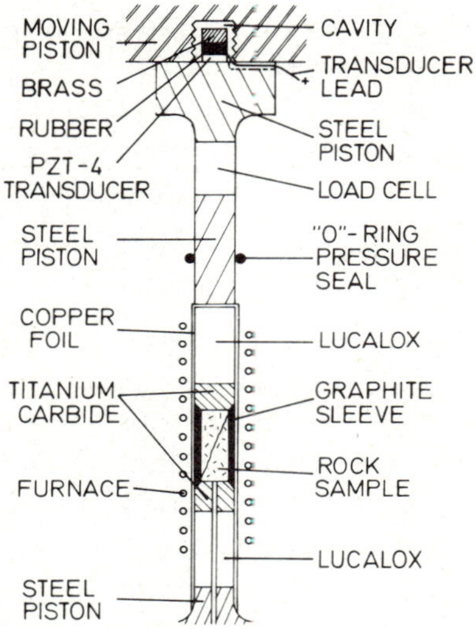

Figure 1
Experimental configuration

The electronics for the signal analysis was similar to that used by SCHOLZ [2]. The emission signal was amplified in three stages using two 40 db amplifiers and the trigger output of the monitoring oscilloscope. Between the first and second stage, the signal was filtered to remove the low-frequency noise picked up by capacitive coupling with the machine, a major source of electrical noise. The high-frequency noise could not be eliminated and limited the sensitivity of the measurements. The acoustic emission pulses, essentially decaying sinusoids, were rectified and shaped [2] and then fed into a multi-channel analyzer (TMC model 102). The analyzer was used in two ways: to count the number of pulses during successive time periods and to count the pulses having peaks within successive amplitude intervals. The latter method determines the *b*-values.

Since the transducer remained outside the furnace and pressure vessel the entire range of pressures and temperatures could be used. This study, however, examined frictional sliding at two pressures, 1 and 2 kbars, and at temperatures from 25°C to 700°C. The procedure used, outlined in detail by STESKY *et al.* [6], was the following. A fault was first formed by fracturing the rock at 500 bars and 25°C. The pressure and temperature were then raised to the desired conditions and the faulted rock loaded until sliding occurred. Rather than have the sliding continue at constant pressure and temperature to examine the effect of displacement, the temperature of sliding with each sample was varied to observe the effect of temperature on acoustic activity, the object of this study. Thus sliding at each set of conditions was continued only until a reasonable estimate of its emission activity could be made. At this point the load was removed, the temperature changed, and the rock reloaded until sliding again occurred. In this way, several measurements could be made with one sample, thus minimizing the effect of fault variability. For clarity, each measurement will be called here a *run* and all measurements for one sample, an *experiment*.

Westerly granite was used in all experiments reported here. Many of the acoustic emission and frictional properties of this rock have been previously studied [2, 3, 6, 7, 9, 17] and its physical characteristics are described elsewhere [2, 6].

3. *Observations during fracture*

The acoustic emissions during fracture at 500 bars, 25°C, were monitored as a check on the technique. SCHOLZ [2] observed that during the fracture of brittle rocks at low pressures two regions of activity occurred: at the low-stress region during initial loading where cracks were closed and at stresses greater than half the fracture strength where new cracks were being formed. As the stress approached the fracture strength the emission rate increased dramatically until fracture. Figure 2 illustrates the emission

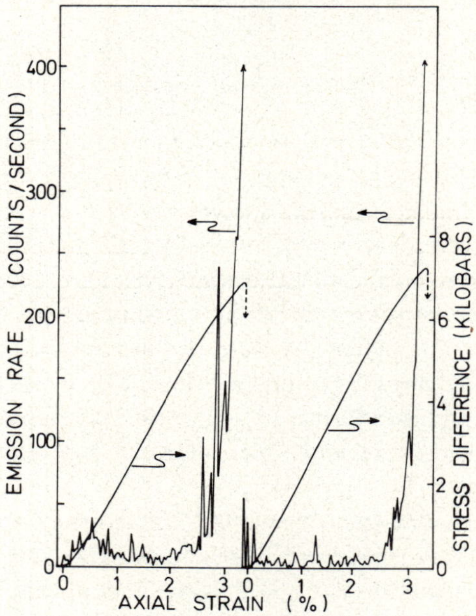

Figure 2
Two fracture experiments in Westerly granite at 500 bars confining pressure and 25°C. Stress and microfracturing rate are plotted against axial strain

rates observed during two typical fracture runs in the present study. The pattern of activity is very similar to that found by Scholz. However, the sensitivity of the apparatus for pulse-height analysis was less than in SCHOLZ's [3] experiments. The number of events detected was lower than Scholz found and, except at stresses greater than half the fracture strength, was insufficient for adequate *b*-value calculation. At stresses above half the fracture strength, all the emissions occurring until just prior to fracture were analyzed together, and thus the stress-dependence of the *b*-value was not determined.

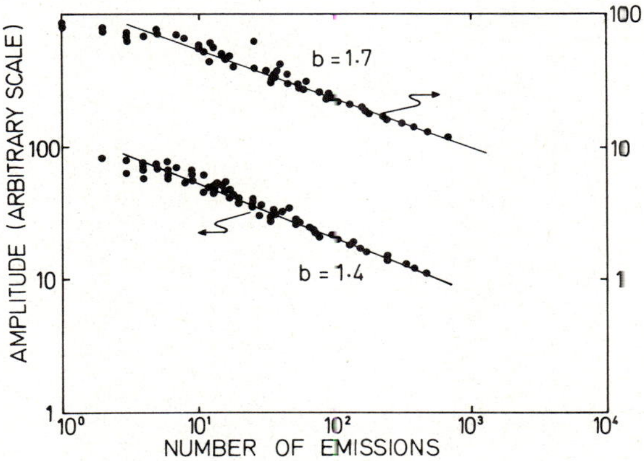

Figure 3

Number of microfracturing events of a given maximum signal amplitude versus amplitude for two fracture experiments in Westerly granite at 500 bars pressure and 25°C. The emissions were accumulated during loading from about half the fracture strength until just prior to failure. The slope of the fitted lines, $1/m$, is calculated from $m = \Delta \log$ (number of events)/$\Delta \log$ (amplitude). The *b*-value is then $b = m - 1$

The results of two fracture runs are shown in Fig. 3. Here we plot the number of events against the maximum trace amplitude on logarithmic coordinates. The low-amplitude range is limited by the ambient noise level. As expected, the measurements follow a power-law relationship and the fall-off in counts at high amplitudes as observed by Scholz is seen here. However, the slope, l/m, of the lines in Fig. 3, or equivalently the *b*-value, defined as $b = m - l$, is consistently higher than in Scholz's study. The *b*-values of four runs ranged from 1.3 to 1.7. SCHOLZ's [3] results suggest that for a rock stressed homogeneously (on a macroscopic scale) the *b*-value should be less than 1.0 and should decrease as the stress level is increased.

The reasons for the high *b*-values in these runs is not known. SCHOLZ [3] suggested that an inhomogeneous stress field could produce *b*-values greater than 1.0. Some stress inhomogeneity was present in our samples. The hole in the end pieces adjacent to the rock likely induced variations in local stress in the region of the rock near the hole. However, whether this anomalous region, representing less than 25% of the total volume of the rock, was sufficient to affect the *b*-values is not known.

4. *Observations during frictional sliding*

Acoustic emission measurements were made at two pressures, one experiment at 1 kbar and several at 2 kbar. The low-pressure emission rates are shown in Fig. 4, along with the stress-shortening curves. Apparently, the acoustic emission response depends considerably on the details of the process, as well as the temperature of sliding. Initially, at 25°C the emission rate was high, with occasional large counts corresponding to small

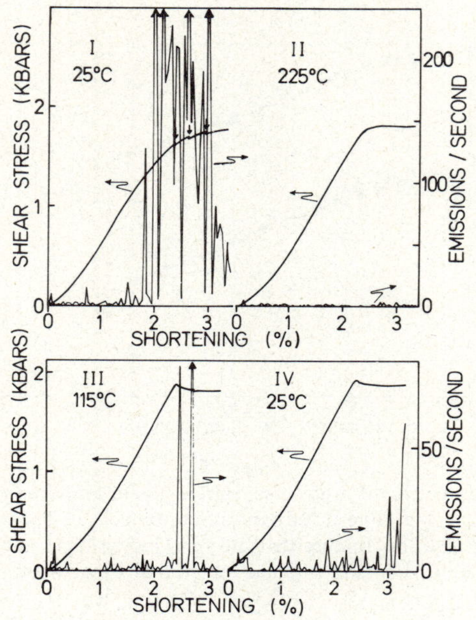

Figure 4
Emission rate and shear stress during frictional sliding in a single faulted sample of Westerly granite at 1 kbar pressure and different temperatures. Note the change in scale of the emission rate between the upper and lower figures. Roman numerals denote the order of the runs. The vertical arrows indicate stick-slips

stick-slips. With displacement, however, the emission rate fell off. At this point the run was stopped and the temperature increased to 225°C. During sliding at this temperature no emissions were detected above the background noise level. The emission rate was slightly higher at 115°C and large counts were detected at two conspicuous small stress-drops. Finally, repeating the room temperature measurement showed that the emission rate was markedly lower than in the first run and that stick-slip did not occur. The emission rate increased with displacement almost to the level at the end of the first run. Unfortunately the experiment was stopped at that point and we do not know whether the increase in emissions was premonitory to a stick-slip event.

Two points are clear from this experiment. The acoustic emission response during frictional sliding at 1 kbar was complex. In general, however, the emission rate was lower during stable sliding and fell off to negligible values at 200°C. To test whether, indeed, acoustic emissions do not occur during high temperature stable sliding or whether the sensitivity of the detecting system was too low, we ran the sliding experiments at 2 kbars. At the higher pressure the local stresses on the fault would be higher and more energy would be available for the emissions.

Figure 5 shows the results of one experiment at 2 kbars. In each run the emission activity began at fairly low stresses and accelerated as the stress increased. Once steady sliding at constant stress began, the emission rate became nearly constant, a behavior

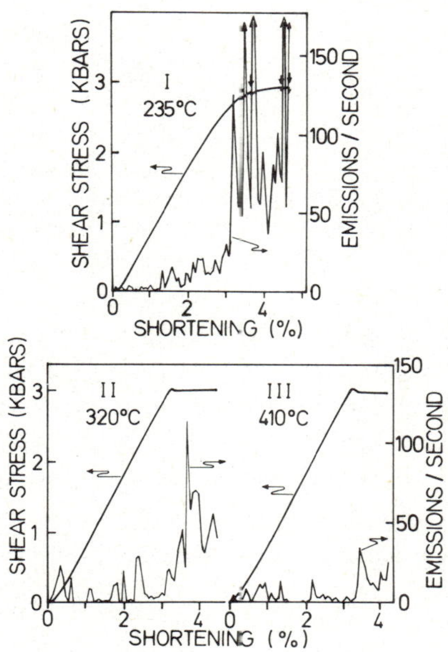

Figure 5
Emission rate and shear stress during frictional sliding in a faulted sample of Westerly granite at 2 kbar pressure. Roman numerals denote the order of the runs. The vertical arrows indicate stick-slips

similar to that observed during the cataclastic deformation of marble [2]. Again, as in the 1 kbar experiment, we see a high emission rate at low temperatures (with large counts during stick-slips) and a decrease in activity with successively higher temperatures. In this case, however, the temperatures ranged upwards from 235°C, the lowest temperature used at 2 kbars. Clearly, emissions do occur at high temperature and during stable sliding.

Figure 6 summarizes the results of several experiments. Here the range of emissions rate during steady sliding at different temperatures are plotted, neglecting the high counts measured during stick-slip events. While there may be some influence of fault geometry and of displacement (as in Fig. 4), the overall trend is a decrease in emission rate with increasing temperature. Even at the highest temperature (660°C) emissions were detected above the background noise level.

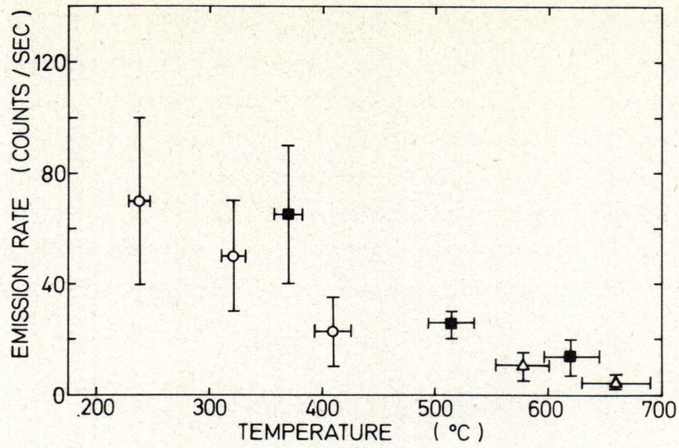

Figure 6
Variation of acoustic emission rate with temperature during stable sliding in Westerly granite at 2 kbar pressure. The vertical bars indicate the range of rates observed and the horizontal bars denote the variation of temperature over the fault surface. Several measurements were made with each sample, the samples being distinguished by different symbols

To test whether the characteristics of the emissions changed with temperature, the relation between emplitude and emission rate during sliding was studied. In many cases, sufficient emissions were produced prior to reaching steady sliding to examine the pre-sliding behavior, although the detailed effect of stress on *b*-value could not be discerned. Figure 7 shows the measurements for one experiment, plotted as in Fig. 3. As for fracturing, a power-law relationship between amplitude and count-frequency seems to hold. The *b*-values, calculated from the slope, seem to be higher than those for fracture.

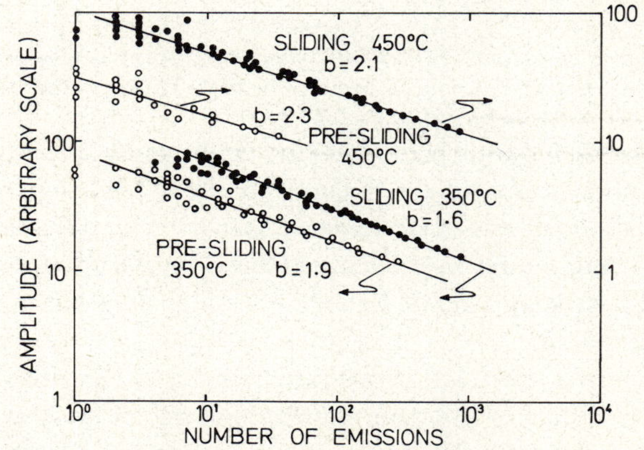

Figure 7
Microfracturing frequency versus maximum signal amplitude for two frictional sliding runs in Westerly granite at 2 kbar pressure. The open symbols indicate the emissions during loading up to stable sliding, while the closed symbols refer to emissions occurring during stable sliding. *b* is calculated as in Fig. 3

The results of several experiments are shown in Fig. 8. Measurements were not made below 300°C because of the unpredictability of the stick-slip events and the influence of machine vibration during the stress drops. For example, note the low *b*-value obtained in a 370°C run where a large stress drop occurred due to jacket failure. Acoustically, stick-slip consisted of a large-amplitude burst of emissions with a total ring-down time (several milliseconds) larger than the dead-time between counts (40 to 1000 μsec). At the highest temperatures the low emission rate prevented separation of pre-sliding and sliding events and all are counted together. Also included in this plot are measurements made prior to stick-slip on sawcut samples at 2 kbar and 25°C [3].

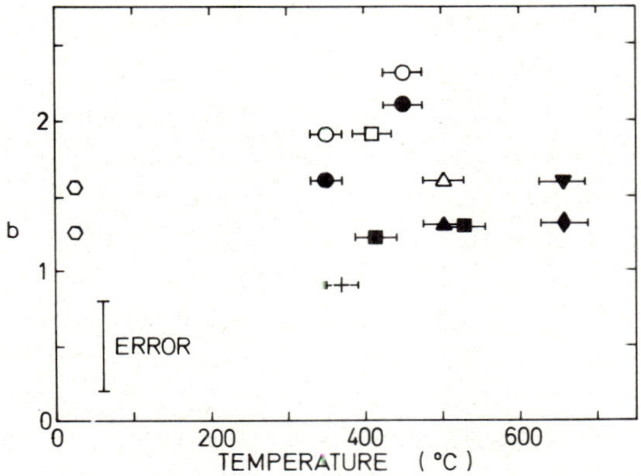

Figure 8

Variation of *b*-value, calculated from the slope of the frequency–amplitude curve, with temperature during frictional sliding of Westerly granite at 2 kbar pressure. Except for the two at room temperature, like symbols indicate measurements made with the same sample. The open symbol denotes *b* for pre-sliding emissions and the closed symbol denotes *b* for stable sliding emissions. The two measurements at 25°C are from SCHOLZ [3]. The vertical error bar has been visually estimated from plots such as in Fig. 7 and may be larger for the pre-sliding emissions. The horizontal bar indicates the range of temperature over the fault surface

Several features are evident in Fig. 8. The *b*-values are high and in general agreement with Scholz's measurements. There seems to be no overall change in *b*-value with temperature within the scatter of the measurements, although there is a weak suggestion, from individual experiments, of an increase in *b* with temperature. Lastly, the emissions during initial loading seem to show a small but consistently higher *b*-value than to those during steady sliding.

5. *Discussion*

Our understanding of the physical mechanisms during high-temperature frictional sliding is meager. BRACE and BYERLEE [7] and, more recently, STESKY *et al.* [6] showed that for all rocks studied the sliding was stable at high temperature. Can the transition from stick-slip to stable sliding at high temperature be attributed to the onset of plastic deformation or, indeed, local melting on the fault surface? At low temperature it is reasonable to assume that most of the emissions in Westerly granite resulted from the brittle cracking of asperities and gouge particles. Such brittle cracking seems to occur even up to 700°C at 2 kbar, as evidenced by the similarity of b-values. That plastic deformation may also have occurred cannot be ruled out. Indeed, BOLAND and HOBBS [8] showed evidence of concurrent fracturing and plastic deformation during high temperature creep of peridotite. Also, acoustic emissions have been observed during the plastic deformation of metals [5]. SCHOLZ [2] similarly observed emissions during the plastic deformation of marble at high confining pressure, although only during yielding and with amplitudes much lower than emissions produced by brittle processes. It seems unlikely that such plastic deformation can account for the emissions found in this study.

The variation of acoustic activity with style of sliding seems to depend on pressure. At 1 kbar stick-slip produced a large number of emissions, while there was almost no activity during stable sliding. That this difference cannot be entirely due to temperature can be seen by comparing the two 25°C runs (Fig. 4). In contrast, at 2 kbar there was little difference in activity (neglecting the high counts during stress drops) between stick-slip and stable sliding other than that due to differences in temperatures as in Fig. 6.

This pressure dependence is puzzling and may be indicative of the mechanisms of gouge deformation. For example, grain crushing may be more significant at higher pressure, producing high emission rates even during stable sliding. At lower pressure, stable sliding may result more from grain rolling or sliding, mechanisms perhaps less conducive to acoustic activity.

The cause of the decrease in emission rate with temperature is not known. During loading of a faulted sample, the emission rate depended on the stress level. But for stable sliding the change in stress with temperature was fairly small (less than 10% between 250° and 650°C) and cannot explain the order of magnitude change in emission rate. Two possible explanations seem most likely: either the decrease in number of emissions is real or the attenuation of the stress waves increases with temperature. If the emissions reflect the occurrence of brittle cracking, then a decrease in the proportion of brittle to plastic deformation will produce a simultaneous decrease in acoustic emission rate. On the other hand, an increase in attenuation with temperature would cause the amplitude of the large number of small events to fall below the noise level so that these events would not be counted. Since all emissions would be attenuated by the same proportion, the b-value would not change with a change in attenuation. At present we cannot distinguish between these possibilities since we have little knowledge of the characteristics of brittle deformation nor of stress-wave attenuation in faulted rock at high temperature.

The pre-sliding behavior is also significant. As shown in Fig. 5, the emissions began early in the loading history and the emission rate increased continually with stress until steady sliding occurred. This behavior differs from that of initial fracturing and of sliding on sawcut samples [2]. In intact rocks the emission rate began to increase at about half the ultimate strength, a higher relative stress level than in faulted rocks (Fig. 2). On the other hand, in sawcut samples the emission rate increased abruptly at a high stress level, above half the ultimate strength, and remained nearly constant until sudden stick-slip [2]. These results indicate fundamental differences between the mechanical behavior of faulted rock and that of intact or sawcut rock.

From an analysis of the time-dependence of high-temperature friction in faulted rocks [9], I observed that transient frictional phenomena occurred when the deformation conditions changed. In that study, changes in the applied boundary displacement rate produced time- and displacement-dependent changes in shear strength and sliding rate. Similarly, HANDIN *et al.* [10] observed transient changes in shear strength following pressure changes. I suggested that during steady-state sliding an equilibrium fault structure is developed that depends on temperature, pressure, sliding rate, and possibly other variables. If conditions change, a new equilibrium structure will be established, but only after some finite displacement. The change in structure may produce transient mechanical behavior. By understanding the fault structure and the factors that affect it, we may be in a better position to explain transient frictional phenomena in the laboratory and in the field. Indeed, stick-slip and perhaps earthquakes may be such transients.

The acoustic emission results give us some insight into the mechanisms of deformation prior to reaching steady-state sliding. In this case, the altered conditions were mainly pressure and temperature. The deformation seems to begin at very low stresses, considerably lower than can be detected by measuring the fault displacement. The intensity of the deformation, as indicated by the emission rate, increased until steady sliding was reached. One would expect then, according to this hypothesis, that if the faulted sample was unloaded after sliding and then reloaded without changing pressure and temperature, little or no emissions would occur prior to reaching steady sliding. Preliminary work by the author indicated that this supposition seems to be correct at room temperature, but this aspect was not examined in the present study. Often a slight peak in shear strength occurred prior to steady sliding, similar to the stress-peaks observed following changes in boundary displacement rate [9] and pressure [10]. However, no significant increase in emission rate corresponding to the stress drop was observed (Fig. 5). The *b*-value analysis indicates that *b* may be slightly greater for the emissions produced during loading than during steady-state sliding. Thus the pre-slip deformation consists of a larger number of small emissions, perhaps related to the readjustments of the fault to a new equilibrium structure.

6. *Conclusions*

Acoustic emissions have been detected and analyzed during high temperature frictional sliding in dry Westerly granite. Although there are ambiguities in the interpretation of the results, the following points seem clear:

1) Emissions were produced during frictional sliding at 2 kbar confining pressure and at all temperatures up to 660°C.
2) During loading, the emissions began at low stresses, below half the ultimate stress, and increased continuously until the sliding stress was reached.
3) During sliding at 2 kbar confining pressure, the emission rate was nearly constant, although in the case of stick-slip sliding, stress drops produced high emission counts.
4) The emission rate during sliding decreased as the temperature was increased. The reason for this decrease is not known but may represent a real decrease in emissions or may result from an increase in attenuation in the fractured rock or piston materials at higher temperature.
5) At 1 kbar confining pressure, the emission rate depended considerably more on the details of the deformation process than at higher pressure. The emission rate decreased rapidly with displacement and was much less during stable sliding than during stick-slip. In contrast, at 2 kbar, there was little difference in emission rate between stick-slip and stable sliding, ignoring the high counts during stress drops.
6) The results of the frequency–magnitude analysis are less reliable. However, the values of b, the slope of the log–log plot of pulse amplitude versus number of counts, are comparable to those obtained in a previous study. They range from about 1.2 to 2.1 and show little change with temperature, within the scatter of the measurements. The b-value for emissions during loading is slightly, but consistently higher than that during steady sliding.

The implications of these results for our understanding of natural faults are two-fold. First, acoustic emissions are produced at high temperature. Their characteristics are similar to those of emissions at low temperature and, by inference, are produced by brittle processes. Thus, we find no evidence that the results of Mogi [11, 12] and of Scholz [2, 3, 13, 14] do not hold at high temperature, providing the deformation processes do not change. Second, unlike sawcut samples, faulted rocks produce emissions at low stresses at a rate which increases continuously with load until sliding occurs. This acceleration of activity prior to slip may be similar to that often observed on natural faults [15], although it is not certain that the mechanisms are the same.

Acknowledgements

This work was carried out under U. S. Geological Survey contracts 14-08-0001-13229 and 14-08-0001-14551. I thank W. F. Brace for his encouragement and supervision and for making his laboratory available to me. Discussions with C. Goetze, K. Hadley, and D. Fitterman were helpful. I also thank P.-Y. F. Robin for reviewing the manuscript.

References

[1] H. R. Hardy, *Application of acoustic emission techniques to rock mechanics research*, Am. Soc. Test. Mat., Spec. Tech. Publ. 505 (1972), 41.

[2] C. H. Scholz, *Microfracturing and the inelastic deformation of rock in compression*, J. Geophys. Res. *73* (1968), 1417.

[3] C. H. Scholz, *The frequency–magnitude relation of microfracturing in rock and its relation to earthquakes*, Bull. Seismol. Soc. Am. *58* (1968), 399.

[4] J. D. Byerlee and L. Peselnick, *Elastic shocks and earthquakes*, Naturwissenschaften *2* (1970), 82.

[5] H. L. Dunegan and A. T. Green, *Factors affecting acoustic emission response from materials*, Am. Soc. Test. Mat., Spec. Tech. Publ. 505 (1972), 100.

[6] R. M. Stesky, W. F. Brace, D. K. Riley and P.-Y. F. Robin, *Friction in faulted rock at high temperature and pressure*, Tectonophysics *23* (1974), 177.

[7] W. F. Brace and J. D. Byerlee, *California earthquakes: why only shallow focus?*, Science *168* (1970), 1573.

[8] J. N. Boland and B. E. Hobbs, *Microfracturing processes in experimentally deformed peridotites*, Int. J. Rock Mech. Min. Sci. *10* (1973), 623.

[9] R. M. Stesky, *Time-dependence of friction in faulted rock at high temperature and pressure*, in preparation (1974).

[10] J. Handin, M. Friedman, J. Logan, G. M. Sowers and D. W. Stearns, *Mechanical properties of rocks affecting earthquake generation*, Ann. Prog. Rept. No. 1, ARPA Contract No. 14-08-0001-12723, Office of Earthquake Res., US Geol. Survey, Menlo Park, California (1972), 147 pp.

[11] K. Mogi, *Study of the elastic shocks caused by the fracture of heterogeneous materials and its relation to earthquake phenomena*, Bull. Earthquake Res. Inst. *40* (1962), 125.

[12] K. Mogi, *Magnitude frequency relation for elastic shocks accompanying fractures of various materials and some related problems in earthquakes*, Bull. Earthquake Res. Inst. *40* (1962), 831.

[13] C. H. Scholz, *Experimental study of the fracturing process in brittle rock*, J. Geophys. Res. *73* (1968), 1447.

[14] C. H. Scholz, *Mechanism of creep in brittle rock*, J. Geophys. Res. *73* (1968), 3295.

[15] R. L. Wesson and W. L. Ellsworth, *Seismicity preceding moderate earthquakes in California*, J. Geophys. Res. *78* (1973), 8527.

(Received 12th November 1974)

Pageoph,
Vol. 113 (1975) Birkhäuser Verlag, Basel

The Study of Precursors of Failure Under Biaxial Compression

By G. A. SOBOLEV[1])

Summary – The process of failure preparation in model material has been studied. To constrain the deformation and fractures to a certain zone within the sample the biaxial compression tests have been chosen. During the deformation the cross-like zone, consisting of a great number of small cracks, appeared followed by the development of macrofracture in the central part of the specimen. The strong minimums in velocity of ultrasonic waves, changes of strain rate, and self-electric potential reliably pointed to the approaching moment of failure.

At the Institute of Physics of the Earth, a programme of laboratory investigations is being carried out to examine the basic laws of the fracturing process for application to the physics of earthquakes. The programme includes investigation of the development of single fractures, interaction of fractures, and formation of macrofractures. Measurements are conducted under different states of stress and with different materials (VINOGRADOV *et al.* [1]; SCHAMINA *et al.* [2]; VOLAROVICH *et al.* [3]). In this paper we consider only one series of experiments made by the author with Z. J. Stakhovskaja and A. V. Kolzov.

To constrain the deformation and fractures to a certain zone within the sample we used the biaxial compression tests. The experimental arrangement consisted of a massive frame and piston. The specimen was tightly adjusted to the mandrel so that under vertical loading in one horizontal direction, displacement was held at zero, leading to the development of stress in this direction. Two faces of the specimen were free. This method of loading, together with friction on the confined faces of the specimen, leads to a complex stress state. Plexiglass was used as a model material, the transparency of the plexiglass permitting the interior of the sample to be observed during an experiment. The specimens had the shape of a parallelepiped and dimensions of $40 \times 40 \times 30$ mm. The tests were conducted under slow loading and lasted from one to two hours in order to prolong all the processes going on within the specimen and to permit termination of the test at the first appearance of visible fractures. Before the experiment the constrained sides of the specimen were marked with a 0.5×0.5 cm^2 grid.

During the deformation two cross-like zones consisting of a great number of small cracks would appear followed by the development of a macrofracture in the central

---

[1]) Institute of the Physics of the Earth, B. Gruzinskaya 10, Moscow, USSR.

part of the specimen. In Fig. 1a one can see the photo of a partially fractured specimen. We noted that macrofracture appears inside the specimen which is, to a certain extent, analogous to processes in the earth. Figure 1b shows full failure of the specimen. From the displacement of grid lines the local deformation can be estimated. The force, total strain, and velocity of longitudinal ultrasonic waves along the axis of main loading were measured during the test. The self-electric potential was measured

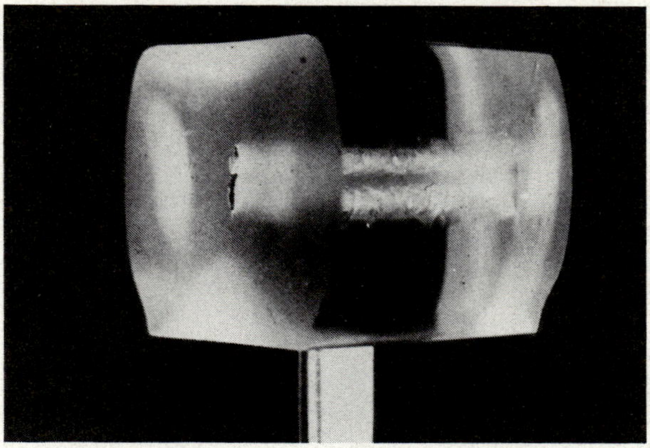

a

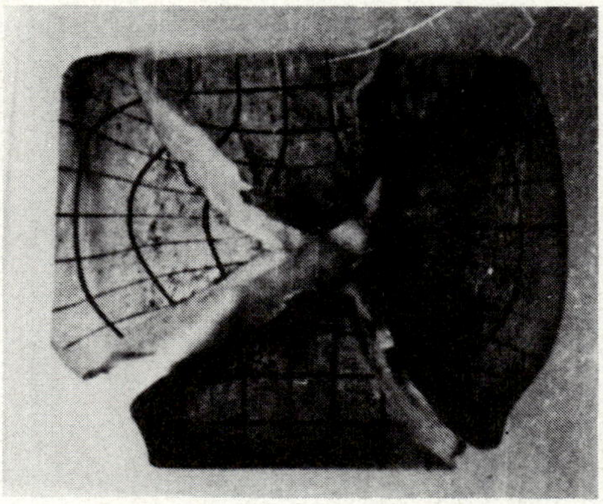

b

Figure 1
The failure of specimens of plexiglass under biaxial compression: a) internal macrofracture; b) the full failure

between an electrode fixed to one of the free faces of the specimen (Fig. 2) and the ground. Longitudinal mode ceramic transducers (ZTS-18), with a natural resonant frequency of approximately 760 KHz, were used for the ultrasonic measurements. Total deformation was registered with the help of deformometers in the form of rings with strain-gauges attached to them.

Figure 2 shows the results of one of the experiments. The loading of the specimen was gradually increased. The strain rate of the specimen in the first part of the experiment was constant 10^{-4} sec^{-1} and then was sharply increased to 10^{-3} sec^{-1}. It is characteristic that in all experiments conducted a period of slower growth of total deformation was observed just prior to complete failure. The distortion of the net on the sides of the partially fractured specimens points to the strong constriction of strain in the zone of future failure.

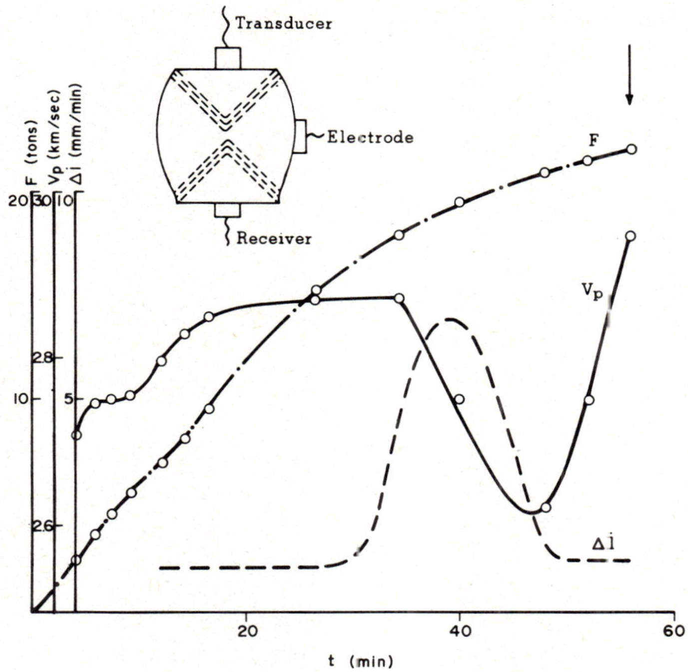

Figure 2
The changes of the load F, total longitudinal strain rate $\Delta \dot{l}$, and velocity of longitudinal ultrasonic wave V_p during experiment under biaxial compression. The arrow points to the moment of macrofracture

The most interesting feature of the change in velocity of ultrasonic waves was the existence of strong minimums at the time of increasing total strain rate. The reduction of velocity was as much as 7%. Just before the moment of failure the velocity increased and exceeded all previous values.

Another experiment is shown in Fig. 3. The changes of velocity, strain, and self-electric potential are plotted as a function of failure force. During the aforementioned variations of velocity and strain the electric potential also changed rapidly. In a number of experiments, high-frequency vibrations of the potential just before the moment of macrofracture formation were also observed.

On the whole, the changes of velocities, strain rate, and self-electric potential reliably pointed to the approaching moment of failure. From comparison of the results

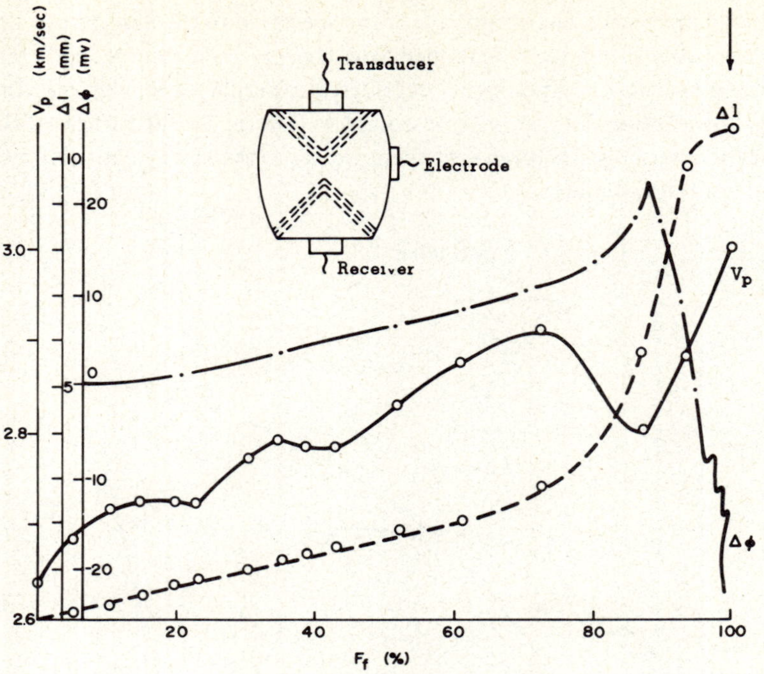

Figure 3
The changes of total longitudinal strain Δl, velocity of longitudinal ultrasonic wave V_p, and electric potential ϕ depending of the per cent of failure force, F_f. The arrow points to the moment of macrofracture

of the measurement of these parameters and visual analyses of specimens, a preliminary conclusion may be drawn that anomalous changes are caused by accelerated development of the process of microfracturing (quasi-plastic deformation). The possible reason for recovery of velocity and strain rate is the process of strain concentration in the cross-like narrow zones of failure and unloading other parts of the sample. This should lead to recovery of the integral properties of the material.

The question is, how are these results applicable to the natural processes going on inside the earth. To a first approximation, the mechanical behaviour of the material of the source zone can be described by a rheological equation which includes at least strain, stress, and rates of strain and stress:

$$P(\varepsilon, \sigma, \dot{\varepsilon}, \dot{\sigma}) = 0 \tag{1}$$

To model such a process in the laboratory it is necessary to satisfy simultaneously the condition of similarity for the rigidity and Young's modulus, and for the coefficient of viscosity (NASAROV [4])

$$E_M = \frac{\beta}{\gamma} \cdot E_N; \qquad \mu_M = \frac{\beta}{\gamma} \cdot \mu_N; \qquad {}_M = \frac{\beta}{\gamma} \cdot \tau\eta_N, \tag{2}$$

where β, γ, and τ are constants of similarity of stress, strain and time, respectively, i.e.,

$$\sigma_M = \beta \cdot \sigma_N; \qquad \varepsilon_M = \gamma \cdot \varepsilon_N; \qquad t_M = \tau \cdot t_N \tag{3}$$

The subscript M means 'model' and N refers to 'nature'. To maintain the similarity during the process of fracturing (which is the most interesting for the physics of earthquakes), it is necessary to satisfy the equations of dynamic similarity

$$\xi_M = \frac{\beta}{\alpha} \cdot \xi_N; \qquad \tau = \alpha \sqrt{\frac{\delta \cdot \sigma}{\beta}} \tag{4}$$

where α is the geometric constant, δ is the constant of density, and ξ is weight. It is also desirable to preserve the similarity conditions on the sides of the fault, taking into account cohesion and friction. Let us especially dwell on the modelling of time phenomena. Laboratory experiments require that such tests should last for not more than several hours or days. This means that the constant of time is $\tau = 10^{-5}$. To observe similarity when displacements are great, the constant of strain γ is to be equal to 1, and high-pressure equipment cannot give $\beta \gg 1$. Therefore, we find that $\tau = \alpha = 10^{-5}$, i.e., the model material should be fine-grained. On the other hand, the viscosity coefficient of the model material η_M should be much lower than the viscosity of natural rocks. Rock tests in the laboratory are of too short duration to demonstrate the rheological behaviour of the same rock in the earth. From this point of view, it is preferable to use model materials equivalent rheological behaviour.

The factor of time has been insufficiently considered in previous laboratory experiments dealing with the study of precursors. Most measurements were made with brittle rocks under a relatively high stress rate, i.e. without satisfying the previously mentioned conditions of similarity. Perhaps the conditions of the experiments did not allow observations of the recovery of physical properties, including the changes of velocities of ultrasonic waves just before failure.

If the fracturing process is sufficiently slow it will be possible for the material to pass through the development of many small cracks in the whole volume and for the eventual concentration of strain to occur in the narrow zone of expected failure. Qualitatively, similar results can be obtained with other materials if the conditions of the experiment (scale of time and material features) are properly chosen. The experiments with plexiglass are not unique in this sense.

References

[1] S. A. Vinogradov *et al.*, Seismicheckaja energija pri rasruschenii obraszov pod postojannoj nagrusko, Izvertija AN SSSR, Physika Zemly. No. 3, 1973.

[2] O. G. Schamina *et al.*, Issledovanie prosessa podgotovky treschin, Izvestia AN SSSR, Physika Zemly, No. 8, 1973.

[3] M. P. Volarovich *et al.*, Physico-mechanicheskije svojstva porod i mineralov pri visokom davlenii i temperature, B Nauka, Moscwa, 1971.

[4] A. G. Nasarov, O mechanicheskom podobii tviordich deformieruemich tel. Izd. AN Arm. SSR, Erevan, 1965.

(Received 22nd November 1974)

Pageoph,
Vol. 113 (1975)

Birkhäuser Verlag, Basel

Magnetism of Rocks and Volumetric Strain in Uniaxial Failure Tests

By RANDOLPH J. MARTIN III and MAX WYSS[1])

Summary – The relation between remanent magnetization and volumetric strain for gabbro samples stressed in uniaxial compression inside a near zero-field μ-metal shield has been examined. For samples with an induced IRM parallel to the axis of compression, remanent magnetization decreased linearly up to the onset of dilatancy. As increased stress produced additional dilatancy, the variation of remanent magnetization became nonlinear, and the stress dependence continually decreased until the rock failed. Stress cycling with the peak stress augmented for each cycle produced a continuous decrease in the zero stress value of the IRM although an appreciable amount of recovery was observed during unloading. When the sample was loaded in constant stress increments after the onset of dilatancy and held for several minutes at each level, time-dependent variations in remanent magnetization coincided with time-dependent increases in inelastic volumetric strain. In general as the inelastic creep rate increases, the rate of change in remanent magnetization increases. These results suggest that dilatancy related effects of the intensity of rock magnetization should be observed in magnetic rocks in epicentral regions prior to earthquakes and may serve as both long- and short-term precursors.

1. Introduction

Precursory changes in the physical properties of rock masses near earthquakes have been reported by many investigators. Perhaps the most widely observed effect is the decrease in V_p/V_s ratio. NUR [1] and others have suggested that the observed premonitory variations in travel time were due to stress-induced increases in crack porosity. SCHOLZ *et al.* [2] developed a dilatancy model for earthquakes based on a correlation between rock properties that showed pronounced changes as dilatancy progressed in laboratory experiments and similar effects observed in tectonic regions prior to earthquakes. For example, BRACE and ORANGE [3] found that the resistivity of water-saturated samples loaded to failure in non-hydrostatic compression decreased several orders of magnitude parallel to the axis of greatest compression. BARSUKOV [4] reports decreases in resistivity prior to several intermediate magnitude earthquakes in Russia. Thus, many of the anomalous changes that precede earthquakes can be related to time-dependent dilatancy of rocks in the epicentral region and serve as precursors to impending seismic events.

[1]) Cooperative Institute for Research in Environmental Sciences, University of Colorado/NOAA, Boulder, Colorado 80302.

If dilatancy produces predictable changes in rock properties prior to earthquakes, then it should be a relatively simple matter to monitor, say V_p/V_s ratio, and thus gain some advance warning of destructive shocks. Obviously it is not that simple, possibly not all earthquakes are accompanied by V_p/V_s anomalies. It seems prudent then to study all the possible effects of dilatant behavior and measure as many parameters in the field as possible to minimize the possibility of an unexpected event in a populated area. One effect that has not received too much attention, in the laboratory at least, is the effect of dilatancy on the magnetism of rocks.

As early as the last century it has been said that geomagnetic field changes precede earthquakes. In his summary of such claims, RIKITAKE [5] showed that after the introduction of precision instruments, the amplitude of the anomalies decreased drastically. Two of the premonitory anomalies measured in Japan are very convincing; that before the Niigata earthquake (FUJITA [6]) and that before the Shizuoka earthquake (TAZIMA [7]). The magnitudes of these anomalies were approximately 10 and 35 gammas. Recently WYSS [8] found that the Sitka earthquake may have been preceded by a 20 gamma anomaly, and JOHNSTON [9] reported a 2 gamma change prior to the 28 November, 1974, earthquake in Hollister, California.

Theoretical models for tectonomagnetic effects (SHAMSI and STACEY [10], NAGATA [11]) are all based on the assumption that magnetization decreases linearly with stress. This approach is adequate to explain tectonomagnetic effects due to elastic rebound following rupture. A good example may be the magnetic step which was caused by the Cannikin explosion (HASBROUCK and ALLEN [12]). If, however, precursors are caused by dilatancy, one has to reconsider the explanation of premonitory magnetic field changes.

OHNAKA and KINOSHITA [13] tested basaltic rocks in uniaxial compression and noted a decrease in magnetization with increasing stress when the magnetic vector was parallel to the axis of compression. These results suggest that local variations in stress will effect a corresponding change in remanent magnetization and corresponding changes in the magnetic anomaly associated with it. No simultaneous strain measurements, however, were made during these tests. Thus, the effect of dilatancy on the characteristic behavior of remanent magnetization is not clear. It seemed worthwhile then to examine changes in the intensity of remanent magnetization during uniaxial compression tests. Furthermore it is of interest to determine if any time-dependent changes in remanent magnetization coincided with time-dependent dilatancy at constant stress. With these problems in mind a set of experiments was initiated.

2. Experimental procedure

Right circular cylinders of Gabbro 2.54 cm in diameter and 7.12 cm long were tested in uniaxial compression. Changes in rock magnetization and volumetric strain were recorded as a function of the applied force until the rock failed. The experimental arrangement is shown schematically in Fig. 1. Magnetic changes were measured with flux gate magnetometers. One instrument had a sensitivity of 10 gammas and the other 1 gamma. The probe detected changes in the component of the magnetic vector parallel to the probe axis. The data are presented in relative magnetic units which is simply the reading in gammas on the flux gate magnetometer 4.5 mm from the surface of the sample. The cylindrical probe, approximately 7 cm long and 1.5 cm in diameter, was taped on to the endcaps of the sample with its axis parallel to the axis of the sample.

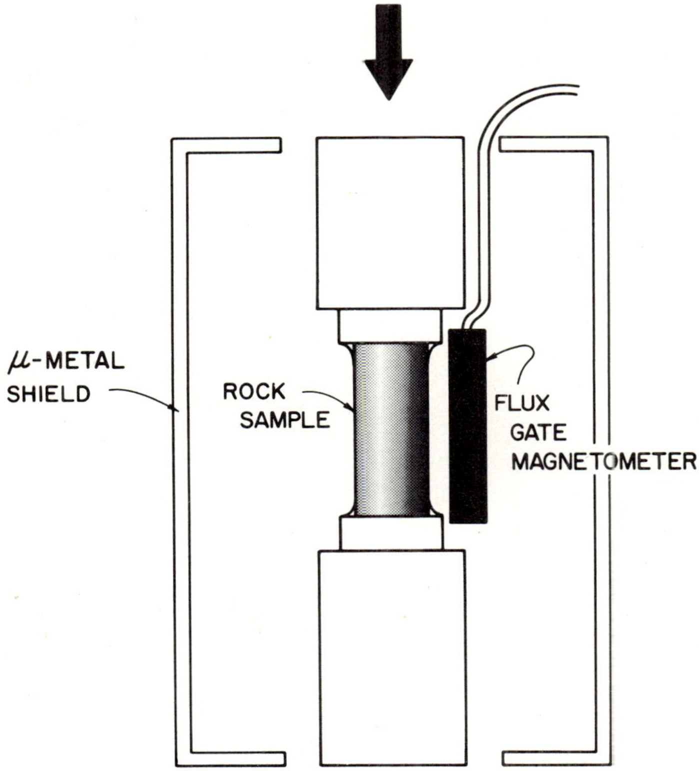

Figure 1
Schematic diagram of experimental setup. The gabbro sample with stainless steel endcaps is placed in a near zero field μ-metal shield. Stainless steel spacers are used to position the sample in the center of the shield. A flux gate magnetometer is positioned with its axis parallel to the axis of the sample. The arrow indicates the direction of loading

All the tests were conducted inside a μ-metal shield that attenuated the external magnetic field. Using this arrangement, only the effect of stress on the remanent magnetization was observed; no susceptibility effects biased the observations. In all but one experiment, a saturation isothermal remanent magnetization (IRM) of intensity 2.3×10^{-1} emu/cc was induced parallel to the axis of the sample. The exception was a specimen with a natural remanent magnetization (NRM) with an intensity of 1.5×10^{-2} emu/cc inclined 39° to the axis of the cylinder.

The sample was compressed between two non-magnetic stainless steel spacers, one resting on the base plate of the press and the other coupled to the piston of the hydraulic cylinder. Dogbone shaped samples were used with stainless steel endcaps and epoxy fillets similar to those described by MOGI [14]. Strain gages were used to measure the axial and circumferential strains. Volumetric strains of 10^{-5} were easily resolved using this arrangement. The stress on the sample was calculated from the pressure in the hydraulic cylinder measured with a pressure gage. The reported absolute stress values are accurate to within 3%.

To ensure that the observed changes in magnetization were not due to the press and stainless steel within the μ-metal shield, two calibration samples were run; one with a lucolux sample and the other with a marble sample. For the calibration runs the fluxgate probe was in the same position as it was for the tests with the gabbro samples. The magnetic field decreased linearly with increasing load by 3 relative magnetic units per 100 bars. This decrease was reversible, and repetition gave the same results. These changes were much smaller than the changes due to the magnetic samples, and therefore the results presented below have not been corrected.

3. Experimental results

The results of a typical experiment are shown in Fig. 2. Here the change in the component of the induced isothermal remanent magnetization parallel to the axis of the sample and the volumetric strain are plotted as a function of stress. The magnetization decreases with increasing stress. The slope of the magnetization versus stress curve is linear to approximately 1200 bars. The slope then becomes nonlinear; the rate of decrease in magnetism with increasing stress was appreciably less than it was below 1200 bars. It is interesting to note that the onset of dilatancy also occurred at 1200 bars.

In several experiments, again with the magnetic vector parallel to the compression axis, the stress was cycled; that is the stress was applied and then removed. During each cycle the peak stress was augmented. Upon unloading, the magnetization never returned to its initial value prior to the experiment. In fact, with each cycle the unstressed intensity continually decreased as shown in Fig. 3. The locus of points for the magnetization at the peak stress of each cycle, however, defines a curve similar to that in Fig. 2. When the maximum stress for each cycle was not augmented, the magnetization appeared to be reversible. For example, Fig. 3 shows five cycles where the stress was varied between 1720 and 2630 bars. The measured values of the magnetization retraced the same line with each stress cycle.

Figure 4 shows the relation between time-dependent inelastic volumetric strain and changes in the NRM vector. The stress was incremented in approximately 175 bar steps and held constant for several minutes at each value. At stresses beyond the onset of dilatancy the inelastic volumetric strain rate increased with each constant stress

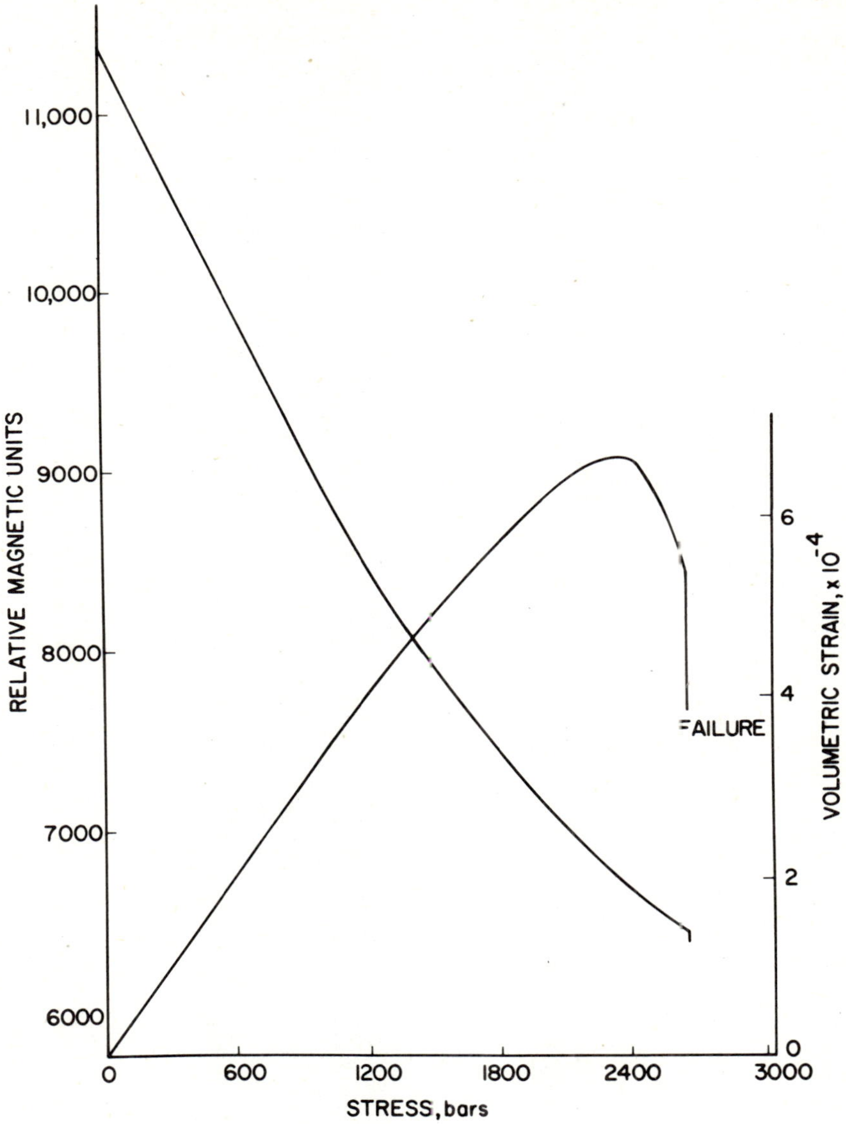

Figure 2
The effect of stress on remanent magnetization initially parallel to the compression axis and volumetric strain is plotted for a uniaxial compression test. Remanent magnetization (IRM) is reported in relative magnetic units which is the reading on the flux gate magnetometer separated 4.5 mm from the surface of the sample. The change in slope of the remanent magnetization vs. stress curve coincides with the point at which the rock ceases to behave linearly elastically at approximately 1200 bars; i.e., the onset of dilatancy. A decrease in volume is positive in this figure

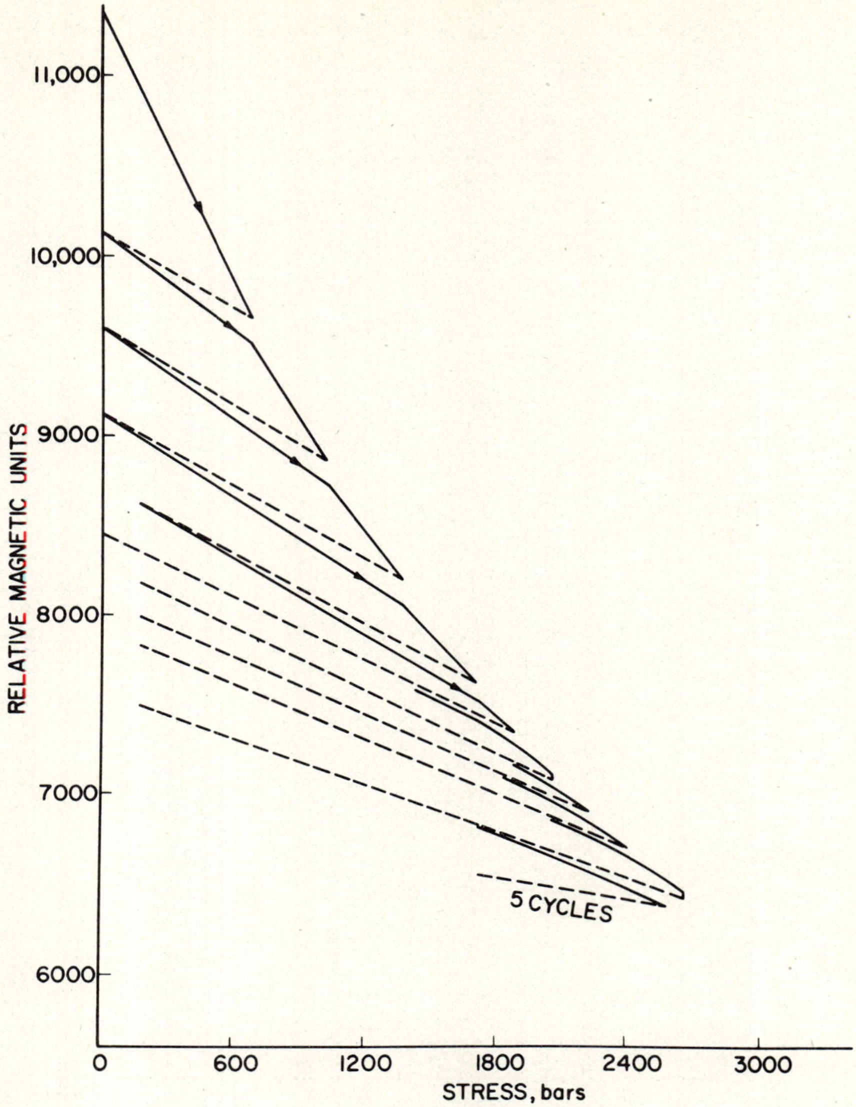

Figure 3
The effect of stress cycling on remanent magnetization is shown for a gabbro sample with an induced IRM parallel to the compression axis. Stress cycling continually reduces the zero stress value of remanent magnetization. Partial unloading and reloading from 2630 bars to 1720 bars was completely reversible for five cycles

increment. Similarly, as the rock deformed at constant stress, the magnetization also continued to increase with time. Furthermore, as the creep rate increased, the rate of change in magnetization increased. All the samples that were studied showed this effect. The data shown in Fig. 4 are from the sample with natural magnetization. For this sample the magnetic vector was inclined 39° to the compression axis. As the stress was applied, the apparent magnetization increased.

The combined effects of stress cycling and creep are shown in Fig. 5. The magnetization was parallel to the axis of compression. Cycling has the effect of reducing the strength of the sample. The last five cycles of a test run are shown. The sample was loaded to 2430 bars and the load maintained for four minutes, during which time the magnetization decreased 35 relative magnetic units (0.5 %). The load was increased to 2500 bars and held constant for 3.5 minutes while the magnetization decreased 30

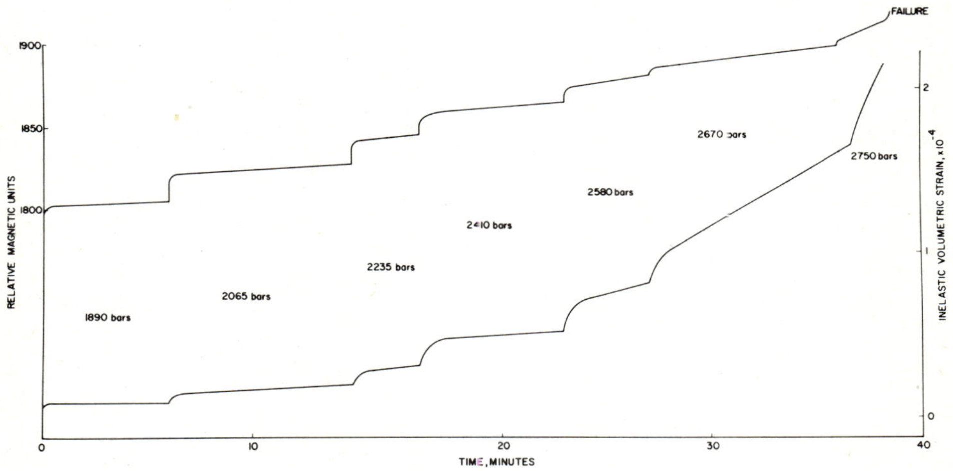

Figure 4
The relation between time-dependent changes in remanent magnetization (NRM) and inelastic volumetric strain at various constant stress levels is shown. The magnetic vector was initially inclined 39° to the compression axis. Time-dependent changes in NRM coincide with time-dependent increases in inelastic volumetric strain. The apparent increase in NRM with stress is probably due to rotation of the magnetic vector into the compression axis. Only the component of magnetization parallel to the sample axis is measured

relative magnetic units. The stress was totally removed and then increased to 2500 bars once again. The magnetization decreased 20 relative magnetic units in several minutes. After two rapid cycles to stresses of 2240 and 1900 bars, a load of 1720 bars was applied. At this point the strength of the sample had been decreased such that, at the comparatively low stress of 1720 bars, rapid inelastic creep occurred. This creep was again accompanied by decrease in magnetization. As one of the strain gages was ruptured, the anelastic volumetric deformation of the sample could no longer be monitored. However, the increasing rate of demagnetization strongly suggested that rupture was imminent, and about one second later the sample ruptured violently. The last value of the magnetic field before rupture showed a change of 70 relative magnetic units (1 %) at constant stress.

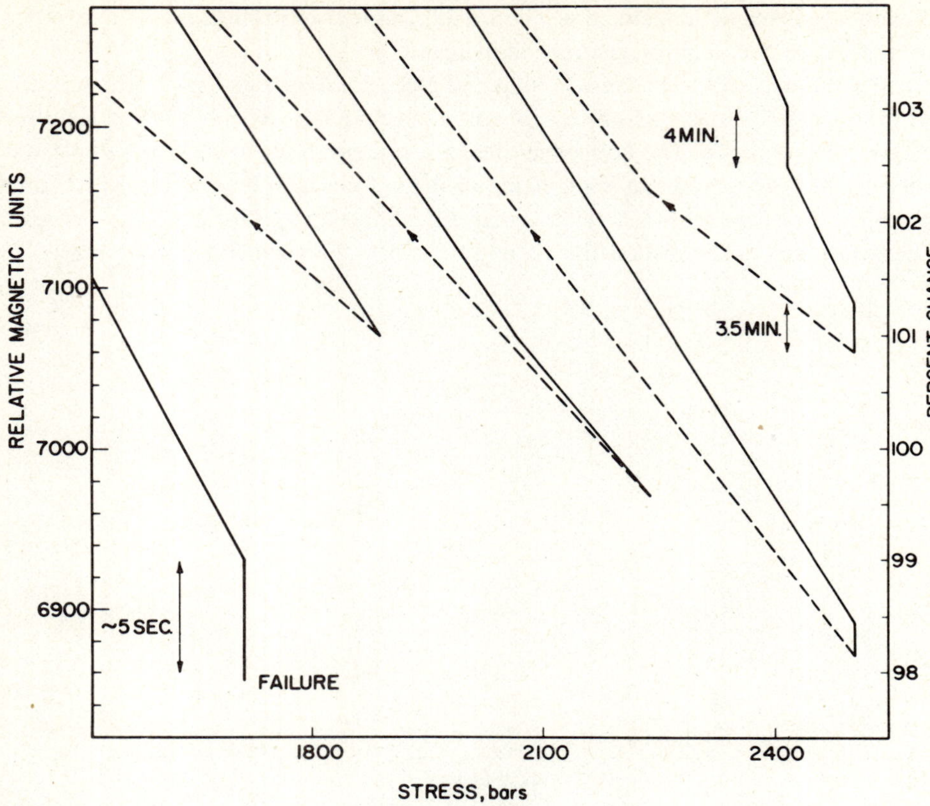

Figure 5
A detailed plot of remanent magnetization (IRM) during stress cycling just prior to failure. IRM was initially parallel to the compression axis. The percent change in IRM, arbitrarily normalized to 100 percent at 7000 relative magnetic units, is shown for reference on the right vertical scale

4. Discussion

Perhaps the most interesting observation in this study is that the onset of dilatancy is associated with a change in the stress-remanent magnetization curve. Furthermore, time-dependent dilatancy at constant stress is also accompanied by a time-dependent variation in IRM and NRM. These two observations imply that measurements of the magnetic field near faults could be useful for earthquake prediction, even though the mechanisms of the effects reported are not fully understood.

BREINER and KOVACH [15] have observed 1 gamma field changes associated with creep events along the San Andreas fault. TALWANI and KOVACH [16] tried to explain these magnetic anomalies by elastic piezomagnetic effects in the crust. In their model they used the calculations by SHAMSI and STACEY [10] and assumed that the creep reduced the stress elastically in the rocks on either side of the fault, and that therefore an elastic piezomagnetic effect took place. Such an effect was observed in laboratory deformation of rocks (e.g., NAGATA and KINOSHITA [17], OHNAKA and KINOSHITA [13]). Under this assumption Talwani and Kovach found that the theoretically expected magnetic anomalies for an infinite fault were about 4 orders of magnitude smaller than the observed ones. They worked around this problem by invoking stress concentrations at the ends of creep dislocations. Alternatively, the magnetic anomalies along the San Andreas fault may be related to dilatant creep similar to that observed in our experiments.

The immediate application of these results to the observed magnetic anomalies in the field seems a bit premature. We have examined the effect of uniaxial stress on the remanent magnetization of gabbro for two orientations of the magnetic vector, parallel to the compression axis and inclined 39° to the axis. In the former case the intensity of magnetization decreased while in the latter it increased. This is most likely due to the fact that only one component of the magnetic vector was monitored and the vector was rotating into the axis of compression. Small rotations of the magnetic vector due to uniaxial stress on rocks have been previously reported (HALL and NEALE [18], DOMEN [19], OHNAKA [20]). Furthermore, CARMICHAEL [21] has theoretically examined the effect of uniaxial stress on the rotation of magnetization from the preferred axis, ⟨111⟩, in magnetite and nickel. As stress is increased, the domain magnetizations will be deflected toward the magnetoelastic minima and away from the magnetocrystalline-preferred ⟨111⟩ orientations. Thus it seems that the effect of uniaxial stress is to cause a reduction in magnetic intensity as well as a rotation of the magnetic vector although the relative importance of each is not readily apparent. In addition, the magnitude of the magnetic effect will be influenced by the chemical composition of the magnetic grains, relation between the remanent magnetic vector and the direction of applied stress, the temperature and the pressure as well as the effect of stress on magnetic susceptibility. What seems to be important then in any estimate of a precursory magnetic anomaly is that once a rock becomes dilatant the stress dependence on the intensity of magnetization decreases; and any calculation based on the magnetic response of a rock in the elastic region must be reconsidered. Hopefully further experiments on a wide variety of rocks and under more general loading and magnetic field conditions will enable us to more accurately estimate the magnetic response of a rock mass as it dilates.

5. *Conclusions*

Uniaxial failure experiments in which rock magnetization and volumetric strain were measured as a function of axial stress furnished the following two critical facts. First, dilatancy decreases the relative effect of stress on remanent magnetization below that observed when the rock behaved elastically; that is the rate of demagnetization decreases with increasing stress after the onset of dilatancy. Second, there is a general correlation between time-dependent dilatancy and time-dependent changes in magnetization; as the creep rate increased, the rate of change of magnetization also increased. These observations suggest dilatancy should produce detectable magnetic anomalies that could serve as long-term precursors, and anelastic creep should produce large magnetic anomalies that could serve as short-term precursors to earthquakes.

Acknowledgements

This work was supported by NSF Grant No. GA-41801 and US Geological Survey Contract No. 14-08-0001-14571. We wish to thank EDWIN LARSON for his assistance and suggestions during the course of these experiments.

REFERENCES

[1] AMOS NUR, *Dilatancy, pore fluids and premonitary variations in t_s/t_p travel times*, Bull. Seism. Soc. Am. *62* (1972), 1217.
[2] C. H. SCHOLZ, L. R. SYKES and J. P. AGGARWAL, *The physical basis for earthquake prediction*, Science *181* (1973), 803.
[3] W. F. BRACE and A. S. ORANGE, *Electrical resistivity changes in saturated rocks during frictional sliding*, J. Geophys. Res. *73* (1968), 1433.
[4] O. M. BARSUKOV, *Variations in the electrical resistivity of rocks and earthquakes*, in *Earthquake Precursors, Sciences USSR*, edited by M. A. SADOVSKII, I. L. NERSESOV and L. A. LATYNINA (Acad. Sciences USSR 1973), p. 198.
[5] T. RIKITAKE, *Geomagnetism and earthquake prediction*, Tectonophys *6* (1968), 59.
[6] N. FUJITA, *The magnetic disturbances accompanying the Niigata earthquake*, J. Geodetic Soc. Japan *11* (1965), 8.
[7] M. TAZIMA, *Accuracy of Recent Magnetic Survey and a Locally Anomalous Behaviour of the Geomagnetic Secular Variation in Japan*, Thesis, Tokyo Univ. 133 pp.
[8] M. WYSS, *A search for precursors to the Sitka, 1972, earthquake: Sea level, magnetic field, and P-residuals*, Pure and Appl. Geophys., this issue.
[9] M. J. S. JOHNSTON, *Tectonomagnetic experiments in Western USA*, Trans. Am. Geophys. Un. *56* (1974).
[10] S. SHAMSI and F. D. STACEY, *Dislocation models and seismomagnetic calculations for California 1906 and Alaska 1964 earthquakes*, Bull Seism. Soc. Am. *59* (1969), 1435.
[11] T. NAGATA, *Tectonomagnetism*, IAGA Bull. *27* (1970), 12.
[12] W. P. HASBROUCK and J. H. ALLEN, *Quasi-static magnetic field changes associated with the Cannikin nuclear explosion*, BSSA *62* (1972), 1479.
[13] M. OHNAKA and H. KINOSHITA, *Effects of uniaxial compression on remanent magnetization*, J. Geomag. Geoelect. *20* (1968), 93.
[14] K. MOGI, *Some precise measurements of fracture strength of rocks under compressive stress*, Rock. Mech. and Eng. Geol. *4* (1966), 41.
[15] S. BREINER and R. L. KOVACH, *Local magnetic events associated with displacement along the San Andreas fault*, Tectonophys. *6* (1968), 69.
[16] P. TALWANI and R. L. KOVACH, *Geomagnetic observations and fault creep in California*, Tectonophys. *14* (1972), 245.
[17] T. NAGATA and H. KINOSHITA, *Magnetization of titaniferous magnitite under uniaxial compression*, J. Geomag. Geoelect. *17* (1965), 121.
[18] J. M. HALL and R. N. NEALE, *Stress effects on thermoremanent magnetization*, Nature *188* (1960), 805.

[19] H. DOMEN, *Piezo-remanent magnetism in rocks and its field evidence*, J. Geomag. Geoelect. *13* (1966), 67.

[20] M. OHNAKA, *Stability of remanent magnetization of rocks under compression – Its relation to grain size of rock forming ferromagnetic minerals*, J. Geomag. Geoelect. *21* (1969).

[21] R. S. CARMICHAEL, *Stress control of magnetization in magnetite and nickel implications for rock magnetism*, J. Geomag. Geoelect. *13* (1968).

(Received 11th February 1975)

Pageoph,
Vol. 113 (1975) Birkhäuser Verlag, Basel

Stable Sliding Preceding Stick-Slip on Fault Surfaces in Granite at High Pressure

By J. D. BYERLEE and R. SUMMERS[1])

Summary – The distance of stable sliding before sudden slip on fault surfaces in granite decreases rapidly as the confining pressure is increased. At a pressure of 6 kb the amount of stable creep is very small or absent. Two orders of magnitude change in strain rate has no effect on the distance of stable sliding. Our results suggest that in the earth, fault creep should predominate in the shallow crust but in the deep crustal layer most of the stresses are probably relieved by sudden earthquake type of motion. Below the crust high temperature would promote stable-slip so in this region creep would once more predominate.

1. Introduction

It has been observed that there are two types of fractional sliding between rock surfaces. One type of sliding occurs smoothly and in laboratory experiments this type of movement is called stable sliding. In the natural situation it is called fault creep. The other type of sliding occurs jerkily; that is, the surfaces suddenly slide forward, lock and then slide again. In laboratory experiments we call this type of motion stick-slip and in the earth sudden movement on pre-existing faults produces earthquakes.

Some of the factors that are known to stabilize sliding in laboratory experiments are the presence of weak alteration minerals such as serpentine (BYERLEE and BRACE [1]), high temperature (BRACE and BYERLEE [2]), high pore pressure (BYERLEE and BRACE [3]), and large thicknesses of fault gouge (BYERLEE and SUMMERS [4]). It has been observed that stable fault creep is occurring on the San Andreas fault in central California (RALEIGH and BURFORD [5]; NASON and TOCHER [6]) and any one of the factors listed above, and possibly others as yet unknown, could be responsible for promoting fault creep in this region. If the fault continues to slide in this way then this region of central California may be free of large earthquakes, but we cannot be sure of this, because sudden slip could commence in the unstable zone either to the north or to the south and propagate through the stable zone.

It has also been observed that in laboratory experiments stable sliding often precedes violent stick-slip (BYERLEE [7]; BYERLEE [8]; DIETERICH [9]; SCHOLZ [10]; SCHOLZ *et al.* [11]) and this observation led SCHOLZ *et al.* [11] to suggest that fault creep could be a warning of an impeding earthquake. In the experiments carried out by SCHOLZ

[1]) US Geological Survey, Menlo Park, California 94025.

et al. [11] stick-slip was always preceded by a small amount of stable sliding but their experiments were carried out with finely ground surfaces of granite at low normal stresses. In the earth, sliding is on irregular-fault surfaces separated by fault gouge, and the normal stresses, at least in the lower crust, could be high. We do not know in this situation whether stick-slip is always preceded by stable sliding. Thus we cannot be sure whether fault creep will always precede a sudden earthquake type of motion.

We have carried out a number of experiments that were designed so that we could measure the distance of stable sliding preceding sudden slip on fault surfaces in granite at confining pressures up to 6 kb and strain rates of 10^{-4}, 10^{-5} and 10^{-6} per second. The purpose of this paper is to document the results and to discuss their significance.

2. *Experimental method and results*

Cylindrical samples of Westerly granite 2.54 cm in diameter and 6.35 cm long were jacketed in polyurethane sleeves and sealed at their ends to hardened steel plugs. The samples were subjected to a confining pressure and loaded at their ends in compression. Both confining pressure and strain rate were kept constant during each experiment. Experiments were run at pressures up to 6 kb and strain rates of 10^{-4}, 10^{-5}, and 10^{-6} per second. These strain rates refer to the elastic strain rate in the specimen and not to the shear strain rate in the fault gouge during sliding, but an order of magnitude change in the elastic strain rate would also correspond to an order of magnitude change in the shear strain rate in the gouge.

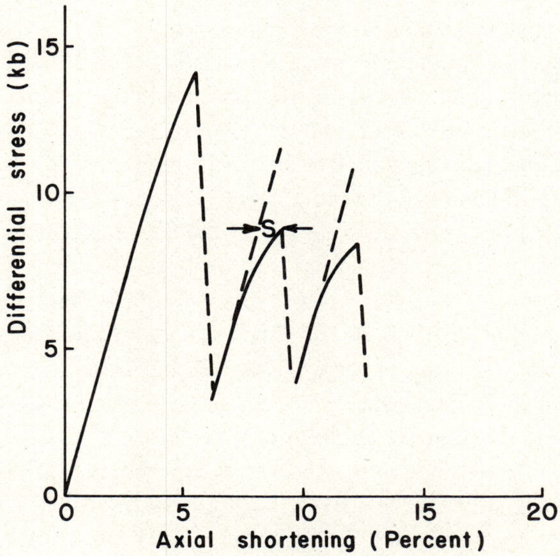

Figure 1
Differential stress as a function of percentage of axial shortening for the deformation of granite at a confining pressure of 2.74 kb and at a strain rate of 10^{-5} per second

A typical stress–strain curve for one of the experiments is shown in Fig. 1. The sample was deformed at a confining pressure of 2.74 kb and at a strain rate of 10^{-5} per sec. The sample failed violently at a differential stress of about 14 kb. After faulting, deformation of the sample was initially elastic until a differential stress of about 6 kb; then a marked non-linearity of the stress–strain curve developed. A small amount of this non-elastic deformation is probably due to compaction of the gouge developed during faulting, but most of the departure, s, from linear elastic strain is, at least at low confining pressures, caused by stable sliding on the newly formed fault surface.

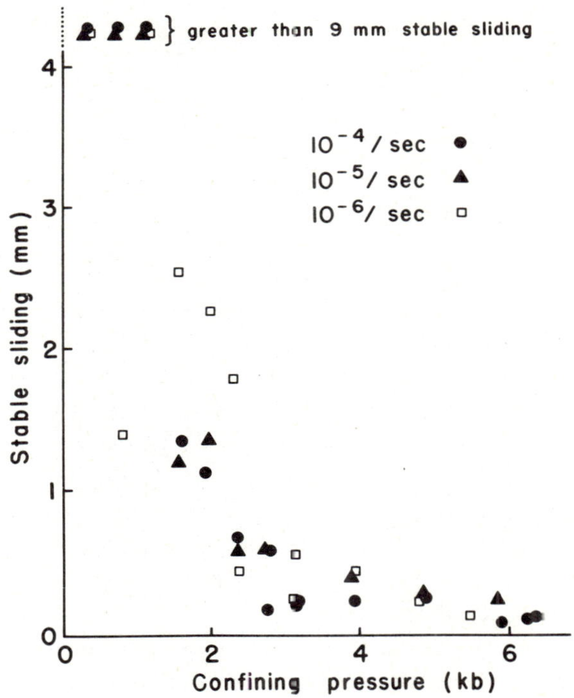

Figure 2
The distance of stable sliding preceding the first sudden slip on fault surfaces of granite at strain rates of 10^{-4}, 10^{-5} and 10^{-6} per second and at confining pressures up to 6 kb

The fault angle was measured after each experiment and the component of s parallel to the direction of movement on the fault plane was calculated. The calculated results for the distance of stable sliding preceding the first and the second sudden slip on the newly formed fault for all the experiments are plotted in Figs. 2 and 3, respectively.

Our experiments were terminated when the distance of slip on the fault surfaces reached 9 mm. We stopped the experiments at this point because we found, by experience, that when the displacement was greater than this the sharp leading edge of the fault ruptured the jacket and allowed the confining pressure medium to enter the rock

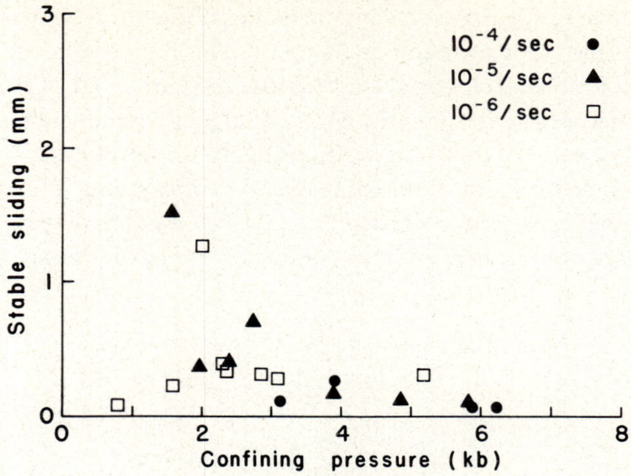

Figure 3
The distance of stable sliding preceding the second sudden slip on fault surfaces of granite

and when this occurs the effective confining pressure drops to zero and further experimental data has little meaning for our purposes.

In all but one of the experiments at pressures of 1.5 kb and below sliding on the newly formed fault surfaces was stable up to the 9 mm limit that we set. At pressures above 1.5 kb the distance of stable sliding preceding sudden slip decreased rapidly as the confining pressure was increased. The similarity between the data shown in Figs. 2 and 3 indicate that the previous sliding history had little effect on the results. At some pressures slower strain rate resulted in a greater distance of stable sliding but at other pressures the reverse was true. Thus at least within the uncertainty of our results there does not seem to be any strain rate effect.

At pressures of 6 kb the distance of stable sliding was reduced to about 0.1 mm. This could be caused by compaction of the gouge and the value of 0.1 mm that we have calculated from our experimental data may not represent stable sliding at all. If the gouge was 1 mm thick, then 10% compaction could account for 0.1 mm of non-elastic deformation of the sample. This amount of compaction could be possible so that even though our results show that non-elastic strain precedes sudden slip at pressures as high as 6 kb we cannot be certain that this is due to stable sliding on the fault.

3. *Discussion*

Our results show that the distance of stable sliding preceding sudden slip on fault surfaces in granite is independent of the strain rate. SCHOLZ *et al.* [11] also found a strain rate independence for the distance of stable sliding preceding sudden slip on finely ground surfaces of granite at low normal loads. Thus in this respect our results are similar to those of SCHOLZ *et al.* [11].

We found, however, that with fault surfaces the distance of premonitory creep decreased rapidly as the confining pressure was increased, whereas SCHOLZ *et al.* [11] found that over the normal stress range of about 100 to 500 bars the distance of premonitory creep with ground surfaces was independent of the normal stress. One of the ways in which our experiments differed from those of SCHOLZ *et al.* [11] was that in our experiments the fault surfaces were irregular and were separated by a large thickness of fault gouge whereas in the experiments carried out by Scholz *et al.* the surfaces were flat and finely ground. In our experiments we were essentially studying the rheological behavior of fault gouge and apparently its behavior is quite different from that of clean flat surfaces. In the natural situation faults contain gouge which may be composed of weak alteration minerals, it may contain fluids at high pressure, it may be at elevated temperature, and these factors may be important in determining whether a fault will slide stably or unstably.

SCHOLZ *et al.* [11] found that in their experiments sudden slip was always preceded by a small amount of stable sliding and they suggested that, if this is universally true, then creep on a fault may be a warning sign of an impending earthquake. Our results shown in Figs. 2 and 3 indicate that at pressures as high as 6 kb a small amount of non-elastic deformation preceded sudden slip on the faults. Thus our results would also suggest that large earthquakes may be preceded by fault creep, but our results indicate that the amount of premonitory creep would decrease as the focal depth of the earthquake increases. For earthquakes that have their focus in the lower crust the amount of premonitory creep may be so small that it could not be detected by changes in strain or tilt measured at the earth's surface.

Our results show that, at least with faulted samples of granite that have undergone the same amount of non-elastic deformation, the amount of energy released by sudden slip increases rapidly as the confining pressure is increased. At pressures below 1.5 kb none of the energy is released suddenly whereas at 6 kb most of the sliding takes place by violent stick-slip. This result would suggest that in the upper crust most of the movement between faults would take place by stable creep but as the depth increases a greater percentage of the movement should take place by sudden earthquake-producing motion.

Analysis of the data collected from the US Geological Survey network of seismic stations in northern California shows that the energy released by earthquakes, over the period from 1971 to 1973, increases rapidly with depth until a maximum is reached at a depth of about 6 km (W. H. K. Lee, oral communication). It is only the unstable deformation that produces earthquakes so that analysis of the energy released in this way in the upper crust will be small if most of the deformation in this region is stable. It should be noted that the energy released by earthquakes in northern California falls off rapidly below a depth of 6 km and this is probably because stable sliding is promoted by an increase in temperature (BRACE and BYERLEE [2]).

References

[1] J. D. Byerlee and W. F. Brace, *Stick-slip, stable sliding, and earthquakes – effect of rock type, pressure, strain rate and stiffness*, J. Geophys. Res. *73* (1968), 6031–6037.

[2] W. F. Brace and J. D. Byerlee, *California earthquakes: why only shallow focus*, Science *168* (1970), 1573–1575.

[3] J. D. Byerlee and W. F. Brace, *Fault stability and pore pressure*, Bull. Seis. Soc. Am. *62* (1972), 657–660.

[4] J. D. Byerlee and R. Summers, *The effect of fault gouge on the stability of sliding on saw cuts in granite*, Trans. Am. Geophys. Union *54* (1973), 1210.

[5] C. B. Raleigh and R. O. Burford, *Tectonics of the San Andreas fault system strain studies*, Trans. Am. Geophys. Union *50* (1969), 380.

[6] R. D. Nason and D. Tocher, *Measurement of movement on the San Andreas fault*, in *Earthquake displacement fields and the rotation of the earth*, edited by Mansinha *et al.* (D. Reidel, Dordrecht, Netherlands 1970).

[7] J. D. Byerlee, *Frictional characteristics of granite under high confining pressure*, J. Geophys. Res. *72* (1967), 3639–3648.

[8] J. D. Byerlee, *The mechanics of stick-slip*, Tectonophysics *9* (1970), 475–486.

[9] J. H. Dieterich, *Time-dependent friction in rocks*, J. Geophys. Res. *77* (1972), 3690–3697.

[10] C. H. Scholz, *Microfracturing and the inelastic deformation of rock*, J. Geophys. Res. *73* (1968), 1417.

[11] C. H. Scholz, P. Molnar and T. Johnson, *Detailed studies of frictional sliding of granite and implications for the earthquake mechanism*, J. Geophys. Res. *77* (1972), 6392–6406.

(Received 26th November 1974)

Pageoph,
Vol. 113 (1975) Birkhäuser Verlag, Basel

The Sliding Characteristics of Sandstone on Quartz Fault-Gouge[1])

By JAMES T. ENGELDER[2]), JOHN M. LOGAN and JOHN HANDIN[3])

Summary – Three types of triaxial compression experiments are used to characterize the frictional processes during sliding on quartz gouge. They are: 1) pre-cut Tennessee Sandstone sliding on an artificial layer of quartz gouge; 2) fractured Coconino Sandstone sliding along experimentally produced shear fractures; and 3) a fine-grained quartz aggregate deformed in compression. The specimens were deformed to 2.0 kb confining pressure at room temperature and displacement rates from 10^{-2} to 10^{-5} cm/sec dry and with water. There is a transition in sliding mode from stick-slip at confining pressures <0.7 kb to stable sliding at >0.7 kb. This transition is accompanied by a change from sliding at the sandstone-gouge contact (stick-slip) to riding on a layer of cataclastically flowing gouge (stable sliding). Quartz gouge between the pre-cut surfaces of Tennessee Sandstone lowers both the kinetic coefficient of friction and the magnitude of the stick-slip stress drops compared to those for a pre-cut surface alone. Stick-slip stress drops are preceded by stable sliding at displacements of 10^{-5} cm/sec. For a decrease in displacement rate between 10^{-3} and 10^{-5} cm/sec, stress-drops magnitudes increase from 25 to 50 bars. Tests on saturated quartz gouge show sufficient permeability to permit fluid-pressure equilibrium within compacted gouge in 10 to 30 seconds; thus the principle of effective stress should hold for the fault zone with quartz gouge. Our results suggest that at effective confining pressures of less than 2.0 kb, if a fault zone contains quartz gouge, laboratory-type stick-slip can be an earthquake-source mechanism only if a planar sliding-surface develops, and then only when the effective confining pressure is less than 0.7 kb.

1. Introduction

Gouge is commonly found between the walls of real faults. Thus, any attempt to characterize or model the faulting process, which may involve creep (aseismic faulting) or unstable slip (seismic faulting), must account for the mechanical properties of the fault with gouge. Most fault zones can be modelled by using three mechanical elements: a gouge zone with wall rock on both sides.

Knowledge of the mechanical properties of fault zones is based primarily on sliding-friction experiments. In contrast to natural faults with long histories of sliding and frictional wear many experimental surfaces have little or no previous frictional wear. In those experiments the frictional force is developed between two intact rocks rather than through a layer of gouge. The purpose of this paper is to report an attempt to simulate sliding after a thick layer of gouge has developed during prior large fault displacements and to determine the frictional properties of rock sliding on fault gouge.

---

[1]) Lamont–Doherty Geological Observatory Contribution No. 2222.

[2]) Presently at Lamont–Doherty Geological Observatory, Palisades, New York 10964 USA.

[3]) Center for Tectonophysics, Texas A&M University, College Station, Texas 77843, USA.

Some of the frictional properties of rock sliding on gouge are known (BYERLEE and BRACE [1]; DIETERICH [2]), but many details are sketchy. To date observations on the changes in frictional properties with the accumulation of gouge have been limited to experiments in which a small amount of gouge had been generated during the limited displacement of an experiment (BYERLEE [3]; COULSON [4]; JAEGER [5]). In Westerly Granite the frictional force reaches a residual value when frictional wear generates enough gouge to completely separate the sliding surfaces (BYERLEE [3]). The accumulation of gouge generated during initial slip causes a decrease in stress drops during stick-slip sliding (JACKSON and DUNN [6]). Some surfaces have a stable sliding to stick-slip transition when sliding on gouge (BYERLEE and BRACE [1]; SCHOLZ *et al.* [7]). At elevated temperatures an indurated gouge or glass may form between the sliding surfaces thus further modifying the frictional force between them (FRIEDMAN *et al.* [8]). Compacted aggregates, which may simulate gouge, fracture and stick-slip at 7 to 10 kb confining pressure (BYERLEE and BRACE [9]). The coefficient of static friction of surfaces with gouge is time dependent (DIETERICH [2]).

The rate of generation of gouge with displacement is a function of several parameters. For example, quartz gouge is generated at a higher rate at higher confining pressure (ENGELDER [10]). JACKSON and DUNN [6] show that the rate of gouge generation depends on the angle of foliation to the sliding surface as well as the confining pressure. For rocks with higher fracture strengths, less gouge is generated per unit displacement between surfaces than for rocks with lower fracture strengths (COULSON [4]). DIETERICH [2] implies that the volume of gouge generated from a rough surface far exceeds that generated from a smooth one, when they are both displaced the same amount at the same normal stress.

Faults may contain any one of a number of gouges depending on the mineralogy of the country rock from which the gouge was generated. Gouge from one section in the San Adreas fault contains predominantly montmorillonite, chlorite and serpentine (NASON and TOCHER [11]). In contrast, some gouge from the Muddy Mountain Thrust contains only quartz particles (ENGELDER [10]). Gouge may also consist of carbonate minerals or of combinations of several different minerals.

2. *Experimental procedure*

This systematic study of the frictional effects of a fault gouge between intact rock is restricted to one type of gouge and, so, does not define the mechanical behaviors of all gouges. In order to document the effects of gouge, the frictional properties of the rock without gouge should be well known, and the gouge generated from that rock should be: 1) common in nature, 2) easy to work with in the laboratory, and 3) amenable to petrographic observation. Quartz fault gouge has these desirable characteristics and the frictional properties of Tennessee Sandstone are well known (LOGAN *et al.* [12]; HUMSTON [13]; BLACKWELL [14]; FRIEDMAN *et al.* [8]).

Experimental quartz fault gouge is generated by shearing a 0.17 cm thick layer of

100 to 250 μm diameter quartz grains between halves of a 5 cm diameter by 10 cm long cylinder of Tennessee Sandstone cut and ground with an 80-grit wheel at $35° \pm 0.1°$ to the cylinder axis (Fig. 1). Pre-cut cylinders are more desirable than mechanically fractured cylinders because the latter have rough sliding surfaces with large asperities which affect sliding characteristics enough to mask the more subtle effects of other experimental parameters. The layer of quartz on the sliding surface is cataclastically reduced to quartz gouge during the initial displacement. The experimental quartz

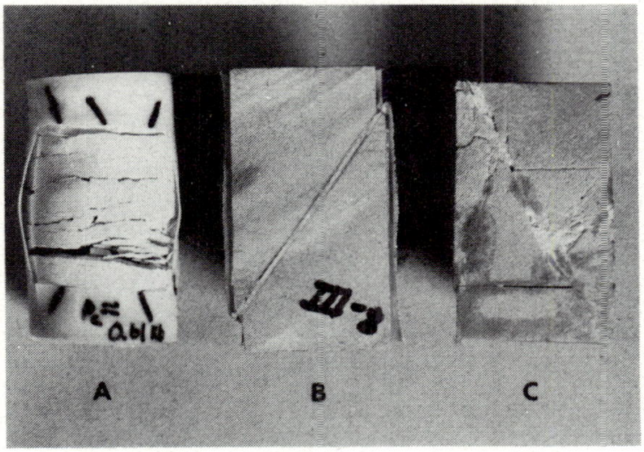

Figure 1

The three types of specimens used in this work. Specimen *a* is a compacted quartz aggregate. Shear fractures which extend between black lines drawn on the polyolefin jacket occur at about 30° to cylinder axis. Large horizontal fractures are extension fractures which resulted from extracting the specimen from the pressure vessel. Specimen *b* is a section through a pre-cut cylinder of Tennessee Sandstone with 0.12 cm thick layer of quartz gouge between sliding surfaces. Specimen *c* is a section through a cylinder of Coconio Sandstone that has been fractured to generate gouge during displacement. Long axis of each cylinder is about 10 cm

gouge has a grain-size distribution comparable to that from natural faults (ENGELDER [10]) and, therefore, is assumed to have the same mechanical properties as the natural counterpart. A typical force-shortening curve and details of the cataclasis accompanying the curve are given by ENGELDER [15].

Experiments are conducted at room temperature, confining pressure to 2.0 kb, pore pressure to 1.0 kb, and slip rates between 10^{-2} and 10^{-5} cm/sec. The apparatus is described by HANDIN *et al.* [16]. The cut and ground cylinders are jacketed with heat-shrink polyolefin sleeves. The quartz grains are spread along the saw-cut by removing one half of the specimen from the sleeve and then distributing the quartz in an even layer on the sliding surface. Once the specimen half is replaced, the ends are carefully aligned. A 0.6 cm thick steel spacer powdered with MoS_2 on both sides is placed between the specimen and the pistons in order to enhance proper alignment and to decrease friction on the ends of the specimen. The reproducibility of these experiments is indicated by a series of four tests at 1 kb confining pressure in which after

0.45 cm of sliding displacement the mean differential stress was 3.00 kb with a standard deviation of 0.05 kb.

Cylinders of fractured or pre-cut Coconino Sandstone are also used to evaluate the frictional properties of quartz gouge (Fig. 1). In addition, 5 × 10 cm cylinders of compacted quartz particles less than 63 μm in diameter are used to study the deformation mechanisms of dry quartz aggregates at strain rates between 10^{-4}/sec and 10^{-7}/sec. The aggregate is prepared by crushing sand and sieving out all particles greater than 63 μm. The remaining particles are poured into a polyolefin jacket that is pre-shrunk to a cylindrical shell 5 cm in diameter × 10-cm long.

3. Experimental results

Friction tests

Pre-cut specimens of Tennessee Sandstone stick-slip with stress drops of more than 1.0 kb at confining pressures greater than 0.5 kb (HUMSTON [13]). Large stress drops are observed even after a film of gouge (<0.01 cm) is generated on the sliding surface. In contrast, a 0.12-cm thick layer of gouge between sliding surfaces causes Tennessee Sandstone to stick-slip with stress drops of less than 10^{-1} kb (Fig. 2). Sandstone sliding

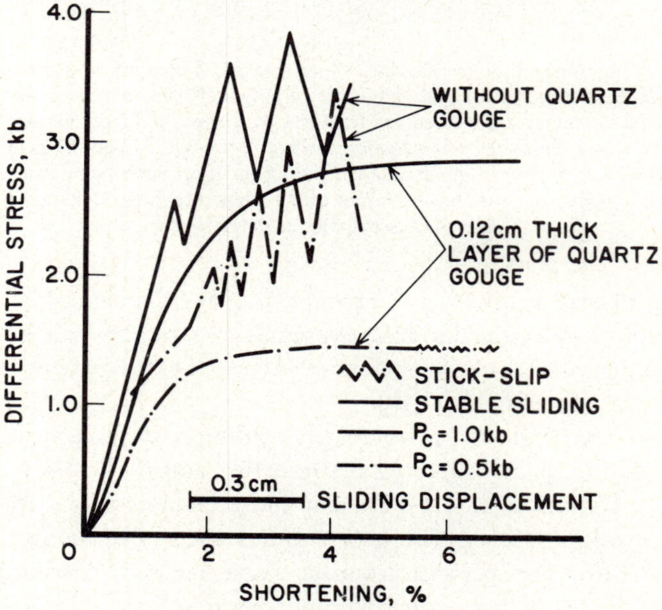

Figure 2
Differential stress versus shortening curves for the sliding of pre-cut Tennessee Sandstone with and without an 0.12 cm thick layer of quartz gouge. Specimens are deformed at room temperature, a constant rate of shortening of 10^{-4}/sec or displacement rate of 10^{-3} cm/sec. Data for Tennessee Sandstone sliding without gouge are from HUMSTON [13]

on the layer of gouge has a lower kinetic coefficient of friction relative to that measured for sandstone without the discrete layer of gouge.

Tennessee Sandstone stick-slips at all confining pressures in the range 0.2 to 2.0 kb (HUMSTON [13]). However, confining pressure in the same range does affect the sliding mode of Tennessee Sandstone on quartz gouge (ENGELDER [15]; Figure 2). With a 0.12 cm thick layer of gouge, Tennessee Sandstone does not stick-slip above 0.7 kb confining pressure. As explained in ENGELDER [15], stick-slip occurs after an initial phase of stable sliding during which gouge is generated on the sliding surface. The onset of stick-slip occurs earlier in the displacement history of the specimen as the confining pressure is increased from 0.14 kb to 0.5 kb. It was expected that stick-slip would occur with even less displacement at 0.7 kb confining pressure, but the sliding is in fact stable at displacements greater than those for which stick-slip occurs at 0.14 kb. Thus, a stick-slip to stable-sliding transition is noted for Tennessee Sandstone sliding on dry quartz gouge.

In order to better understand this transition, a sequentially loaded experiment is used to investigate sliding characteristics of a single specimen both above and below 0.7 kb confining pressure. During this type of experiment the confining pressure is changed after the specimen has slid a short distance; the specimen is unloaded prior to each change in confining pressure. At the onset of a sequentially loaded experiment, gouge is generated by the initial sliding. Then the confining pressure is readjusted to the desired level. The specimen in Fig. 3 was initially loaded at 1.0 kb to generate

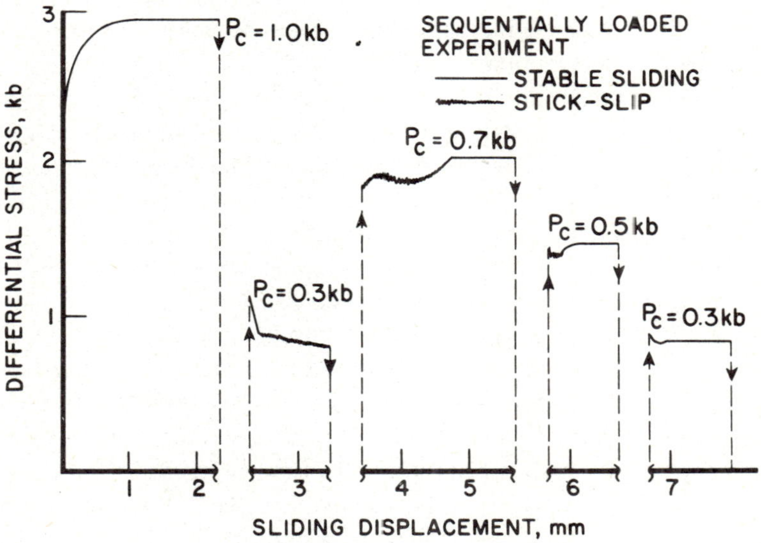

Figure 3
Effect of sequential loading on frictional sliding of a pre-cut specimen of Tennessee Sandstone with an 0.12 cm thick layer of dry quartz gouge. Confining pressure adjusted five times during the same experiment; for each adjustment the differential stress is reduced to zero and then reapplied (dashed lines)

gouge from sand; stable sliding occurs after the gouge is generated. Whenever the specimen is reloaded at a confining pressure lower than 0.7 kb, stick-slip begins after an initial large stress drop. The sliding behavior during the first two loadings illustrated in Fig. 3 corresponds with that discovered by conventional testing where a stick-slip to stable sliding transition occurs at 0.7 kb confining pressure. This fundamental change in the sliding process occurs regardless of the previous sliding history of the specimen, provided that the confining pressures during sliding are never held between 0.7 and 0.8 kb.

The sliding of the sequentially loaded specimens at 0.7 kb confining pressure is atypical. Stick-slip occurs initially, but is then followed by stable sliding. This behavior occurs only during sequentially-loaded tests; the specimen slides stably at 0.7 kb during the entire conventional experiment. At 0.7 kb the stick-slip to stable-sliding transition seems to be a function of displacement in that a certain amount of stick-slip precedes stable sliding. Also the kinetic coefficient of friction is lower during stick-slip than stable sliding.

During the fourth and fifth cycles (Fig. 3) sliding is stable at confining pressures of 0.5 and 0.3 kb, for which stick-slip is expected. This occurs only during sequential-loading experiments. Stable sliding occurs at confining pressures of less than 0.7 kb only when preceded by slip at between 0.7 and 0.8 kb confining pressure. Slip at 0.7 kb confining pressure results in a fundamental change within the gouge that affects subsequent experiments. The gouge, deformed at 0.7 kb confining pressure during stable sliding, suppresses stick-slip at lower confining pressures, whereas the gouge deformed at 1.0 kb does not. Thus there is more than one process for stable sliding and accompanying gouge deformation.

DIETERICH [2] observes that the static coefficient of friction of rock in contact with gouge increases with the logarithm of the time of contact. A similar behavior is found for stick-slip sliding at 0.5 kb confining pressure on 0.12 cm thick quartz gouge (Fig. 4). At 10^{-5} cm/sec a small amount of stable sliding precedes the stress drop. Stable sliding may precede the stress drop at faster displacements, but the apparatus is not sensitive enough to detect it. SCHOLZ *et al.* [7] have observed stable sliding and then episodic sliding prior to stick-slip if gouge is between sliding surfaces.

An increase in gouge thickness to 0.36 cm results in a decrease in magnitude of stress drops in the stick-slip regime of Tennessee Sandstone on quartz fault gouge. At 1.0 kb this rock slides stably on all thicknesses of gouge between 0.03 and 0.36 cm. The shear stress necessary to cause stable sliding at 1.0 kb confining pressure on all thicknesses of gouge does not vary by more than the experimental error determined from four tests for sliding on 0.12 cm thick gouge. The effect of gouge thickness on the transition from stick-slip to stable sliding was not studied.

The stick-slip to stable sliding transition occurs even when the 0.12 cm thick layer of gouge is water saturated. The principal effect of the saturation is an increase of the magnitude of the stick-slip stress drops with a small amount of preceding stable sliding (Fig. 4). Saturating the gouge has about the same effect as does decreasing the displacement rate from 10^{-3} to 10^{-5} cm/sec. The kinetic coefficient of friction for Tennessee

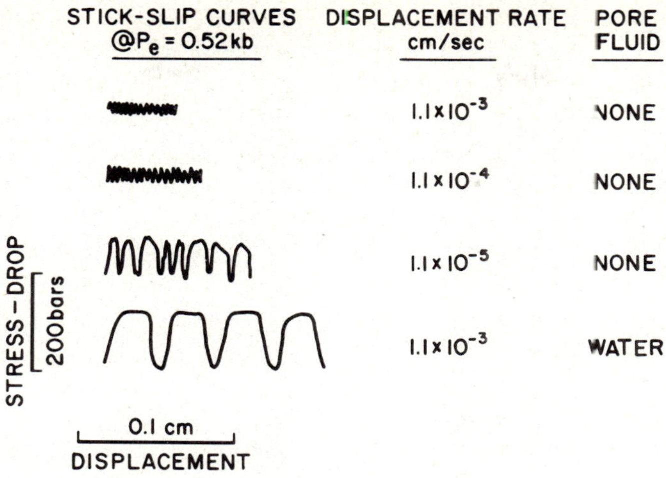

Figure 4
Diagram of the character and magnitude of stress drops for stick-slip of pre-cut specimens of Tennessee Sandstone sliding on a compacted 0.12 cm thick layer of quartz gouge. The effects of displacement rate and pore water are illustrated

Sandstone sliding on saturated quartz gouge at 0.5 kb effective confining pressure is higher than for sliding on dry gouge (Fig. 5). The two curves for saturated tests are within experimental reproducibility of each other but the curve for this dry sample is not within experimental reproducibility of the others.

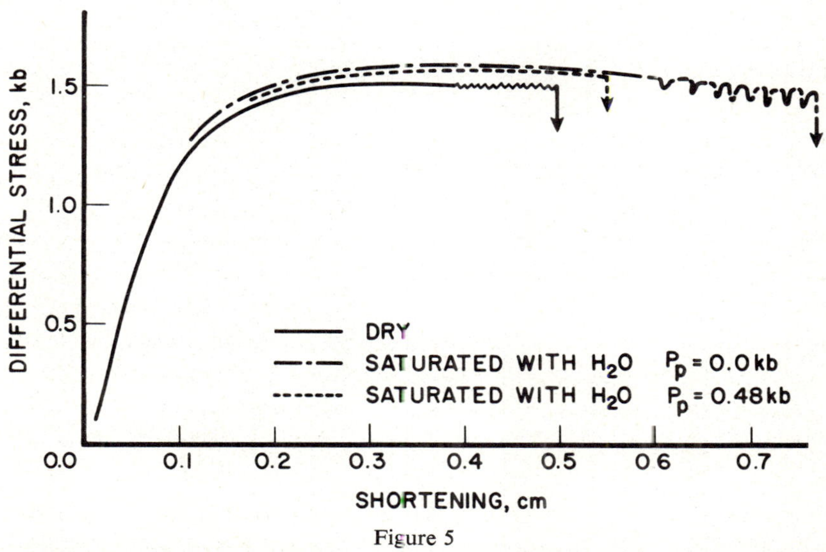

Figure 5
Differential stress-shortening curves for pre-cut specimens of Tennessee Sandstone sliding on either dry or saturated, compacted quartz gouge at an effective confining pressure of 0.5 kb. Gouge thickness 0.12 cm. Displacement rate 0.7×10^{-3} cm/sec

Changes in porosity of the layer of gouge during a sliding friction experiment can be estimated by measuring the change in volume of the pore-pressure reservoir necessary to re-establish the original pore pressure after a certain amount of sliding has occurred. Because Tennessee Sandstone has a low permeability (0.7 millidarcys) a hole is drilled through the center of the specimen to allow easy access of water to the gouge zone. All changes in porosity are assumed to occur only within the gouge and not in the intact rock. The only detectable change in porosity occurs during the generation of the quartz gouge from quartz sand. Here the porosity is reduced from about 30%, typical of sand compacted at 0.5 kb confining pressure, to 11%. It remains about the same throughout the remainder of an experiment.

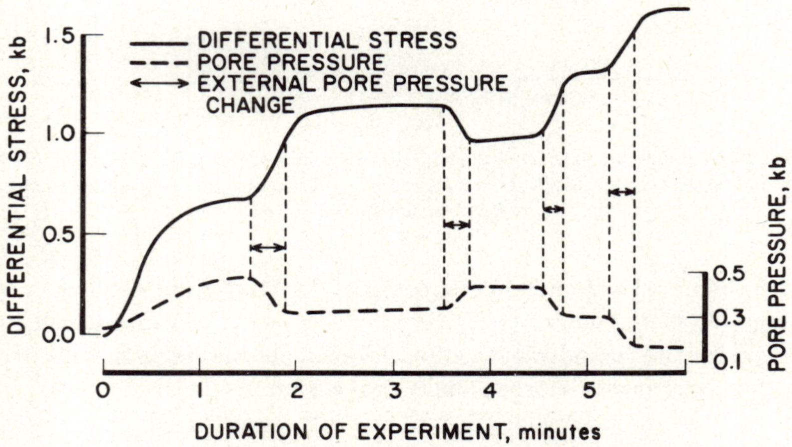

Figure 6

Curves of differential stress and pore-fluid pressure versus time for pre-cut Tennessee Sandstone sliding on an 0.12 cm thick layer of quartz gouge. Confining pressure was 0.8 kb and displacement rate was 10^{-3} cm/sec

Some idea of the effectiveness of the pore-fluid pressure and the relative short-term permeability of the quartz gouge is gained by varying the pore pressure during sliding (Fig. 6). During the initial differential loading of the specimen, the increase in pore pressure is due to the collapse of pores in the sand layer which is cataclastically deforming into a finer-grained gouge. An external pore-pressure adjustment then decreases the pore pressure to 0.2 kb. The result is an increase in the effective pressure within the specimen with a corresponding increase in differential stress necessary to maintain sliding on the quartz gouge. Likewise, an increase in pore pressure causes a decrease in both effective confining pressure and differential stress necessary to maintain sliding. Following each decrease in pore pressure of 0.2 kb over 20 sec, the differential stress continues to increase for an additional 10 sec. This indicates a small lag between the external pressure and pore pressure within the layer of gouge, but permeability is sufficient to allow rapid fluid migration within the gouge.

Compacted aggregates

Compacted aggregates of quartz particles, all less than 63 μm in size, are deformed to simulate the mechanical behavior of the quartz fault-gouge and thus help to explain the cause for the transition in the sliding mode. Each specimen is precompacted at 2 kb confining pressure by shortening the sample 3%. The arbitrary pre-compaction is necessary to obtain a specimen with about the same cohesiveness as that of the dry gouge after removal from the gouge zone after an experiment. These experiments are undoubtedly history-dependent, and the results should be considered with this in mind. During loading the specimen compresses nearly linearly then yields and flows by pervasive cataclasis, and finally fractures at all confining pressures (Fig. 7). During

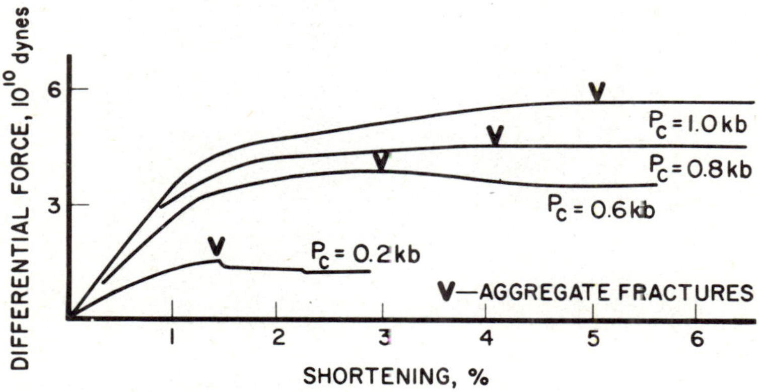

Figure 7
Effect of confining pressure on the mechanical behavior of compacted quartz aggregates composed of particles less than 63 μm

the quasi-linear portion of the stress-shortening curve, non-elastic compaction by pore collapse takes place. A positive slope after the 'yield point' always indicates pervasive cataclasis which causes the specimen to bulge with no evidence of macroscopic fracturing. Flat or negative slopes indicate fracturing along 5 to 10 conspicuous macrofractures (Fig. 1), a larger number than in intact sandstone deformed under similar conditions. More cataclastic flow precedes fracture at the higher confining pressures, as is indicated by the larger amount of shortening before fracture. The specimens fail with a slight loss of load only below 0.8 kb confining pressure. With a large number of fracture surfaces involved, slip on a single surface for a prolonged distance and the resulting macroscopic failure of the sample do not occur.

A cursory effort was given to determining the effect of strain rate on the deformation of the precompacted aggregates. The load necessary to shorten the sample 5% at 0.76 kb confining pressure decreases about 5% for each decade of decrease in strain rate between 10^{-4} and 10^{-7}/sec.

4 Microscopic observations

In order to further clarify the deformation mechanisms which influence the sliding of sandstone on quartz fault-gouge, thin sections of the gouge are examined petrographically. Typical gouge has a median grain size of about 25 μm and is not intact. ENGELDER [10] describes the relevant observations. 1) Cataclasis is more intense at higher confining pressures; thus the layer of quartz gouge is finer grained relative to those in lower confining-pressure tests. 2) Quartz gouge is much finer grained when it is generated in the presence of water at the same effective confining pressure.

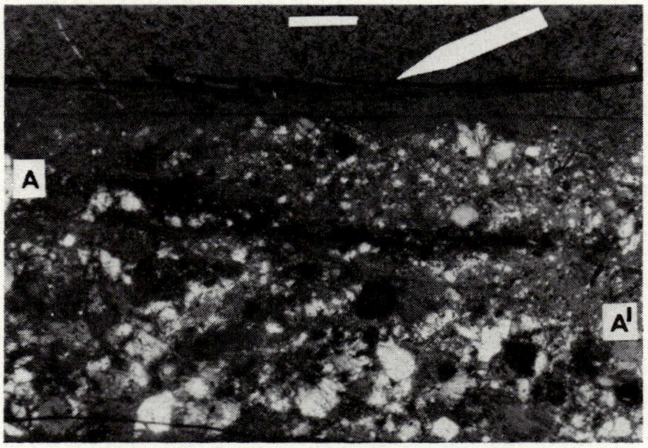

Figure 8
Photomicrograph of indurated gouge next to typical gouge. The indurated gouge (arrow) is preserved in epoxy which is used to impregnate the gouge during preparation of thin sections. The boundary (A–A') between fine and coarse-grained gouge is used to infer a sense of shear of the intact sandstone on the layer of gouge. In this case the upper surface has moved to the right. Scale line is 0.02 cm; partially crossed polarized light

At confining pressures below 1.0 kb the gouge shows incipient banding with layers of coarse grained material between layers of fine-grained gouge (Fig. 8). Banding occurs at angles of 20° to 30° to the sliding surface. When banding occurs at >1.0 kb confining pressure, it tends to be parallel with the sliding surface.

A 0.006 cm thick layer of indurated quartz forms between the gouge and pre-cut surfaces of the sandstone during stable sliding at 0.7 kb confining pressures (Figs. 8 and 9). The criteria for indurated gouge are: 1) it can easily be flaked from the sliding surface, 2) it can withstand small loads without breaking up, and 3) it can support brittle fracture (FRIEDMAN *et al.* [8]). This indurated material consists of 2 to 5 μm quartz particles embedded in a matrix which appears to be isotropic and which has a refractive index of 1.518 ± 0.002. The apparent isotropy may be due to the very fine grain size and lack of birefringence or to amorphous silica which has been fused by heat generated during sliding. Transmission electron micrographs of the indurated gouge appear in ENGELDER and MCKEE [17].

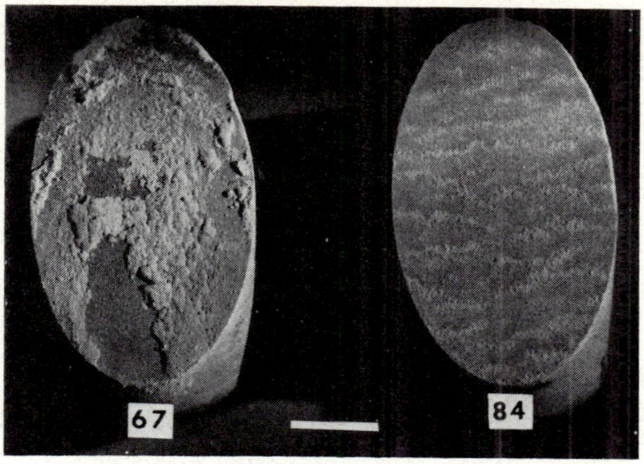

Figure 9
Photograph of indurated gouge on the pre-cut surface of Tennessee Sandstone experiment 67 formed upon continuous displacement at a rate of 10^{-3} cm/sec at 0.7 kb confining pressure. The indurated material partially spalls off the surface in large flakes. All specimens deformed at other conditions have surfaces similar to experiment 84 which is covered with patches of compacted gouge. The white banding on the surface of experiment 84 is gouge from the fine-grained bands formed within the gouge at angles 20°–30° to the sliding surface. Scale line is 2.0 cm

5. *Discussion of experimental results*

Stick-slip to stable-sliding transition

Thin section studies of the quartz gouge together with force-shortening records give clues to the mechanisms of stick-slip and stable sliding. The idea is that during stable sliding the sandstone rides on a layer of gouge which is flowing cataclastically, whereas during stick-slip the intact rock slips along the contact with the gouge layer.

In all experiments, stable sliding first occurs as the initial layer of sand is cataclastically deformed. Because cataclasis pervades this layer, shear deformation is probably distributed throughout, although banding indicates that the intensity of cataclastic deformation varies somewhat from point to point. At this stage no discrete sliding plane has developed either within the gouge or at the gouge-rock interface. Sliding involves only the scale of individual grains moving past each other. Essentially, stable sliding occurs because the intact sandstone is carried on a layer of gouge which is flowing internally by pervasive cataclasis.

As displacement progresses, stick-slip becomes the mode of sliding at confining pressures less than 0.7 kb. Stick-slip occurs when the gouge is reduced to a median grain-size of about 25 μm. It starts at smaller displacements as the confining pressure increases from 0.14 to 0.5 kb because cataclastic deformation per unit displacement is more intense at the higher confining pressure.

Most curves from stable sliding experiments have a peak which represents the ultimate strength of the specimen, followed by a negative slope which results from decreasing load due to a decreasing contact area. If stick-slip follows stable sliding, the force-shortening curves at confining pressures of less than 0.8 kb have inflection points which occur at the onset of stick-slip (Fig. 10). The inflection point becomes more striking as the confining pressure increases.

At 0.7 kb confining pressure the inflection point represents a shift in the manner by which the rock slides on the layer of gouge. Initially, the rock is riding on a layer

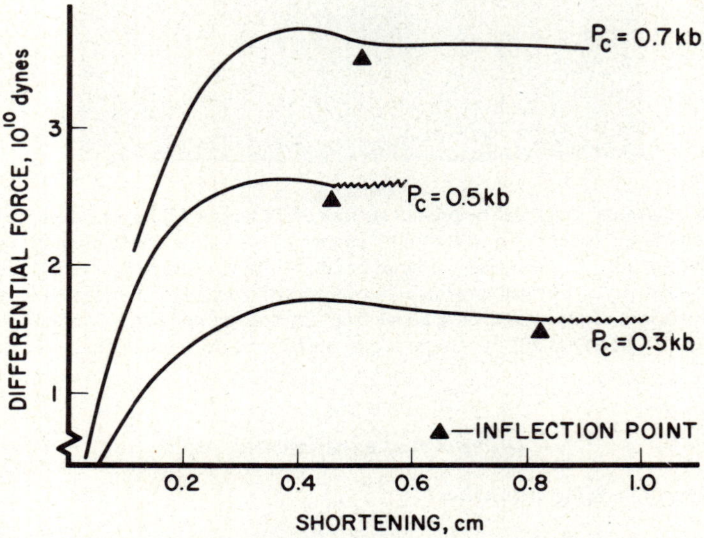

Figure 10

Differential force-shortening curves which have inflection points. The curves represent the sliding of pre-cut Tennessee Sandstone on an 0.12 cm thick layer of dry quartz gouge at a displacement rate of 10^{-3} cm/sec

which is flowing by pervasive cataclasis. Beyond the inflection point the shear zone becomes restricted to the rock–gouge interface. This is indicated by the formation of indurated gouge within a restricted zone at the contact with intact sandstone (Figs. 8 and 9). Indurated gouge is found only in specimens which have been displaced beyond the inflection point at 0.7–0.8 kb confining pressure and have slid stably. The indurated gouge is a manifestation of intense cataclasis due to the sliding along the gouge–sandstone interface.

At confining pressures of less than 0.7 kb, the sliding mode changes from stable to stick-slip at the inflection point. It is assumed that as in the 0.7 kb confining-pressure tests, the inflection point represents a shift from sliding on cataclastically flowing gouge to sliding along the interface. However, for unknown reasons, indurated gouge does not form at <0.7 kb confining pressure.

Other experiments also indicate that stick-slip may occur when sliding is restricted to a planar interface. When 0.1 cm deep grooves are cut into the surface of Tennessee

Sandstone before sliding on quartz gouge, planar shear surfaces do not develop at 0.5 kb confining pressure and only stable sliding occurs. HUMSTON [13] observed that stick-slip is repressed for sliding of mismatched surfaces of Tennessee Sandstone and BLACKWELL [14] says that it occurs only when the fit between planar surfaces is good. JAEGER and COOK [18] observe that stick-slip is more likely for pre-cut surfaces than fracture surfaces. Their observation is in agreement with some tests we have conducted on Coconino Sandstone (Fig. 1). Fractured Coconino Sandstone does not stick-slip at 0.5 kb confining pressure apparently because the rough fracture surface prevents the formation of the necessary smooth sliding surfaces. However, pre-cut Coconino Sandstone does stick-slip on a 0.12 cm thick layer of gouge.

At confining pressures above 1.0 kb, the indurated gouge does not form, yet the sliding is stable. It has been established that an inflection point in the force-shortening curve indicates a change in the intensity of cataclastic deformation at the gouge-rock interface at 0.7 kb confining pressure. At 1.0 kb confining pressure the force-shortening curve has a constant negative slope, which suggests a single mode of deformation throughout the experiment. At small displacements stable sliding occurs because the intact rock is carried on a layer of gouge which is flowing by pervasive cataclasis. It follows that pervasive cataclasis probably continues to carry the intact specimen throughout the experiment. That cataclasis is pervasive above 1.0 kb is supported by the observations that cataclasis is more intense at higher confining pressures (ENGELDER [10]) and banding of gouge into coarse and fine layers is less common than at lower pressures.

BRACE and BYERLEE [19] observed a stable-sliding to stick-slip transition for faulted specimens of Westerly Granite with 0.05 cm of gouge at 1.5 kb confining pressure. Stick-slip might not have occurred below 0.7 kb because of the poor fit between the surfaces of fractured Westerly Granite. It is also possible that this stable sliding observed at 1 kb confining pressure is caused by the same mechanism which is suggested for precut specimens of Tennessee Sandstone. Owing to the necessity of using smaller specimens and the difficulties of sample preparation and jacketing, the effects of confining pressures above 2.0 kb have not been investigated systematically for Tennessee Sandstone on quartz fault gouge. BYERLEE and BRACE [9] observed stick-slip for a compacted aggregate of sand at 7 to 10 kb confining pressure. We can speculate that Tennessee Sandstone would also stick-slip on quartz gouge at some unknown higher confining pressure.

Mechanical properties of fine-grained aggregates

Fine-grained quartz aggregates have been deformed in triaxial compression to document the behavior of simulated quartz gouge without the large shear tractions which were applied during the friction tests. By our definition a rock exhibits ductile behavior when it undergoes more than 5% permanent strain before faulting. Therefore, the pre-compacted quartz aggregates have a macroscopic brittle–ductile transition below 1.0 kb confining pressure and near the stick-slip to stable-sliding transition for quartz fault-gouge. The mechanism for stable sliding as outlined above required that quartz fault-gouge be able to sustain large 'macroscopic' shear strains by pervasive cataclasis, and this is true of the precompacted aggregates at the higher confining pressures.

It may also be significant that at confining pressures below 0.7 kb, a partial drop in load follows the development of fractures in compacted aggregates. The suggested mechanism for stick-slip requires that slip occurs along a discrete plane at the rock-gouge interface. The stick-slip stress drops below 0.7 kb confining pressure and the stress drops below 0.7 kb confining pressure due to fracture of the aggregate may be manifestations of the same process. Likewise, no stress drop occurs above 0.7 kb in either the friction experiments or in the compacted aggregates which fracture only after large 'macroscopic' strains.

Effect of pore water

The presence of water increases the kinetic coefficient of friction for Tennessee Sandstone sliding on quartz gouge (Fig. 5). A saturated gouge is finer grained compared to that of a dry gouge subjected to the equivalent deformation (ENGELDER [10]). The generation of the finer-grained material is probably due to stress corrosion of the quartz by the water. An effect of the finer grain-size is that the actual contact area between gouge and intact sandstone is greater. Because a larger shear force is necessary for sliding on a larger contact area, the measured kinetic friction is greater.

Energy considerations for the generation of quartz gouge

About 2.7×10^9 ergs is dissipated during sliding of pre-cut Tennessee Sandstone 0.1 cm on the layer of gouge at 0.7 kb confining pressure (Fig. 11). This is the work required to overcome frictional resistance in the gouge–sandstone system where fric-

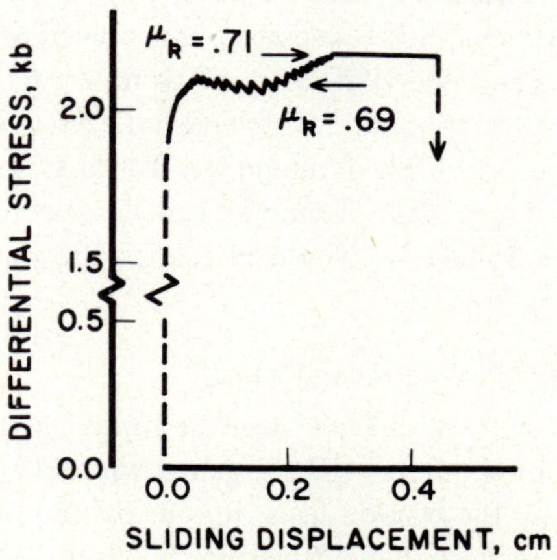

Figure 11
Differential stress-displacement curve illustrating the change in sliding mode and kinetic coefficient of friction during the generation of the indurated gouge. This is a drawing of cycle 3 of Fig. 3

tional heating, fracturing of grains, and radiation of seismic waves are three principal energy sinks. Creation of new surface area by brittle fracture during cataclasis of the entire layer of gouge requires only a small fraction of this work, but the additional work expended for the creation of the bands of indurated gouge is of the same order of magnitude as that necessary to create new surface area.

Part of the work of cataclastic deformation is spent in breaking atomic bonds and creating new surfaces during fracturing of the initial sand aggregate. The work done in breaking atomic bonds (i.e., fracture–surface energy) is equivalent to the increase in total surface (Helmholz) free energy, if the surface is created by an isothermal and reversible process (INMAN and TIPLER [20]). The amount of work (W) expended to create the large surface area of the fine-grained gouge is calculated by multiplying the estimated increase in area (A) by the surface energy per unit area (γ) of quartz:

$$W(\text{ergs}) = \gamma(\text{ergs/cm}^2) \times A(\text{cm}^2).$$

In order to estimate this increase in surface energy, it is assumed that the initial sand layer contains perfectly spherical quartz grains, all 0.02 cm in diameter. The layer is initially 0.17 cm thick on 28 cm^2 of sandstone, and it has 30% porosity. If it is assumed that the gouge generated from 0.02 cm spheres consists of spherical quartz particles 2.5×10^{-3} cm in diameter (ENGELDER [10]), then the increase in surface area is about 7.7×10^3 cm^2. The surface energy for quartz as determined by the Oberimoff–Gilman method ranges from about 400 ergs/cm^2 for (1011) to 1000 ergs/cm^2 for (1010) (BRACE and WALSH [21]). Taking 500 ergs/cm^2 as a representative value, we calculate the work required to generate the new surface area as 3.9×10^6 ergs. This is only a small fraction of the total work of 2.7×10^9 ergs.

Indurated gouge is generated during the 0.1 cm of displacement beginning at about the point where the magnitudes of the stick-slip stress drops begin to decrease (i.e., at a displacement of 0.2 cm in Fig. 11), and ending at the onset of stable sliding (i.e., at a displacement of 0.3 cm). During this 0.1 cm of sliding, the shear stress increases by 50 bars. Integrating the shear force (1.4×10^9 dynes) over 0.1 cm of displacement gives an increase in work of 7×10^7 ergs. About 5% of the gouge with a median size of 2.5×10^{-3} cm is reduced to indurated gouge, assumed to consist of 2.5×10^{-5} cm spheres (Fig. 8). About 2×10^7 ergs is required to create this new surface area. This is the same order of magnitude as the increase in work (7×10^7 dyne-cm) necessary to slide the sandstone 0.1 cm during the generation of the indurated gouge. Our point is that only a small fraction of the total work expended during the whole of the frictional sliding goes into fracturing of quartz to create new surface area. However, the small increase in work just prior to stable sliding (Fig. 11) can be accounted for by the large increase in surface area during the generation of the indurated gouge.

Natural faulting process

What, if anything, do these experiments tell us about natural faults? The best field analogs that we have found are faults like the Muddy Mountain Thrust Fault that have 3 to 30 cm of quartz gouge between sliding surfaces that are as planar as those in our experiments (ENGELDER [10]). The Muddy Mountain Thrust carries a section 4 to 5 km thick, and with a normal pore pressure, the effective confining pressure would have been 0.5 kb (BROCK [22]). This is in the stick-slip regime for sandstone on quartz gouge. However, our experiments would predict that thrust faults carrying thicker sections at normal pore pressure would slide stably.

Earthquakes are known to occur at depths in the crust where our experiments suggest that stable sliding should occur. Earthquake generation may well depend on untested parameters, including time of loading, type of gouge, temperature and surface roughness. Or it may be that seismogenic fault zones are under water pressures higher than hydrostatic.

HOSKINS *et al.* [23] and SCHOLZ *et al.* [7] observe small amounts of stable sliding prior to stock-slip stress drops. This also occurs in our tests at a rate of sliding of 10^{-5} cm/sec. These studies all suggest that if laboratory stress drops are really analogous to earthquakes, then stable sliding prior to a stress drop may be the same as aseismic creep which has preceded some large earthquakes.

6. Conclusions

1) For pre-cut surfaces of Tennessee Sandstone sliding on quartz gouge, the mode of sliding changes from stick-slip to stable sliding at effective confining pressures between 0.7 and 0.8 kb. This change in mode correlates with the macroscopic brittle–ductile transition in compacted quartz aggregates.

2) The stick-slip to stable-sliding transition is accompanied by a change from sliding of intact sandstone along its planar contact with the compacted gouge to riding on a layer of cataclastically flowing gouge.

3) The sliding-mode transition for quartz fault–gouge would eliminate stick-slip as a valid mechanism for 'locking' a fault zone in sandstone between effective confining pressures of 0.8 kb and at least 2.0 kb.

4) Compared to bare rock surfaces sliding without gouge, stress drops during stick-slip are drastically reduced by the presence of quartz gouge.

5) For pre-cut surfaces of sandstone sliding on compacted gouge at three different displacement rates at 0.5 kb confining pressure, the stress drop increases 10–15 bars for each decade of decrease in displacement rate.

6) An indurated film of gouge forms between the sandstone and an 0.12 cm thick layer of compacted quartz gouge only at about 0.7 kb confining pressure.

7) From considerations of the total work required to create new fracture surface during grain-size reduction, it is reasonable to conclude that cataclastic deformation of the gouge requires only a small fraction of the total work needed to slide pre-cut sandstone on gouge. Work expended solely for the creation of the indurated-gouge layers is of the same order of magnitude as that necessary to create the new surface area.

8) Quartz gouge is sufficiently permeable to permit fluid-pressure equilibrium throughout the compacted gouge in 10 to 30 sec. Thus the principle of effective stress should hold for the fault zone.

9) The fact that shallow earthquakes in pre-existing fault zones do occur at depths where effective confining pressure would exceed 0.7 kb for hydrostatic water pressure may imply that pore pressure in these zones are well above hydrostatic, although many other untested parameters may also affect the stick-slip to stable-sliding transition in nature.

Acknowledgement

Support from the Advanced Research Projects Agency of the Department of Defense through the US Geological Survey is acknowledged. This work benefited from discussions with M. FRIEDMAN, A. GANGI, G. SOWERS and D. STEARNS.

REFERENCES

[1] J. D. BYERLEE and W. F. BRACE, *Stick-slip, stable sliding, and earthquakes – Effect of rock type, pressure, strain rate, and stiffness*, J. Geoph. Res. *73* (1968), 6031–6037.

[2] J. H. DIETERICH, *Time-dependent friction in rocks*, J. Geoph. Res. *77* (1972), 3690–3697.

[3] J. D. BYERLEE, *Frictional characteristics of granite under high confining pressure*, J. Geoph. Res. *72* (1967), 3639–3648.

[4] J. H. COULSON, *The effects of surface roughness on the shear strength of joints in rocks*, Technical Report MRD-2-70, Missouri River Division, Corps of Engineers, Omaha, Nebraska (1970), 283 pp.

[5] J. C. JAEGER, *Friction of rocks and stability of rock slopes*, Geotechnique *21* (1969), 97–134.

[6] R. E. JACKSON and D. E. DUNN, *Experimental sliding friction and cataclasis of foliated rocks*, Int. Jour. Rock Mech. Min. Sci. *11* (1974), 235–249.

[7] C. H. SCHOLZ, P. MOLNAR and T. JOHNSON, *Detailed studies of frictional sliding of granite and implications for the earthquake mechanism*, J. Geoph. Res. *77* (1972), 6392–6406.

[8] M. FRIEDMAN, J. M. LOGAN and J. A. RIGERT, *Glass indurated quartz gouge in sliding-friction experiments on sandstones*, Geol. Soc. Am. Bull. *85* (1974), 937–942.

[9] J. D. BYERLEE and W. F. BRACE, *High-pressure mechanical instability in rocks*, Science *164* (1969), 713–715.

[10] J. T. ENGELDER, *Cataclasis and the generation of fault gouge*, Geol. Soc. Am. Bull. *85* (1974), 1515–1522.

[11] R. D. NASON and D. TOCHER, *Stone Canyon observatory: guidebook to the Gabilan Range and adjacent San Andreas Fault*; Amer. Assoc. Petr. Geol., Pacific Section, (1967), 105.

[12] J. M. LOGAN, T. IWASAKI, M. FRIEDMAN and S. A. KLING, *Experimental investigation of sliding friction in multilithologic specimens*, in *Geological Factors in Rapid Excavation*, edited by PINCUS, Geol. Soc. Am. Eng. Case History *9* (1973), 55–67.

[13] J. A. HUMSTON, *Experimental study of stick-slip in Tennessee sandstone* (M.S. Thesis), College Station, Texas A&M University (1972), 68 pp.

[14] M. L. BLACKWELL, *The influence of pore fluids on the frictional properties of quartzose sandstone* (M.S. Thesis), College Station, Texas A&M University (1973), 69 pp.

[15] J. T. ENGELDER, *Coefficients of friction and their values for sandstone sliding on quartz gouge*, 3rd Int. Cong. Int. Soc. Rock Mech., v. II (1974), 499–504.

[16] J. HANDIN, M. FRIEDMAN, J. LOGAN, L. PATTISON and H. SWOLFS, *Experimental folding of rocks under confining pressure: buckling of single-layer rock beams, in flow and fracture of rocks*, Geophys. Monograph Series *16* (1972), 1–28.

[17] J. T. ENGELDER and T. R. MCKEE, *Electron microscopical study of indurated quartz gouges*, Proc. 31st Ann. Mtg. Electron Microscopy Soc. Am. (1973), 214–215.

[18] J. C. JAEGER and N. G. W. COOK, *Fundamentals of Rock Mechanics* (London, Methuen, 1969) 513 p.

[19] W. F. BRACE and J. D. BYERLEE, *California earthquakes: why only shallow focus*, Science *168* (1970), 1573–1575.

[20] M. C. INMAN and H. R. TIPLER, *Interfacial energy and composition in metals and alloys*, Metal Rev. *8* (1963), 105–166.

[21] W. F. BRACE and J. B. WALSH, *Some direct measurements of the surface energy of quartz and orthoclase*, Am. Mineralogist *47* (1962), 1111–1122.

[22] W. G. BROCK, *Characterization of the Muddy Mountain–Keystone Thrust Contact and related deformation* (Master's Thesis), College Station, Texas A&M University (1973), 179 pp.

[23] E. R. HOSKINS, J. C. JAEGER and K. J. ROSENGREN, *A medium-scale direct friction experiment*, Int. J. Rock. Mech. Min. Sci. *5* (1968), 143–153.

(Received 28th October 1974)

Pageoph,
Vol. 113 (1975) Birkhäuser Verlag, Basel

Clay Gouges in the San Andreas Fault System and their Possible Implications

By F. T. Wu, L. Blatter and H. Roberson[1])

Summary – In the northern and central sections of the San Andreas Fault Zone, and along Calaveras and Hayward faults, clay gouges have been found to occur on the surface and at shallow depths.

It is consistent with the available geochemical data that such gouges can exist at depths down to 10 km. If extensive gouge materials exist in a fault zone then their properties will determine, to a large extent, the behavior of the fault. From known properties of clays in the presence of water we can infer that, in such cases, the tectonic stress and the stress drops for earthquakes will be low and substantial creep will take place before earthquakes.

1. Introduction

Recent work on the physical mechanism of large, shallow, earthquakes led to the conclusions that faulting takes place along pre-existing weak planes and that friction is the controlling factor (Brace and Byerlee [1]). These conclusions solved a number of conceptual difficulties such as the small stress drops involved in earthquakes (Brune [2]) and the total deviatoric stress needed to initiate the stick-slip process (Brune *et al.* [3]). Laboratory experiments are designed to further the understanding of the process (Johnson *et al.* [4]; Wu *et al.* [5]) and thoughts concerning earthquake predictions and control are based on these concepts. Quantitatively, however, stresses involved in this model are not compatible with currently accepted data. Stesky and Brace [6] conclude from their laboratory data that the frictional stresses in the San Andreas fault would range from about 750 bars to 2000 bars. These values are much too high compared to those (~100 to 200 bars) independently estimated by Brune *et al.* [3] from heat flow data. Stesky and Brace [6] suggest that frictional stresses may be reduced by the presence of crushed and hydrothermally altered materials or extremely high pore pressures.

In the laboratory a fault is often modeled as two surfaces with various size asperities in contact. Both natural or saw-cut surfaces are used (Brace and Byerlee [1]; Scholz *et al.* [7]); in either case some 'gouge', a product of mechanical wear, is sometimes present after stick-slip has taken place. Such gouge seems to affect the behavior of stick-slip in a significant way (Scholz *et al.* [7]). The San Andreas fault in California

[1]) Department of Geological Sciences, State University of New York, Binghamton, New York, 13901.

has been active for millions of years and the two sides of the fault have been displaced for hundreds of kilometers (CROWELL [8]), and the accumulation of such material must be extensive.

Three types of intrafault materials have been found in the San Andreas system (including the Howard and Calaveras faults). In central California, along the San Andreas fault, from San Juan Bautista to Parkfield and along the Calaveras and Hayward faults, clay gouges are exposed (NASON [9, 10]). North of San Juan Bautista, near Crystal Spring Lake, mylonites have been reported (WATERS and CAMPBELL [11]). Ground-up rocks have been found north of San Francisco and in the transverse range along the San Andreas fault, while in San Jacinto and Elsinore faults there are zeolites and other evidence of hydrothermal alterations (MANUEL N. BASS, personal communication, 1974).

This distribution of intrafault material is obviously related to two factors: 1) the country rock that provides the ground-up minerals, and 2) the availability and the composition and temperature of aqueous solutions in the fault zone. From Pt. Arena to San Luis Obisbo, the San Andreas fault is bordered on the east by the Franciscan rocks and on the west by granitic rocks and Cenozoic sedimentary rocks. South of San Luis Obisbo, the faults are mostly within the granitic rocks. The Hayward and Calaveras faults on the other hand are within the Franciscan rocks. Cataclastic feldspar, quartz, serpentine and other minerals from Franciscan rock may react with aqueous solutions to form clay.

That clay gouge appears in the same fault zone, not too far from a zone where mylonites with the same basic parent materials but in an entirely different mineralogic and physical state (WATERS and CAMPBELL [11]), seems to argue against the surface origin of the clay and support the thesis that the depth of formation of these materials is not shallow. The Franciscan rocks, which seem to be necessary for the formation of clay gouge, extends to a depth of about 12 km on the eastern side of the San Andreas Fault (STEWART [12]). We shall discuss the geochemical condition under which Franciscan rocks can convert to clays of various composition. There is some other circumstantial evidence, however, that the clay gouge we observe on the surface is formed at depth: 1) considering that the San Andreas fault system has been active and exposed for >50 my (CROWELL [8]) and that a reasonable estimate of erosion rate is about 0.1 mm/yr, what we see on the surface may have been at a depth of 5 km, and 2) clay gouge has been observed to occur in deep mines in the fault zone down to a depth of several kilometers.

If the system in the depth range of interest (3–12 km) is dry then the products of grinding is cataclastic rocks of various sorts or possibly glass (MCKENZIE and BRUNE [13]; FRIEDMAN *et al.* [14]), even though the gouge near the surface may be clayey. But it is likely that assorted clay minerals form in the zone in the presence of water.

There are many types of clay minerals. Due to their different chemical and structural characteristics they may have diverse mechanical properties. For example, montmorillonite has the tendency to adsorb cations and, of special interest to the study of earthquake generation, to adsorb water, dilate and lose its strength. On the other hand, illite does not adsorb as much water and tends to be somewhat more stable mechanically.

If, as we propose, the clay gouge or mylonite does extend down to 12 km then the mechanical behavior of the fault will be almost completely controlled by the physical and chemical properties of the gouge. One clear example of gouge-zone controlled faulting is the 1966 Parkfield earthquake, where the displacement occurred on two sides of the faulted graben, which is filled with gouge (OAKSHOTT [15]; also Geologic Map of California, USGS Map G67017). Although much is to be learned about the formation of clay and the behavior of clay and cataclastics under shear, moderate confining pressure (up to 3–4 kb), and moderate temperature (up to 300°C), we must turn to the existing extensive body of soil mechanics literature to make an assessment of the influence of gouge such as in the Parkfield example.

We should emphasize again that at this stage there are a large number of unknowns in this problem, such as the real nature of gouge material, the amount and the chemistry of the pore water, the fabric of the gouge, the mechanical properties of gouge under the proper range of pressure and temperature, etc. These can only be known by further laboratory experiments and eventually, deep drilling into the fault zones. The purpose of this paper is to look into the possible influence of clay gouge on the mechanical behavior of faults based on available data. The importance of knowing the details of the material in the gouge zone is obvious when the earthquake prediction and control problems are considered.

2. *Possible existence of clay gouge down to 12 km*

To look into the problem of whether clay gouge can be generated down to a depth of 12 km, we shall use a simplified model of the country rock for San Andreas in Northern and Southern California (Table 1) as a 'starting' condition. The Franciscan rock contains graywacke (~80%) (with a high soda to potash ratio), shale, altered mafic volcanic rock, chert and minor limestone (BAILEY *et al.* [16]).

Table 1

Simplified crustal model of California in the vicinity of San Andreas fault system

Depth (km)	San Andreas				Calaveras and Hayward Faults	
	Northern		Southern			
	East	West	East	West	East	West
0–12	'Franciscan'	Granitic Rocks	Granitic Rocks	Granitic Rocks	'Franciscan'	'Franciscan'
>12	Basic Rocks					

Formation of clay minerals is initiated by the cataclasis of the wall rock during repeated fault movement. We may expect the product to be mixed in size at first, but during each episode of sudden movement, the crushed rocks are further reduced in size. In between the episodes, creeping of the fault may also cause cataclasis. The pulverization process is of course facilitated by the presence of water.

Clay minerals could be expected to form quite rapidly as an alteration of the pulverized rocks due to leaching by percolating fluids (either meteoric or hypogene). However, with increasing depth, and hence temperature, the rate of clay formation increases 3–4 fold with each 10°C increase in temperature (MILLOT [17]). The nature of the hypogene solutions is a crucial factor in the formation of various clay minerals since the clay mineral developed will depend on the pH and composition of the solutions. The nature of the solutions is determined by its origin (e.g., fresh vs. marine) and the composition of the rock they invade and react with.

The fault zone, as well as the repeatedly strained rocks near the fault zone, form a natural conduit for migrating solutions. Recent observations of precursory velocity changes (e.g., SCHOLZ *et al.* [18]) in the fault zone may indicate the mobility of fluids in the vicinity of the fault.

The types of clay minerals that can form depend upon: 1) the composition of the country rock; 2) the composition of the solutions; 3) the temperature and pressure. Because of a paucity of data on the second factors we cannot be certain of the exact mineralogy of the assemblage that might develop. We shall discuss in general terms what clay minerals could be formed under sets of possible conditions.

In a sodium-, potasium-, magnesium- or calcium-rich environment, experimental data have shown montmorillonite (or more generally, smectite, see GRIM [19]) to be the stable clay mineral phase up to temperatures of about 400°C (MUMPTON and ROY [20]; AMES and SANDS [21]). In Northern California, where the San Andreas fault cuts predominantly sodium–calcium-rich graywacke, ultrabasic igneous rocks, volcanic ash and metamorphic Franciscan rock assemblages (BAILEY *et al.* [16]), smectite is expected to develop in the fault zone.

The mineralogy of the clays would of course change if the concentration of potassium in the pore solution or the pH increased to the point that a more illitic (non-expandable) phase became the stable clay mineral. Thus, assuming that the pore solutions were dilute, highly expandable clays are expected to form from the Franciscan fault gouge and persist to depths corresponding to ~400°C. However, if the pore waters were not dilute, but rather were saline at depth, then the presence of potassium may promote the development of a mixed-layer illite/smectite phase. The rate of development of a mixed-layer illite/smectite phase has, however, been found to be greatly retarded if the K/cation rate is low (~0.5) (BLATTER [22]). Therefore, even if the pore waters were marine in origin, the sodium and calcium released from the country rock during alteration may keep the K/cation ratio lower than that required for illite to be the stable clay mineral and hence only expandable clays would form.

In Southern California where the fault cuts more potassium-rich granitic rocks

(BAILEY *et al.* [16]), the type of clay which could be expected would depend more heavily on the origin of the pore solutions than was the case in the sodium rich rocks of the Franciscan assemblage. Experimental studies have shown that a Na–K–feldspar immersed in sea water at 200°C altered to an illite after six weeks. However, the same feldspar immersed in distilled water under the same conditions formed an expandable phase (BLATTER [22]). Therefore, it could be expected that an expandable smectite would again develop from the alteration of the granitic rocks as long as the pore solutions remained relatively K^+-poor. The leaching of the granite by meteoric waters may result in an accumulation of potassium to the point that its activity became large enough for a mixed-layer illite/montmorillonite phase to form. If, on the other hand, the pore solutions were K^+-rich, the more illitic phase may be expected to be the direct alteration product. In this case, the pressures that could be generated due to the hydration of the clays would be significantly less than if montmorillonite were the dominant phase. However, in either case, if an abundance of either of these clays were formed, the friction of the fault would be reduced as a result of the weak bonding strength of both clay minerals.

The above discussion has been restricted to depths up to approximately 12 km. Below this depth, under a normal geothermal gradient, clay minerals start to become unstable and the minerals comprising the greenschist facies become stable. At depths at which the greenschist facies become the stable assemblage, the clay minerals begin to alter to chlorite, pyrophyllite and micas (muscovite and biotite). While these minerals are not hydratable, they do have a platy morphology and a weak bonding strength. Therefore, while expandable clay minerals would not be expected at great depths, weakly bonded platy minerals which would also offer little resistance to the fault movement would be expected to replace them.

3. *Mechanical behavior of clay gouge*

Most of the experimental studies pertaining to the mechanical behavior of clays have been conducted under very low confining pressure of a few bars (see e.g., LEE and INGLES [23]; a few experiments are performed under moderate pressures of up to 70 bars (e.g. OLSON and PAROLA [24]). Although these pressures correspond to shallow depths (up to about 300 m), judging from trends of those data, some of the general conclusions from these experiments are most probably valid[2]) even though the absolute magnitude of the quantities measured cannot be extrapolated.

Under the confining pressure of 1–3 kb and perhaps shear stress of up to 200 bars (BRUNE *et al.* [3]), the gouge can be expected to be densely packed. The platelets are parallel to each other locally, though on a larger scale they may appear to have clumps with different dominant directions and slickensides resulting from repeated internal deformation under shear. Because of the confining pressure, it is very likely that the

[2]) In the interpretation of soil mechanics data, the initial compaction has to be taken into consideration. The heavily compacted clays are closer to gouge than the 'remodeled' clays.

adsorbed water has been partially driven out. However, this depends very much on the permeability of the system. A poorly drained system may cause the gouge to remain saturated.

An undersaturated, and densely packed gouge under shear will experience a *dilatancy* effect. This effect is produced by the rearrangement of the individual platelets in the dense structure. The rearrangement may involve rotation and translation and the result is that a large number of voids are created (see, for example, SCOTT [25], p. 51). Decrease in pore pressure will cause the effective stress to increase, a strengthening effect, much like that in rocks (NUR [26]; SCHOLZ *et al.* [18]). Thus, an originally almost impervious material may now be open for water to enter.

Water will then be drawn into the dilatant gouge by: 1) the presence of a pressure gradient; and 2) the adsorption potential of clay. The first factor is self-evident. The second factor is complicated and highly variable from clay to clay. The controlling factors include: the mineralogy, the exchangeable ions, electrolyte content of the aqueous phase, confining pressure, void size and distribution, etc. (MIELENZ and KING [27]). The mechanisms of adsorption according to MIELENZ and KING [27] are: 1) a reduction of effective compressive stress due to an enlargement of capillary film; and 2) osmotic imbibing. The capillary compressive force increases as the radius of capillary decreases, therefore, clays of small grain size (e.g., montmorillomite) have high capillary pressure and low effective confining pressure. Sodium montmorillonite is one of the most adsorbent clay minerals and compact montmorillonite has been found to exert a high swelling pressure when it is confined. Potassium, calcium, magnesium montmorillonite, and illitic minerals are found to be somewhat less adsorbent and kaolinite is non-adsorbent (*ibid.*). The relaxation of effective compressive stress will, of course, decrease the strength of the clay. When dry or undersaturated, clays can have quite high strength; however, with variations of $6\frac{1}{2}\%$ to 18% in water content for some clay, the strength can decrease by two orders of magnitude (OLSON and PAROLA [24]).

For the San Andreas fault system most of the water must come from the country rock or percolate along the fault zone. Depending on the rate of intake of water by the clay and the permeability of the country rock, we may have the area immediately surrounding the fault (where the porosity and permeability can be expected to be higher than some distance away from the fault due to jointing) drained of most of its water and gradually filled again when the water slowly returns to the area. The gouge, if the shear strength is still greater than the shear stress, will continue to take up water and swell until it is weakened further and the shear stress exceeds the strength of the clay; whence a rapid failure of the gouge will create stress drops and therefore earthquakes. The amount of water needed and the extent of total volume change of the gouge depend on the species and the volume of clay taking part in the dilatancy process. It is clear, however, that the participation of only a very narrow region near the gouge country rock contact is necessary to decrease 'frictional strength' of the fault (cf. STESKY and BRACE [6]).

Before the failure, clay may exhibit a large strain. OLSON and PAROLA [24] report 20–30% of axial strain before failure. This phenomenon could explain the precursory creep observed for several Central Californian earthquakes (NASON [28]). Also, the *en echelon* cracks in the Parkfield area (ALLEN and SMITH [29]) could very well be related to extensive creeping before the catastrophic failure of the gouge material. This mechanism is quite different from observed accelerating creep before a stick-slip event in laboratory experiments (SCHOLZ *et al.* [7]; WU *et al.* [5]).

For some clays, when subjected to repeated shear, the angle of internal friction becomes successively lower, i.e., it is more and more prone to slippage. This happens when the interlocking grains are rearranged so that the scale-like and flake-shaped clay particles adjust themselves parallel to the shear plane until they form a continuous, shiny shear surface (slickensided clays) (BOROWICKA [30]). This effect tends to delineate the slip-plane and subsequent fault motion may occcur along it. Because of the low strength, progressive rupture or very small earthquakes may take place in a zone where this has happened. This may account for the more or less continuous creep with or without small earthquakes along the section of San Andreas between San Juan Bautista and Cholaine.

We have based our discussions so far on the premise that the system is drained, i.e., the water leaves the gouge after shear stress is removed or has decreased to a level below that necessary to create dilatancy. If the system is undrained then the clay may not regain its strength. In this situation we may also expect constant creep.

It should be pointed out that in this model with substantial clay gouge, a water-controlled precursory velocity change will be effected if the dilatancy volume is large enough so that the water taken into the clay gouge is more than the reservoir in the country rock close to the fault zone (where the crack porosity and permeability are expected to be higher than those further out) can supply. Then, initially, when dilatancy starts we may expect V_p to decrease in a narrow zone next to the fault, and V_p will resume the original value when water fills the evacuated cracks.

4. *Discussion*

The motivation for proposing the foregoing model for the San Andreas fault system stems from the observations that, for a fault that has been active for millions of years with displacement along the fault for a few kilometers to hundreds of kilometers, the amount of gouge material in the fault zone must be extensive and that with extensive gouge the rock-on-rock stick-slip model of faulting (BYERLEE and BRACE [31]) may not be applicable.

The alteration of gouge, with the composition of the participating wall rocks, to that of clay is suggested by works on the hydrothermal alteration of country rocks at depth, as well as the occurrence of clay gouge at tens of feet to a few kilometers below the surface in tunnels and wells near the surface. Only deep drillings can confirm the exact nature of the gouge and the pore water. These factors will allow us to know more precisely the mechanical behavior of the fault system. Simultaneously, data on the equilibrium conditions of clays with various solutions under various temperature and pressure conditions[3]) are needed to extend the knowledge gained from very few deep holes that we can afford to drill.

The general behavior of clay gouge at depth can be extrapolated from existing soil mechanics experimental data and theories. However, experiments under much higher confining pressure and temperature and with various pore solution pressure and composition are needed so that the exact amount of dilatancy swelling, the amount of yielding before failure, the failure strength, and premonitoring phenomena can be determined.

We may summarize our discussions as follows: 1) for a fault with repeated movements we may expect the accumulation of cataclastic rocks to be extensive and under favorable conditions, such material may alter to clay gouge; 2) in this case we may expect the faulting to follow the failure behavior of clay gouge – thus, low 'tectonic stress' needed to activate the fault, low stress drop and substantial precursory creep prior to earthquakes; 3) the strain is constantly being released; and 4) we may observe velocity anomalies in the region surrounding the fault due to the dilantancy–water adsorption properties of the clay gouge.

Acknowledgement

Discussions with Drs. Robert Nason, Rosemary Vidale, and Marc Bodine were helpful. We wish to thank Thomas Flynn for the hydrothermal run of the gouge samples and Thomas Fernalld for the X-ray identification of the samples.

The first author (FTW) was supported by NSF Grant A038712 while carrying out this research.

References

[1] W. F. Brace and J. D. Byerlee, *California earthquakes: why only shallow focus?* Science *168* (1970), 1573–1575.

[2] J. N. Brune, *Tectonic stress and the spectra of seismic shear waves from earthquakes*, *75* (1970), 4997.

[3] J. N. Brune, T. L. Henyey and R. F. Roy, *Heat flow, and rate of slip along the San Andreas Fault, California*, J. Geophy. Res. *14* (1969), 3821–3827.

[4] T. Johnson, F. T. Wu and C. Scholz, *Source parameters for stick-slip and earthquakes*, Science *179* (1973), 278–280.

[5] F. T. Wu, K. C. Thomson and H. Kuenzler, *Stick-slip propagation velocity and seismic source mechanism.* Bull. Seism. Soc. Am. *62* (1972), 1621–1628.

[6] R. M. Stesky and W. F. Brace, *Estimation of frictional stress on the San Andreas Fault from Laboratory Experiments*, Proc. of the Conf. on Tectonic Prob. of the San Andreas Fault system, School of Earth Sciences, Stanford University, California, (1973), 206–214.

[7] C. H. Scholz, P. Molnar and T. Johnson, *Detailed studies of frictional sliding of granite and implications for the earthquake mechanism*, J. Geophys. Res. *77* (1972), 6392–6406.

---

[3]) A preliminary experiment has been performed by subjecting two clay gouge samples with montmorillonite composition) collected by Robert Nason in Central California to a pressure of 3 kb and temperature at 300°C. After three months, the mineralogy of the samples was not significantly altered.

[8] J. C. CROWELL, *Displacement along the San Andreas fault, California*, Special GSA paper, No. 71 (1962).

[9] R. D. NASON, *Investigation of fault creep slippage in northern and central California*, Ph.D. Thesis, UCSD (1971), 232 pp.

[10] R. D. NASON, *Propagation of fault creep events*, NOAA Technical Report ERL 236-ESL21 (1972).

[11] R. C. WATERS and C. D. CAMPBELL, *Mylonites from the San Andreas fault zone*, Am. Jour. of Science *29* (1935), 473–503.

[12] S. W. STEWART, *Preliminary comparison of seismic travel times and inferred crustal structure adjacent to the San Andreas fault in the Diablo and Gabilan ranges of Central California*, Proc. Conf. on Geologic Prob. of San Andreas Fault System, Stanford University, (1968), 218–230.

[13] D. P. MCKENZIE and J. N. BRUNE, *Melting on fault planes during large earthquakes*, Geophy J.R.A.S. *29* (1972), 65–78.

[14] M. FRIEDMAN, J. M. LOGAN and J. A. RIGERT, *Glass-indurated quartz gouge in sliding friction experiments on sandstone*, Bull. Geol. Soc. Am. *85* (1974), 937–942.

[15] G. B. OAKSHOTT, *Parkfield earthquakes of June 27–29, 1966, Monterey and San Luis Obispo Counties, California – Preliminary Report, geologic features*, Bull. Seismol. Soc. Amer. *56* (1966), 961–971.

[16] E. H. BAILEY, W. P. IRWIN and D. L. JONES, *Franciscan and related rocks, and their significance in the geology of western California*, Calif. Div. of Mines & Geology Bull. *83* (1964).

[17] MILLOT, *Geology of Clays* (Springer-Verlag, New York 1970).

[18] C. SCHOLZ, L. SYKES and Y. R. AGGARWAL, *The physical basis for earthquake prediction*, Science *181* (1973), 803–810.

[19] R. E. GRIM, *Clay Mineralogy* (McGraw-Hill Book Co., 1968), p. 596.

[20] F. A. MUMPTON and R. ROY, *The influence of ionic substitution on the hydrothermal stability of montmorillonites*, National Res. Council Pub. *456* (1956), 337–339.

[21] L. L. AMES and L. B. SANDS, *Factors effecting maximum hydrothermal stability in montmorillonites*, Am. Mineralogist *43* (1958), 641–649.

[22] C. L. BLATTER, *The interaction of clay minerals with distilled and saline solutions at elevated temperatures*, Unpublished Ph.D. Thesis, State University of New York at Binghamton, New York (1974).

[23] I. K. LEE and O. G. INGLES, *Strength and deformation of soils and rocks*, in *Soil Mechanics, Selected Topics* (American Elsevier Publishing Co., Inc., New York 1968).

[24] R. E. OLSON and J. F. PAROLA, *Dynamic shearing properties of compacted clay*, Int. Symp. on Wave Propagation and Dynamic Properties of Earth Material, Univ. of N. Mexico Press (1961), 173–182.

[25] C. R. SCOTT, *An introduction to Soil Mechanics and Foundations* (McLaren, London 1969).

[26] A. NUR, *Dilatancy, pore fluids, and premonitory variations of t_s/t_p travel times*, Bull. Seismol. Soc. Amer. *62* (1972), 1217.

[27] R. C. MIELENZ and M. E. KING, *Physical–chemical properties and engineering performance of clays, clays and clay technology*, California Div. of Mines Bull. *169* (1965), 196–266.

[28] R. D. NASON, *Fault creep and earthquakes on the San Andreas Fault*, Proc. of the Conference on Tectonic Problems of the San Andreas Fault System, edited by R. L. KOVACH and A. NUR, Geol. Soc. vol. XIII, Stanford University (1973), 275–285.

[29] C. R. ALLEN and S. W. SMITH, *Pre-earthquake and post-earthquakes surficial displacements*, Bull. Seism. Soc. Am. *56* (1966), 966–967.

[30] H. BOROWICKA, *The influence of the colloidal content on the shear strength of clay*. Sixth Int. Conf. on Soil Mech. and Foundation Engineering, Vol. 1 (University of Toronto Press, 1965).

[31] J. D. BYERLEE and W. F. BRACE, *Stick-slip, stable sliding and earthquakes – effects of rock type, pressure, strain rate and stiffness*, J. Geophys. Res. *73* (1968), 6031–6037

[32] B. VELDE, *The compositional foin muscovite-pyrophyllite at moderate temperatures and pressures*, Bull. Soc. Franc. Miner. Crist. *92* (1969), 360–368.

(Received 4th November 1974)

Pageoph,
Vol. 113 (1975) Birkhäuser Verlag, Basel

An Analysis of Local Changes in Gravity Due to Deformation

By J. B. Walsh[1])

Summary – Deformation causes changes in gravity even though the measuring station is fixed relative to the center of the earth and the total mass in the vicinity of the station does not change. Deformation has two effects: the density at a point changes in response to changes in local stresses and the point itself moves relative to the station. The general expression derived here shows that the gravity change depends on the uplift immediately beneath the station and the deformation of cavities at depth – cracks, pores, magma reservoirs, etc. The approximation is made that density is uniform except for cavities, but no assumptions about the constitutive behavior of the medium are necessary.

The gravity change studied here shows interesting analogies with the gravity effect due to buried bodies with anomalous density. For example, I find that the net volume change due to deformation can be found by integrating the gravity anomaly over the earth's surface, in analogy with the calculations for excess mass.

The theory is especially useful when both deformation and net fluid flow into or out of the system occurs. The results are applied to uplift and gravity changes observed during the great Alaskan earthquake, the Matsushiro earthquake swarm, and a draw-down operation at the Wairakei geothermal field. The data at Wairakei are used to estimate the net amount of water taken from the aquifer. The data at Matsushiro is not sufficiently precise to discriminate between the various mechanisms that have been proposed, but the Alaskan earthquake data suggest that no dilatancy or fluid flow occurred between the survey in 1962 and the resurvey in 1964 just after the event.

1. Introduction

Gravity measured near volcanoes and earthquake zones changes in response to the deformation which accompanies activity. Deformation has two direct effects upon gravity. First, the density of a point in the rock matrix changes when the local stress distribution changes. In addition, the point itself may move, carrying an element of mass closer to or farther from the measuring station. Although the density changes and displacements are small, the deformed region is large, and the effect upon gravity is found in the analysis here to be appreciable.

Deformation is the source of other changes in gravity, as well, and the anomaly is the net result of all such changes. Gravity typically is measured at a station on the earth's surface. Therefore, the station moves away from the center of the earth when uplift occurs, and gravity decreases accordingly. Also, deformation may open new fractures or change the size of existing ones; fluids – water or magma, for example – flowing into or out of the cavities change the mass beneath the station and produce a

[1]) Department of Earth and Planetary Sciences, Massachusetts Institute of Technology, Cambridge, Massachusetts 02139.

change in gravity. Both of these effects can be calculated following well-established procedures, however, and the results can be applied to the problem here without further analysis.

Measurements of deformation and changes in the gravity effect have been found to be especially useful for intepreting events where fluid flow may have occurred. BARNES [1] correlated uplift due to the Alaskan earthquake in 1966 with changes in gravity and suggested that flow of rock had occurred at depth, increasing the mass beneath the survey area. NUR [2] and KISSLINGER [3] studied gravity and uplift measurements spanning the Matsushiro earthquake swarm and interpreted the results in terms of the flow of ground water into a dilatant zone. Similar techniques have been applied to volcanic eruptions by IIDA *et al.* [4] and the drawing of liquids and gases from a geothermal field by HUNT [5]. I show here that a different interpretation of the data is possible in some of these studies when the gravity anomaly due to deformation found in the analyses is considered.

To analyze the anomaly due to deformation *per se* (i.e., the reference point is fixed in space and the mass of the system does not change), we assume that deformation occurs over a region which is small enough that the earth can be approximated by the half-space $z \geqslant 0$, with z taken positive downward (see Fig. 1). The medium is assumed

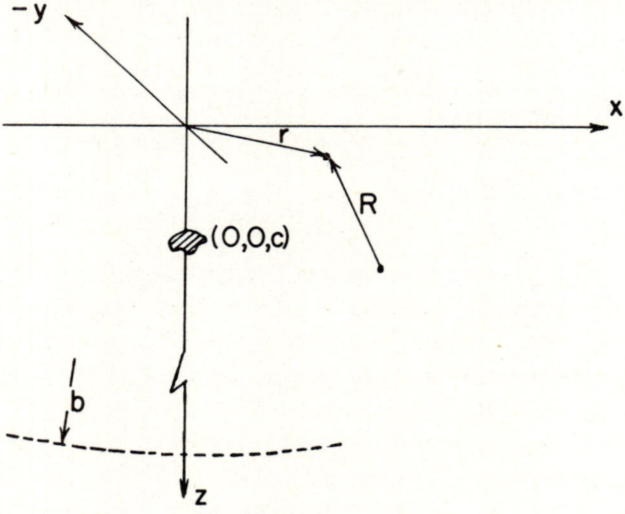

Figure 1

Coordinate system. The earth is assumed to be a half-space with z positive downward. **R** is a vector from an arbitrary point in the interior to an arbitrary reference point on $z = 0$ located a distance **r** from the origin. The dotted line is a hemisphere having radius b, which is much larger than the size or depth of the source. Shown in the figure is a point source at depth c, but the analysis can be applied to sources of any size or depth

to be homogeneous with uniform density. The analysis is carried out without specifying any particular constitutive behavior, and so the expression for the gravity effect can be applied to any material whether elastic, plastic, viscous, etc.

2. Analysis

The gravitational potential U due to a distribution of particles having mass ρdV is

$$U = -\int_V (G/R)\,\rho dV \tag{1}$$

where G is the gravitational constant, R is the distance between a point and the reference, and the integration is over the volume V. The change dU in potential when the density and distribution of the particles is changed is

$$dU = -G\int_V (d\rho/R - dR/R^2)\,dV \tag{2}$$

where $d\rho$ is the change in density. Let $\mathbf{u}$ be the displacement of a point and $\mathbf{R}$ a vector from the point directed toward the reference. The change $d\rho$ in density and the increase dR of the radius can be expressed in terms of the displacement $\mathbf{u}$, as follows:

$$\begin{aligned} d\rho &= -\rho\nabla\cdot\mathbf{u} \\ dR &= -\mathbf{u}\cdot\mathbf{R}. \end{aligned} \tag{3}$$

Substituting (3) into (2), we find

$$dU = G\rho\int_V \left(\frac{\nabla\cdot\mathbf{u}}{R} - \frac{\mathbf{u}\cdot\mathbf{R}}{R^2}\right) dV$$

or, equivalently,

$$dU = G\rho\int_V \nabla\cdot(\mathbf{u}/R)\,dV. \tag{4}$$

The divergence theorem is now applied to (4), giving

$$\begin{aligned} dU &= G\rho\int_S \mathbf{n}\cdot(\mathbf{u}/R)\,dS \\ &= G\rho\int_S (\mathbf{n}\cdot\mathbf{u})(1/R)\,dS \end{aligned} \tag{5}$$

where the surface S includes both interior and exterior surfaces in V, and $\mathbf{n}$ is the outward-drawn unit vector on S. We see in (5) that the change in potential depends only on the mass flux, given by $\rho(\mathbf{n}\cdot\mathbf{u})\,dS$ at a point on S and the distance R from that point to the reference.

The gravity effect Δg is found by differentiating (5), giving the expression,

$$\Delta g = -\partial/\partial z(dU) = -G\rho\int_S (\mathbf{n}\cdot\mathbf{u})(z/R^3)\,dS. \tag{6}$$

Integration of (6) over interior surfaces S_c is straightforward. Integration over the surface S_0 defined by the plane $z=0$ must be separated into two parts because of the singularity at $R=0$. The procedure is the same as that used for deriving the expression for the 'equivalent stratum' (see GRANT and WEST [1], p. 214), and we find that the contribution from the surface S_0 to the gravity effect depends only on the mass flux immediately beneath the station, i.e.,

$$\Delta g = 2\pi\rho G h + G\rho \int_{S_c} (\mathbf{n}\cdot\mathbf{u})(z/R^3)\, dS \tag{7}$$

where h, taken positive upwards, is the vertical displacement of the reference point on S_0. The gravity effect due to a point mass dm is $Gdm(z/R^3)$ in the notation here. The integral in (7), therefore, can be considered to be the gravity effect due to a distribution of points having mass $\rho(\mathbf{n}\cdot\mathbf{u})\, dS$ (where the mass is positive or negative depending on the sense of $\mathbf{u}$) distributed on S_c. Thus, the gravity anomaly due to deformation is the same as would be found if no deformation occurred, but material having density ρ was added to or excavated from all surfaces to simulate the actual deformation. Note, however, that this process would generally require a net increase or decrease in the mass of the system, whereas, in actuality only the volume, and not the mass, changes during deformation as the problem is formulated here. The density ρ in (7) must be uniform, but the medium need not be elastic or linear, because no assumptions about constitutive behavior were made in the derivation.

Spatial anomalies, i.e., differences in gravity in an area relative to values in the surrounding 'normal' region, indicate the presence of high or low density material. The theory needed for interpreting these anomalies is well developed, and much of it can be adapted to the problem here. Consider, for example, the spatial anomaly due to a buried mass of volume v having density $\Delta\rho$ greater than the surrounding material. The 'excess mass' $\Delta\rho v$ associated with this body can be calculated by integrating the gravity anomaly over the surface $z=0$ (see GRANT and WEST [6], p. 227), as follows:

$$\int_{S_0} \Delta g dS = 2\pi G v \Delta\rho. \tag{8}$$

An analogous relationship can be found in the present problem by adapting the procedures presented by GRANT and WEST [1]. First, note that the potential U in (1) satisfies Poisson's equation, and so the differential potential dU in (2) must also; that is,

$$\nabla\cdot\nabla(dU) = 4\pi G d\rho. \tag{9}$$

Equation 9 is integrated over V giving

$$\int_V \nabla\cdot\nabla dU = 4\pi G \int_V (d\rho)\, dV. \tag{10}$$

Equation 3 is introduced into the right-hand side of (10), and the divergence theorem is applied to both integrals, giving

$$\int_S (\mathbf{n} \cdot \nabla dU)\, dS = -4\pi G\rho \int_S (\mathbf{n} \cdot \mathbf{u})\, dS. \tag{11}$$

The integral on the right-hand side of (11) is the increase ΔV in the volume of the half-space. The left-hand side is separated into integration over the surface $z = 0$ and over a hemisphere S_b in $z \geqslant 0$ with radius b (see Fig. 1) which is much larger than the dimensions of the region where most of the deformation occurs. Equation 11 becomes

$$-\int_{S_0} [\partial(dU)/\partial z]\, dS + \int_{S_b} [\partial(dU)/\partial R]\, dS = -4\pi G\rho \Delta V. \tag{12}$$

The first integral in (12) is the gravity effect integrated over $z = 0$. The second integral is found by differentiating (2):

$$\partial(dU)/\partial R = G \int_V [d\rho/R^2 - 2(\mathbf{u} \cdot \mathbf{R})/R^3]\, dV,$$

where V is the volume of the deformed region. The radius of the hemisphere is very large, and so we can assume that the density increase and displacements can be replaced by average values $-\rho\Delta V/V$ and $\langle \mathbf{u} \cdot \mathbf{R} \rangle$ for the deformed region. Equation 13 can be now written in the form

$$\partial(dU)/\partial R = -(G\rho/R^2)(\Delta V + 2\langle \mathbf{u} \cdot \mathbf{R} \rangle V/R). \tag{13}$$

Equation 13 is introduced into (12), integrated over S_b, and evaluated in the limit as $b \to \infty$. The contribution of $\langle \mathbf{u} \cdot \mathbf{R} \rangle$ is found to be negligible, and (12) becomes

$$-\int_{S_0} \Delta g dS - 2\pi G\rho \Delta V = -4\pi G\rho \Delta$$

or

$$\int_{S_0} \Delta g dS = 2\pi G\rho \Delta V. \tag{14}$$

Thus, the gravity anomaly integrated over the surface is proportional to the increase in volume of the half-space. Note the similarity between (8) and (14).

The expressions used to calculate the center of mass of buried bodies also have analogies in the present problem. The analysis is nearly the same as the standard treatment (GRANT and WEST [1], p. 228) and need not be presented here. One finds that the coordinates $(\bar{x}, \bar{y})$ of the center of gravity of the change ΔV in volume can be found from the distribution of Δg on S_0; that is,

$$2\pi G\rho\bar{x}\Delta V = \int_{S_0} x\Delta g\ dS$$

$$2\pi G\rho\bar{y}\Delta V = \int_{S_0} y\Delta g\ dS. \tag{15}$$

3. Free-air correction and the effect of fluid flow

I assumed in deriving the expression for the gravity effect that the reference point is fixed in space and that no mass enters or leaves the system. Gravity measurements typically are measured with the meter fixed relative to the earth's surface, however, and so they must be corrected for the free-air effect, which accounts for movement of the reference away from the center of the earth. This correction, as discussed in standard texts (e.g., see GRANT and WEST [6], p. 237) depends to a small extent on latitude. Here, we neglect this term because of the uncertainties inherent in the method and the data, and write the free-air correction as

$$\Delta g_{FA} = -(8\pi/3)\, G\rho_E\, h \tag{16}$$

where ρ_E is the average density of the earth (5.52 g/cm³).

The increase in mass beneath the reference is determined by the amount of fluid which flows in to fill existing cavities or to fill the voids produced as cavities change size. The effect Δg_F upon gravity in either case can be written as

$$\Delta g_F = G\rho_F \int_{S_c} (\mathbf{n}\cdot\mathbf{u}_F)(z/R^3)\, dS \tag{17}$$

where $\rho_F(\mathbf{n}\cdot\mathbf{u}_F)\,dS$ is the mass flux of fluid past a point on the surface of a cavity. The net effect upon gravity when (16) and (17) are added to the intrinsic effect of deformation given by (7) is

$$\Delta g/G = 2\pi h(\rho - \rho_E) + \int_{S_c} \mathbf{n}\cdot(\rho\mathbf{u} + \rho_F\,\mathbf{u}_F)(z/R^3)\, dS \tag{18}$$

Note that $\mathbf{n}\cdot\mathbf{u}$ is negative for an expanding cavity, and $\mathbf{n}\cdot\mathbf{u}_F$ is positive for an influx of fluid.

4. Point sources

Calculating the gravity anomaly is particularly simple when the size of the source region is small compared with its depth c, because we can make the approximations that $z \simeq c$, and $R^2 \simeq c^2 + r^2$, where r is the radial coordinate of the reference point on $z = 0$. Integrating (7) for this case gives immediately the following expression, valid for any point source:

$$\Delta g/G\rho = 2\pi h - \Delta V_c \frac{c}{(r^2 + c^2)^{3/2}}, \tag{19}$$

where an increase ΔV_c in the volume of the source is taken as positive. We see in (19) that the gravity anomaly depends on ΔV_c and not on the individual components of displacements at the source, except as these affect deformation on the surface S_0.

As an example, consider the gravity anomaly due to a point source of uniform dilatation. The surface uplift, in the notation here, is [7]

$$h = (\Delta V_c/\pi)(1 - \nu)\, c/(c^2 + r^2)^{3/2}. \tag{20}$$

Substituting (20) into (19), we find that the gravity anomaly is

$$\Delta g = G\rho \Delta V_c(1 - 2\nu)\, c/(r^2 + c^2)^{3/2}. \tag{21}$$

Now consider the anomaly due to dilatancy in a horizontal penny-shaped region of radius a located at depth c. The vertical displacement at the surface is found by integrating the effect of a single source over the surface of the penny-shaped region. The differential uplift dh due to a dilation over a differential area dS_c on the penny-shaped region is found from (20) to be

$$dh = (u_c/\pi)(1 - \nu)\, c\, dS_c/(c^2 + r^2)^{3/2}, \tag{22}$$

where u_c is the vertical displacement of a point on S_c. Integrating (22) over the surface of the source gives

$$h = \int_0^a 2\pi u_c(1 - \nu)\, c[c^2 + (r - r')^2]^{-3/2}\, r'\, dr', \tag{23}$$

where r' is the radial coordinate of a point on the source. Carrying out the integration in (23), and evaluating the resulting expression for $a \ll c$ gives

$$h = 2a^2 u_c(1 - \nu)\, c/(c^2 + r^2)^{3/2}$$

or

$$h = (\Delta V_c/\pi)(1 - \nu)\, c/(c^2 + r^2)^{3/2}, \tag{24}$$

where $\Delta V_c = 2\pi u_c a^2$.

The surface displacement field in (24) due to dilation in a small penny-shaped region is identical with that given in (21) for a point source of uniform dilatation. The gravity anomaly for these two sources is therefore the same. The surface displacements due to dilatancy in a short vertical column is found also to be the same as (20) and (24). Apparently, to first order, the shape of the source has no effect upon the observed gravity anomaly. On the other hand, uplift is not the same as (20) if deformation at the source is not a uniform dilatation or if the size of the source is not small compared with depth (see MARUYAMA [8], for examples of such sources).

5. *Applications*

Consider a simple faulting event which involves no dilatancy or fluid flow. The integral in (18) is zero in this case, and

$$\Delta g = 2\pi h G(\rho - \rho_E). \tag{25}$$

The gradient $\Delta g/h$ in (25) is called the 'Bouger' gradient by BARNES [1], NUR [2], and KISSLINGER [3], and we shall use that terminology here. Deviations from the Bouger gradient are due to the contribution from the integral in (18) and indicate the opening or closing of porosity at depth or flow of fluid into or out of the system.

BARNES [1] presents uplift and changes in gravity from observations made before and after the Alaskan earthquake of 1964. The initial leveling survey was carried out in 1922, and the gravity measurements were made in 1962. The surveys were repeated in 1964 several months after the earthquake. Barnes found that his most reliable data followed the Bouger gradient. This result suggests that either negligible dilatancy and fluid flow was involved in this earthquake or that conditions had returned to the 1962 level by the time the bench-marks were reoccupied several months after the event. The second explanation seems less likely when one considers the time constant likely to be associated with an earthquake of this magnitude.

Uplift and measurements of gravity for the Matsushiro earthquake swarm are close to the limits of resolution. Data presented by NUR [2] and a corrected version of more or less the same observations presented by KISSLINGER [3] show similar trends: the decrease in gravity during uplift is greater than calculated from the Bouger gradient, and recovery during subsidence occurs at a rate greater than that during uplift. Nur and Kisslinger interpret this behavior as indicating an increase in dilatancy during uplift and a decrease in dilatancy and in inflow of water during subsidence. This interpretation is in harmony with the analysis here.

STUART and JOHNSTON [9] contend that uncertainty in the data is so large that subtleties such as dilatancy and fluid flow cannot be resolved. They suggest that the uplift is caused by the intrusion of magma. Assuming that the density of magma is about equal to that of rock and that magma fills all the crevices which are formed, we can make the simplification that $\rho\mathbf{u} = -(\rho\mathbf{u})_F$ in (18). Changes in gravity therefore should follow the Bouger gradient. Gravity, if we accept the uncertainties cited by Stuart and Johnston, as corrected by KISSLINGER [3], falls near the Bouger gradient, but generally below it in NUR's [2] version. Therefore, we cannot choose between magma intrusion and dilatancy as the source of uplift at Matsushiro without first establishing the reliability of the observations.

We can use as the final example an investigation by HUNT [5] of gravity changes and subsidence at the Wairakei, New Zealand, geothermal field as water was drawn from the borehole. Hunt, following the standard procedure for calculating excess mass, calculated the draw-off using gravity observations corrected only for the free-air effect. We find by combining (8) and (14) that the gravity effect with the free-air correction integrated over $z = 0$ depends on the volume v of water drawn from the well and the volume change ΔV of the rock matrix:

$$\int_{S_0} (\Delta g + \Delta g_{FA})\, dS = 2\pi G(\rho_F v + \rho\,\Delta V). \tag{26}$$

Hunt included only the first of the terms in parentheses in (26) in his calculation. Let us make the approximation that the aquifer is small enough that drawing water from the field can be considered to be a uniform negative dilatation of a point source. The water level in wells remained approximately constant, and so we can assume a direct relationship between the volume v of water removed from the aquifer and the net change ΔV in the volume of the half-space. For an elastic medium, the relationship between ΔV and v is found by integrating (20) over $z = 0$, with the result

$$\Delta V = (1 - 2\nu)\, v. \tag{27}$$

Equation 27 is substituted into (26), giving

$$\int_{S_0} (\Delta g + \Delta g_{FA})\, dS = 2\pi G \rho_F v[1 + \rho(1 - 2\nu)/\rho_F]. \tag{28}$$

Note that the bracketed term in (28) gives the error when the free-air effect rather than the Bouger effect is used for correcting the observations. One finds from (28), for $\nu = 1/3$ and $\rho = 2.7$, that the amount of water drawn from the well was about half the value calculated by HUNT [5].

Note added in proof.

The uplift and gravity changes associated with the San Fernando earthquake in 1971 are remarkably similar to those during the Alaskan earthquake which are discussed above. The San Fernando measurements, where changes as great as 2 m and –0.5 milligals are found, also plot along the Bouger gradient. This result suggests that the porosity and fluid content of the region beneath the stations did not change significantly between the times of measurement for either the Alaskan or San Fernando shocks. The existence of the San Fernando data was brought to my attention by Sean Solomon, and H. W. Oliver generously sent me a reprint of his report (H. W. OLIVER, S. L. ROBBINS, R. B. GRANNELL, R. W. ALEWINE and SHAWN BIEHLER, in press, *Surface and subsurface movements determined by remeasuring gravity*, Chap. 10 (pp. 192–411) *in* G. B. OAKSHOTT (Ed.), *San Fernando Earthquake of February 9, 1971*; Calif. Div. Mines and Geology Bull. *196*) on the measurements.

Acknowledgements

Discussing the analysis with KEIITI AKI, RAUL MADARIAGA, and ROBERT BURRIDGE got me restarted on several occasions when I was bogged down. LEE AAMODT gave me the reference to the data for the Wairakei geothermal field, and preprints of their Matsushiro earthquake studies were kindly sent by CARL KISSLINGER and by WILLIAM STUART and MALCOLM JOHNSTON. The work was supported by the US Geological Survey under Contract No. 14-08-0001-14551 and by the Defense Nuclear Agency under Subtask Y99 QAX-SB101 – Site Characterization.

REFERENCES

[1] D. F. BARNES, *Gravity changes during the Alaskan earthquake*, J. Geophys. Res. *71* (1966), 451.

[2] A. NUR, *Matsushiro, Japan, earthquake swarm: Confirmation of the dilatancy–fluid diffusion model*, Geology *2* (1974), 217.

[3] C. KISSLINGER, *Processes during the Matsushiro, Japan, earthquake swarm as revealed by leveling, gravity, and spring-flow observations*, Geology *3* (1975), 57.

[4] K. IIDA, M. KAYAKAWA and R. KATAYOSE, *Gravity survey of Mihara volcano, Ooshima Island, and changes in gravity caused by eruption*, Geol. Soc. Japan Rept. *152* (1952).

[5] T. M. HUNT, *Gravity changes at Wairakei geothermal field, New Zealand*, Geol. Soc. America Bull. *81* (1970), 529.

[6] F. S. GRANT and G. F. WEST, *Interpretation Theory in Applied Geophysics* (McGraw-Hill, New York 1965).

[7] J. B. WALSH and R. W. DECKER, *Surface displacements associated with volcanism*, J. Geophys. Res. *76* (1971), 3291.

[8] T. MARUYAMA, *Statical elastic dislocations in an infinite and semi-infinite medium*, Bull. Earthquake Res. Inst. *42* (1964), 289.

[9] W. D. STUART and M. J. S. JOHNSTON, *Intrusive origin of the Matsushiro earthquake swarm*, Geology *3* (1975), 63.

(Received 18th February 1975)

Pageoph,
Vol. 113 (1975)

Birkhäuser Verlag, Basel

Mean Sea Level Before and After Some Great Strike-Slip Earthquakes

By Max Wyss[1])

Summary – The annual mean sea levels of all permanently recording tide gauge stations along the North American west coast were inspected for evidence of elevation changes before and after large strike-slip earthquakes. The Yakutat earthquake, 1958 with $M = 7.9$ was preceded by 3 ± 3 cm of uplift 50 km west of the central part of the 400 km rupture. The Sitka earthquake, 1972 with $M = 7.1$, was preceded by 3 ± 3 cm of uplift at a station 40 km east of the center of the 170 km rupture. The Queen Charlotte earthquake, 1949 with $M = 8.1$, was followed by 5 ± 1 cm of subsidence 200 km east of the center of the 300 km rupture. This subsidence is interpreted as dilatancy recovery. In this case the uplifted area had a radius of approximately 200 km, and the precursor time may have been about 30 years. The San Francisco earthquake, 1906 with $M = 8.3$, was preceded by no uplift during seven years before it and was followed by no elevation change during 20 years after the event. The measurements were taken at the Presidio 6 km from the fault trace. The Long Beach earthquake, 1933 with $M = 6.3$, was preceded by 1 ± 1 cm of uplift at a point 7.4 km from the 50 km long aftershock zone.

These tide gauge data do not indicate any uplift during recent years (up to 1973) except for the Los Angeles area which started to rise during the 1950s and seems to now be elevated by approximately 5 cm with respect to the San Diego area.

1. Introduction

Annual mean sea levels have long been used in Japan to estimate local uplift and subsidence associated with large earthquakes (e.g. Yamaguti [23]). Vertical movements are of primary interest in Japan because most large earthquakes there are thrusts associated with considerable vertical components of displacement. The hypothesis of pre-earthquake dilatancy in the crust (Nur [11]) implies that thrust as well as strike-slip earthquakes may be preceded by local uplift. It becomes important, therefore, to study vertical movements near all active faults, including strike-slip faults. The amount and character of premonitory vertical movement has to be one of the most important constraints for any dilatancy earthquake source models. In addition, if vertical movements of the crust precede strike-slip earthquakes, the uplift may be used as a premonitory indicator. The purpose of this paper is to determine how many continuously operated tide gauges along the North American west coast are located near the major active fault systems, and to see whether any of the recent large earthquakes along the west coast have been preceded or followed by vertical displacement.

---

[1]) Cooperative Institute for Research in Environmental Sciences, University of Colorado/NOAA, Boulder, Colorado 80302.

Dilatancy, volumetric increase under uniaxial stress, has been documented in the laboratory (REYNOLDS [15]; BRACE *et al.* [1]). SCHOLZ *et al.* [16] have outlined the expected vertical motion as a function of time if dilatancy precedes crustal earthquakes. SCHOLZ [17] later proposed that pre-earthquake uplift caused by dilatancy may take a considerable length of time (~1 year) to disappear if the porosity caused by dilatancy became filled with pore fluids.

If we hypothesize that dilatancy precedes strike-slip earthquakes, we may expect a time history of vertical displacements as sketched in Fig. 1. During the strain build-up no vertical displacements are expected while shear strain is accumulating in a direction

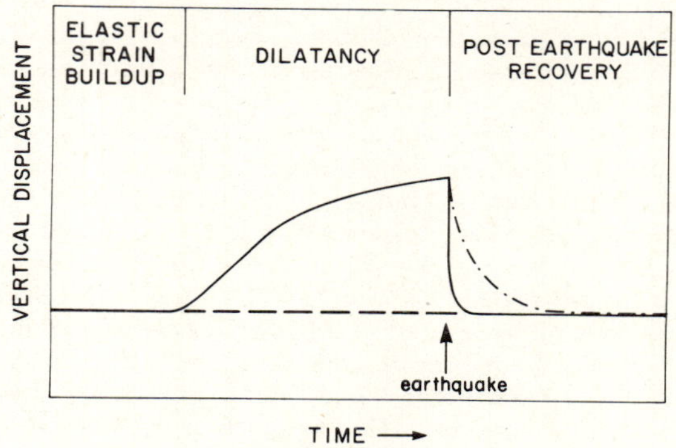

Figure 1
Sketch of uplift versus time as it may be observed if dilatancy occurs before a strike-slip earthquake. If the intermediate principal stress is much larger than the least principal stress, no uplift may result (dashed line). If fluid filled the dilatancy pores, it may take considerable time for the dilatancy uplift to disappear (dotted line)

perpendicular to the fault. When the critical stress for dilatancy is exceeded, porosity will open in the direction of the least principal stress, which is horizontal in the case of strike-slip tectonics. If the least principal stress is much smaller than the intermediate one (vertical), then no uplift of the surface may result (dashed line in Fig. 1). If the intermediate and least principal stresses are approximately the same, dilatancy volume increase should become evident as surface uplift (solid line, Fig. 1). After the earthquake, the dilatancy uplift should disappear slowly or fast, depending on whether or not the porosity caused by dilatancy became filled with ground fluid. According to the inclusion theory by BRADY [2], dilatancy uplift could start to disappear before the earthquake occurs, depending on the geometrical relationship between the fault and the observation point.

The occurrence of dilatancy in the crust could be proven if regional uplift before a strike-slip earthquake and subsequent subsidence could be found. In the case of thrust faults, the usually geographically restricted record of uplift will allow a number of different interpretations as in the case of the Niigata earthquake where the data could be explained without as well as with dilatancy. The main problem with thrust earthquakes is of course that the elastic and permanent tectonic displacements are in part also in the vertical direction obscuring the displacement caused by dilatancy. For vertical strike-slip events, this is not the case It would therefore help the understanding of the earthquake source process if a vertical displacement pattern like that shown in Fig. 1 could be found in the case of a strike-slip earthquake. Unfortunately the cause of uplift at Matsushiro during the famous earthquake swarm is still a matter of controversy (NUR [12]; STUART and JOHNSTON [19]; KISSLINGER [10]). As HANKS [6] pointed out, there are so many unknown parameters that we do not know what order of magnitude of uplift to expect before a given earthquake.

2. *Data*

A map of continuously operating tide gauges in the Alaska panhandle and Canada is shown in Fig. 2. The annual mean sea level of the Alaskan stations (HICKS and CROSBY [8]) is shown in Fig. 3. The data from the station Cordova (COR) are not shown because only six annual means are available. The continuous rapid land uplift at Skagway (SKA) and Juneau (JUN) is evidently due to glacier unloading (HICKS and SHOFNOS [7]). During the early part of the record, Yakutat (YAK) and Sitka (SIT) also seem to have been affected by the ice unloading. Further south at the station Ketchikan (KET) and all the Canadian stations (DEPARTMENT OF THE ENVIRONMENT, OTTAWA [5]) the annual mean sea levels appear to be comparatively stable since the beginning of the record in early 1900 (Fig. 4). The standard deviations of the entire record of SIT, KET, PRU and VIC are respectively 3.7, 3.7, 4.0 and 3.3 cm. It is clear from Figs. 3 and 4 that none of the tidal stations showed appreciable vertical

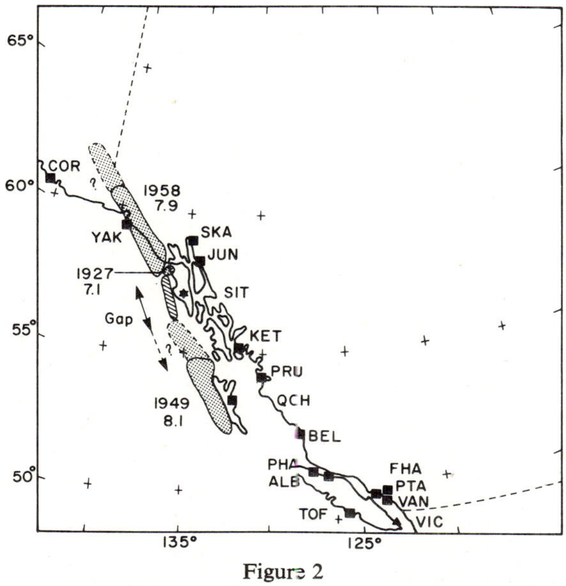

Figure 2
Map showing tide gauge stations in Alaska and Canada. Aftershock areas of large earthquakes shown stippled and dashed for the 1972 Sitka earthquake

movement in relation to the nearby earthquake, except for PRU where the sea level rose about 5 cm in the average during the three years following the Queen Charlotte earthquake.

The continuously recording tide gauge stations along the United States west coast with records longer than 20 years are shown in Fig. 5. The annual mean sea levels of these stations are plotted in Fig. 6. Again it is immediately evident that no appreciable vertical movements were associated with either the San Francisco or the Long Beach earthquake.

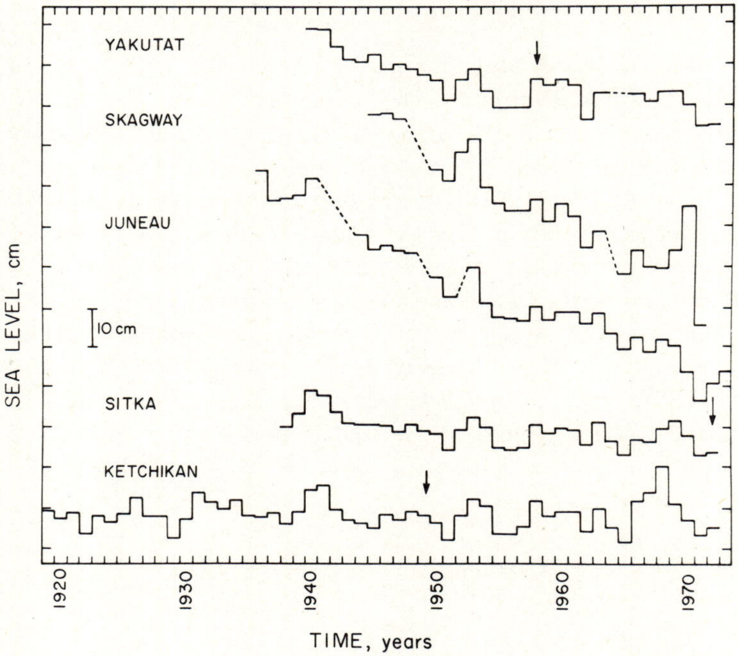

Figure 3
Annual mean sea level as a function of time for Alaskan stations. Arrows indicate occurrence of the earthquakes whose aftershock zones are shown in Fig. 2

The resolution of the data can be improved by taking the difference between the sea level at a nearby reference station and the station of interest (Figs. 7 and 8). The station KET for instance has a standard deviation of 3.7, whereas the differences KET–PRU have standard deviations of less than half that (Table 1). When the stations are farther apart, only little improvement results from comparison (Table 2) because the meteorological conditions which cause part of the sea level fluctuations will be somewhat different at the two stations. Tables 1 and 2 show the limit of resolution of the data. The sea level for approximately 10 year intervals can be estimated with a standard deviation of about 1.5 cm if reference stations are used.

Yakutat. The sea level at Yakutat has been dropping over the entire record (Fig. 3). HICKS and SHOFNOS [7] estimated a rate of 0.7 cm/year, and they suggested that the land at Yakutat was rising because of glacier unloading in the area. However, a clear change in uplift rate occurred around 1954. Before that time the rate of uplift was rapid; after it, the rate of uplift was small and could be zero. This pattern is evident in the direct data (Fig. 3) as well as in the difference SIT–YAK (Fig. 7).

One interpretation of these data is that all uplift at YAK was caused by retreating glaciers, including the change in rate. A second interpretation is that the uplift caused

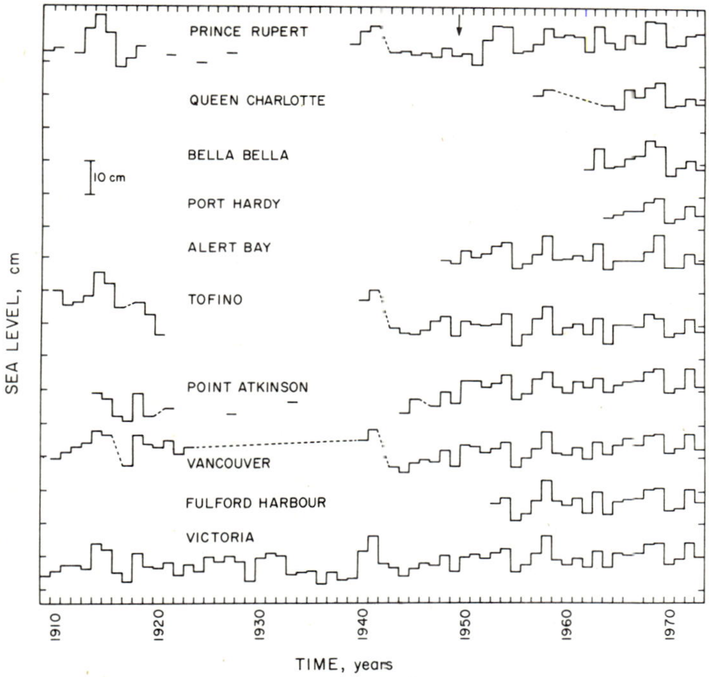

Figure 4

Annual mean sea level as a function of time for Canadian stations. Arrow indicates occurrence of the Queen Charlotte eathquake

by retreating glaciers was constant, and that this rate should be estimated from the post earthquake record. In this case an excess uplift of 3 to 5 cm before the earthquake could have been caused by dilatancy. A third possibility is to assume that 10 cm pre-earthquake uplift occurred, related to dilatancy; and that after the event, no recovery was made because of a permanent uplift due to the rupture. However, no matter which interpretation is preferred, the possible uplift had to be small. The best estimate may be 3 ± 3 cm of pre-earthquake uplift.

The Yakutat earthquake (Fig. 2) broke a 400 km to 600 km segment of the Fairweather fault (SYKES [20]). The tide station is located near the center of the aftershock zone shown by SYKES [20] at an approximate distance of 50 km from the mapped fault trace (Tocher [21]). Along the southern part of the surface rupture the maximum right lateral strike-slip was 6.5 m (Tocher [22]). The fault plane solution also indicated a strike-slip displacement along the Fairweather fault (Stauder [18]).

The absence of uplift at YAK indicates that either no dilatancy expansion occurred in the vertical direction or that such an expansion was restricted to a strip of less than 100 km width but over 400 km length.

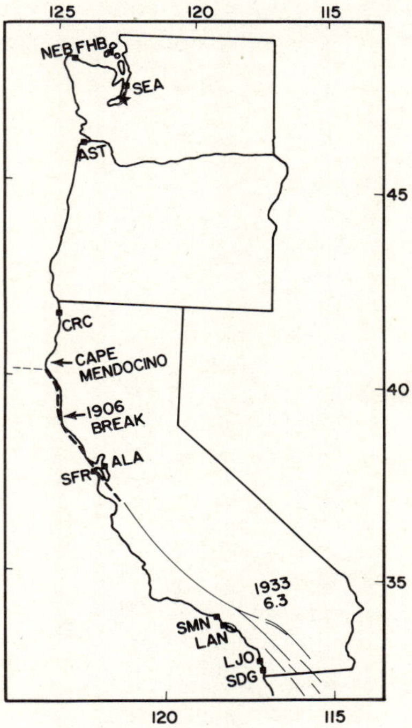

Figure 5
Map showing the absence of tide gauge stations along the United States west coast. A star marks the epicenter of the 1949 Puget Sound earthquake. Aftershock zone of Long Beach earthquake stippled

Sitka. The tide gauge station Sitka was located 40 km to the east of the center of the 1972 Sitka break (Fig. 2), which was 170 km long. The uplift in this case could not have exceeded 3 cm, and it could have been zero (Figs. 3 and 7). Because of the distance to the fault trace, the uplift along the trace could have been an order of magnitude larger. Precursors to this magnitude 7.1 earthquake are discussed in more detail by Wyss [25].

Prince Rupert. The Queen Charlotte earthquake (Fig. 2) was 300 to 500 km long according to Sykes [20]. This event broke along the Queen Charlotte fault which accommodates right lateral strike-slip between the Pacific and American plates (Chandra [4]). The tide station Queen Charlotte (QCH) unfortunately did not exist

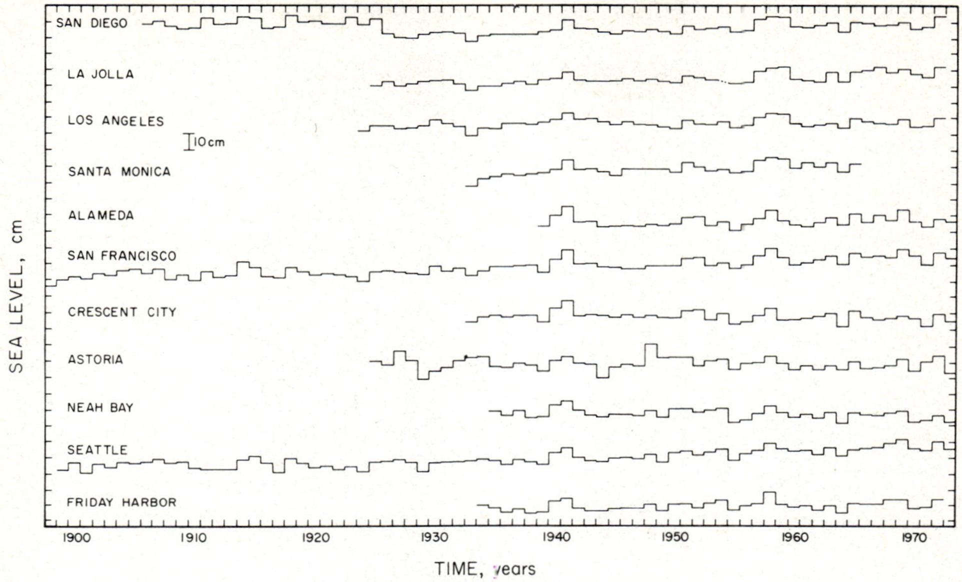

Figure 6
Annual mean sea level as a function of time for western United States stations

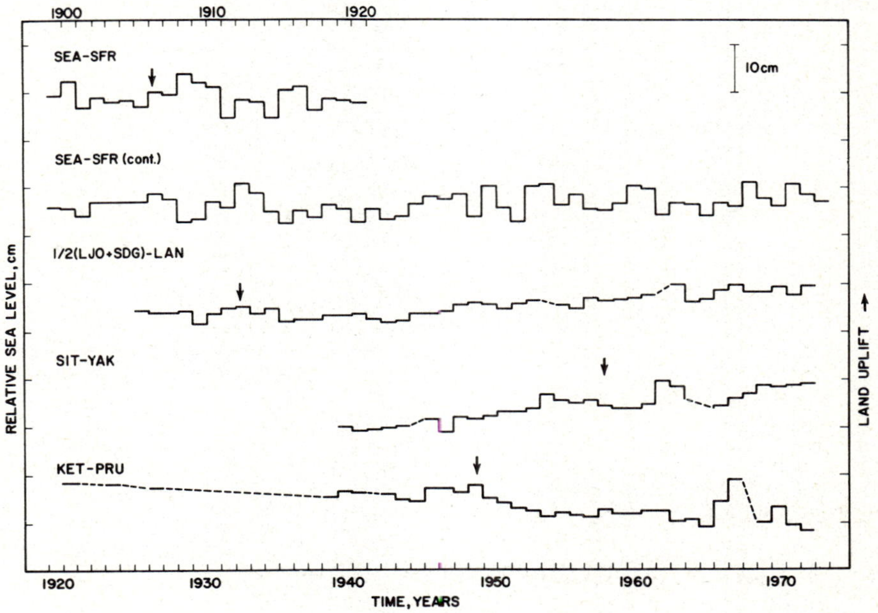

Figure 7
Relative sea level changes between reference stations and stations near earthquakes. Because the latter are subtracted, land uplift at the stations near earthquakes is upwards on this graph. Arrows indicate occurrence of earthquakes

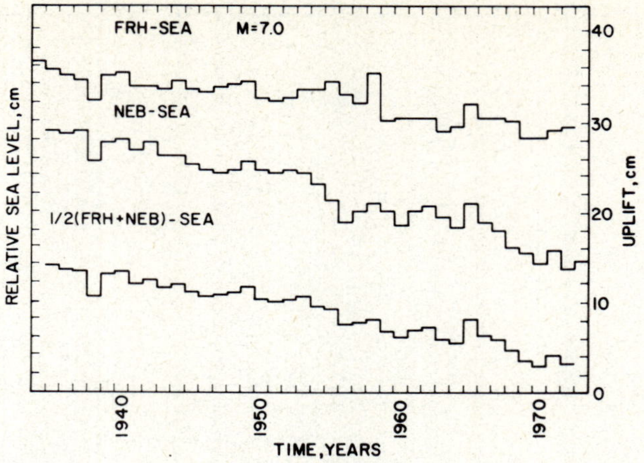

Figure 8
Relative sea level at Seattle. Uplift at Seattle upwards on graph

Table 1
Prince Rupert

Period Beginning	End	Number of years	Sea level cm	Standard Deviation cm	Standard Deviation of the mean cm
		Ketchikan minus Prince Rupert			
1921	1949	13	5.5	1.2	
1953	1972	17	0.0	1.5	
		Victoria minus Prince Rupert			
1909	1917	8	−6.6	3.4	1.2
1921	1951	15	−3.9	3.0	0.8
1952	1973	22	−6.0	2.1	0.5

Table 2
San Francisco

Period Beginning	End	Number of years	Sea Level cm	Standard Deviation cm
1898	1906	9	9.7	3.3
1907	1913	7	9.7	2.7
1914	1926	13	11.8	3.2
		Seattle minus San Francisco		
1899	1906	8	7.3	1.8
1907	1913	7	7.9	3.0
1914	1926	13	6.4	1.8
1928	1939	12	5.8	2.6
1940	1957	18	7.3	4.0
1958	1973	16	7.3	2.2

at the time of this earthquake. KET and PRU are at 200 km distance, the closest tide gauges to this earthquake.

The sea level at KET remained constant through the entire record length, 1919 to 1972 (Fig. 3). On the other hand, PRU showed a clear increase in sea level following the earthquake. The differences between KET and PRU show that the land at PRU sank by 5.5 cm during the early 1950s (Figs. 4 and 7 and Table 1). Considering the vertical movement at PRU, it is surprising that KET remained stable even though it was located at approximately the same distance from the fault trace. The explanation may be that the part of the aftershock zone outlined by a dashed line and provided with a question mark by SYKES [20] was not a part of the main rupture. Then KET would be located perpendicular to the end and PRU perpendicular to the center of the event.

One might argue that the increase of sea level at PRU could have been caused by settling of the station foundation due to the shaking by the earthquake. This seems unlikely in view of the fact that it took several years after the event for the sea level to change. This observation is rather in accord with the idea of SCHOLZ [17] that dilatancy uplift may disappear only gradually because of the damping effect by added pore fluid (dot-dashed line, Fig. 1).

I consider this change in elevation at PRU to be real and related to the Queen Charlotte earthquake. In this case the conclusion that dilatancy has preceded this great strike-slip rupture seems inescapable. The absence of an elevation change at KET indicates that the radius of the dilatant area was 200 km, and it suggests that the fault length was 300 km. This is also the length preferred by SYKES [20].

It is hard to pin down the beginning of the uplift because of the sketchy record from 1919 to 1939 at RUP. Also before 1919 there is no record at KET for comparison, and one is forced to compare with Vancouver (VAN), Victoria (VIC) and Tofino (TOF). The comparison with these stations does not improve the standard deviation, evidently because they are located too far away from PRU. The three available yearly means during the 1920s indicate by themselves and in comparison with KET as well as VIC that the land was uplifted at PRU at that time. The data from 1909 to 1917 compared with VIC (Table 1) and VAN suggest that sea level at PRU during the earliest part of the record was the same as that after the earthquake.

This suggests a precursor time of approximately 30 years. The standard deviations of these comparisons, however, are large (Table 1), and if one uses TOF for a reference station, one would conclude that sea level at PRU was lower during 1910–1917 than at any other time. Also the direct averages at PRU (Fig. 4) do not seem to indicate higher sea level during the earliest part of the record. The suggested precursor time of 30 years is therefore not determined for certain.

Puget Sound. The Puget Sound earthquake of 1949 had a magnitude of 7.3 and was a strike-slip event (NUTTLI [13]). The tide gauge station Seattle (SEA) was located at a distance of 50 km from the epicenter. Thanks to the nearby reference stations Friday Harbor (FHB) and Neah Bay (NEB), the resolution at SEA can be improved considerably. The sea level at SEA appears to be rising at a more or less constant rate (Fig. 8).

If a rising trend of 1 cm/year is assumed for the interval 1935 through 1955, the standard deviation from this trend is 0.7 cm for the curve $\frac{1}{2}$(NEB + FHB) − SEA in Fig. 8. This figure shows that at SEA no elevation changes larger than 1 cm preceded or followed the Puget Sound earthquake.

San Francisco. The San Francisco 1906 earthquake, one of the most discussed events in seismological history, left the sea level at the station San Francisco (SFR) unchanged (Fig. 6). Strike-slip displacements of approximately 5 m occurred along a rupture of about 450 km, and the magnitude was 8.3 (Reid [14]). The tide gauge was located at the Presidio approximately 6 km from the fault trace. The closest station with a record old enough for comparison is Seattle (SEA). Because of its large distance from SFR, the differences of sea levels do not appreciably improve the scatter in the data (Fig. 7, Table 2).

Only seven years of the record predate the earthquake. If the dilatancy diffusion model as proposed by Nur [11], Whitcomb *et al.* [24] and Scholz *et al.* [16] is correct, we would expect an uplift pattern as sketched in Fig. 1. That is, dilatancy uplift should disappear only after the earthquake. As no such recovery is observed (Figs. 6 and 7), one would have to conclude that no dilatancy uplift (resolution 2 cm) occurred prior to the San Francisco earthquake. The inclusion model by Brady [2], on the other hand, predicts that dilatancy near the fault should disappear before the event. Even if Brady's model is correct, it would be surprising if dilatancy related uplift would disappear seven years before the earthquake.

Long Beach. The Long Beach earthquake of 1933 had a fault length of approximately 50 km and $M_L = 6.3$ (Hileman *et al.* [9]). The Los Angeles tide gauge station (LAN) was located about 7.4 km from the closest point of the 50 km long aftershock zone. No substantial sea level change preceded or followed this strike-slip earthquake (Fig. 6).

The reference sea level in this case was the average sea level between the stations San Diego (SDG) and La Jolla (LAJ) (Fig. 7). It appears that with respect to this reference, the land was 1.4 cm higher during 10 years before the Long Beach earthquake than during 9 years after it (Fig. 7, Table 3). However, the difference is marginal.

Table 3
$\frac{1}{2}$(SDG + LAJ) − LAN

Period		Number of years	Sea Level cm	Standard Deviation cm
Beginning	End			
1926	1935	10	0.8	1.0
1936	1944	9	−0.6	0.6
1948	1960	12	2.8	0.6
1961	1972	11	5.2	1.2

Although the sea level at LAN has been increasing continuously (Fig. 6), the level relative to $\frac{1}{2}$(SDG + LAJ) has decreased by about 5 cm. This apparent uplift of LAN started during the 1950s, and it may be related to the general trend of uplift observed for the Los Angeles and Transverse Range area (CASTLE *et al.* [3]). If this elevation change is due to dilatancy uplift, an earthquake with precursor time exceeding 25 years may be approaching. This precursor time would correspond to an earthquake of approximately magnitude eight (SCHOLZ *et al.* [16]; WHITCOMB *et al.* [24]).

3. Conclusions and suggestions

Along the North American west coast there are about a dozen tide gauge stations located near major active strike-slip faults. The annual mean sea levels of these stations have standard deviations of approximately 3 cm. If the sea levels near faults are referenced to sea levels away from faults, the standard deviation of the estimated relative elevations is approximately 1.5 cm. It follows that elevation changes of a few centimeters can be detected using annual mean sea level data, without correction for water temperature and atmospheric pressure.

Mean sea levels near three great strike-slip ruptures indicate: 1) The Queen Charlotte earthquake, 1949, was followed by about 5 cm of subsidence at 200 km distance which is interpreted as dilatancy recovery. 2) The Yakatat earthquake may have been preceded by 3 ± 3 cm of uplift at a distance of 50 km from the fault. 3) The San Francisco, 1906, earthquake was not preceded by uplift for at least seven years before it at 6 km from the trace. The Sitka, 1972, earthquake may have been preceded by 3 ± 3 cm of uplift. The Puget Sound, 1949, and the Long Beach, 1933, earthquakes were not preceded by uplift more than 1 cm at stations 50 and 7.5 km from the epicenter areas.

The only uplift indicated by the recent record (through 1972) is occurring in the Los Angeles area. Since the 1950s, Los Angeles has been uplifted by about 5 cm with respect to San Diego. On the other hand subsidence seems to occur in the area of Seattle, Victoria, and possibly Vancouver. These three stations all indicate a sea level elevation of 5 to 10 cm whereas the surrounding stations show no change. This subsidence seems to be a tectonic movement affecting an area of about 200 km length.

It is clear that tide gauges can provide a relatively cheap estimate of elevation changes of a few centimeters. The coastal areas where vertical strike slip faults cross from the ocean on land or where they parallel the coast are ideal locations for cheaply testing the dilatancy hypothesis. The two most promising places to study are the San Andreas fault north of San Francisco and the Fairweather fault south of Yakutat. At present there are no tide gauges in these areas. A network of well-located tide gauges along these faults (and perpendicular to the faults in the case of the Fairweather fault), could show conclusively whether dilatancy exceeding 1 cm in the vertical direction will occur before the next large earthquakes along these faults.

Acknowledgements

I thank S. D. HICKS for sending me the data on US tide gauges and R. KIND for critically reading the manuscript. This study was sponsored by the US Geological Survey Contrast No. 14-08-0001-14571.

References

[1] W. F. Brace, B. W. Paulding, Jr. and C. Scholz, *Dilatancy in the fracture of crystalline rocks*, J. Geophys. Res. *71* (1966), 3939.

[2] B. T. Brady, *Theory of earthquakes, Part I, A scale independent theory of rock failure*, Pure and Appl. Geophys., this issue, 1975.

[3] R. L. Castle, J. N. Alt, J. C. Savage and G. I. Balazs, *Elevation changes preceding the San Fernando earthquake of February 9, 1971*, Geology *2* (1974), 61.

[4] U. Chandra, *Seismicity, earthquake mechanisms, and tectonics along the Western Coast of North America from 42°N to 61°N*, Bull. Seism. Soc. Am. *64* (1974), 1529.

[5] Department of the Environment, Ottawa, Canada, *Monthly and yearly mean water levels*, Marine Environmental Data Service, Ottawa.

[6] T. C. Hanks, *Constraints on the dilatancy–diffusion model of the earthquake mechanism*, J. Geophys. Res. *79* (1974), 3023.

[7] S. D. Hicks and W. Shofnos, *The determination of land emergence from sea level observations in South East Alaska*, J. Geophys. Res. *70* (1965), 3315.

[8] S. D. Hicks and J. E. Crosby, *Trends and variability of yearly mean sea level 1893–1972* (NOAA Technical Memorandum NOS13 National Ocean Survey 1974.

[9] J. A. Hileman, C. R. Allen and J. M. Nordquist, *Seismicity of the Southern California region, 1 January 1932 to 31 December 1972* (Seismological Laboratory, California Institute of Technology 1973.

[10] C. Kisslinger, *Processes during the Matsushiro earthquake swarm as revealed by leveling, gravity, and spring-flow observations*, Geology *2*, in press.

[11] A. Nur, *Dilatancy, pore fluids and premonitory variations of t_s/t_p travel times*, Bull. Seism. Soc. Am. 62 (1972), 1217–1222.

[12] A. Nur, *Matsushiro, Japan, earthquake swarm: Confirmation of the dilatancy–fluid flow model*, Geology *2* (1974), 217.

[13] O. W. Nuttli, *The western Washington earthquakes of April 13, 1949*, Bull. Seism. Soc. Am. *42* (1952) 21.

[14] H. F. Reid, *The mechanics of the earthquake, volume II of the report of the state earthquake investigation commission on the California earthquake of 1906* (Carnegie Institute, Washington 1910).

[15] O. Reynolds, *Dilatancy*, Nature *33* (1886), 429.

[16] C. H. Scholz, L. R. Sykes and Y. P. Aggarwal, *The physical basis for earthquake prediction*, Science *181* (1973), 803.

[17] C. H. Scholz, *Post-earthquake dilatancy recovery*, Geology *2* (1974), 551.

[18] W. Stauder, *The Alaska earthquake of July 10, 1958: Seismic studies*, Bull. Seism. Soc. Am. *50* (1960), 293.

[19] W. D. Stuart and M. J. S. Johnston (1974), personal communication.

[20] L. R. Sykes, *Aftershock zones of great earthquakes, seismicity gaps, and earthquake prediction for Alaska and the Aleutians*, J. Geophys. Res. *76* (1971), 8021.

[21] D. Tocher, *The Alaska earthquake of July 10, 1958*, Bull. Seism. Soc. Am. *50* (1960a), 217.

[22] D. Tocher, *The Alaska earthquake of July 10, 1958: Movement on the Fairweather fault and field investigation of southern epicentral region*, Bull. Seism. Soc. Am. *50* (1960b), 267.

[23] S. Yamaguti, *On the changes in the heights of yearly mean sea levels preceding the great earthquakes*, Bull. Earthq. Res. Inst. *46* (1968), 1269–1273.

[24] J. H. Whitcomb, J. D. Garmany and D. L. Anderson, *Earthquake prediction: variation of seismic velocities before the San Fernando earthquake*, Science *180* (1973), 632.

[25] M. Wyss, *Sea level, magnetic field, and P-residuals before the Sitka earthquake 1972*, Pure Appl. Geophys., this issue, 1975.

(Received 11th December 1974)

Pageoph,
Vol. 113 (1975) Birkhäuser Verlag, Basel

Dilatancy and Crustal Uplift

By JOHN R. BOOKER[1])

Summary – Opening of cracks before an earthquake (dilatancy) causes inelastic crustal uplift. Raising the pore pressure in these cracks as pore fluid diffuses into the dilatant zone produces further elastic uplift. Since the time between resaturation of the dilatant zone and the earthquake is shorter than the life of the v_p/v_s anomaly, the elastic uplift is less than the inelastic. The magnitude of the v_p/v_s anomaly gives a lower bound on the compressibility of the dilatant zone. Velocity and uplift measurements lead to the conclusion that if dilatancy occurred before the 1971 San Fernando earthquake, the pore pressure in the hypocentral region at the time of the event must have been at least an order of magnitude below hydrostatic.

1. Introduction

Dilatancy, which is the swelling of a rock at deviatoric stresses approaching fracture, has become a unifying hypothesis in earthquake prediction research. A wide range of earthquake forerunners can apparently be explained if under-saturated cracks open near the fracture stress and are subsequently resaturated by diffusion of pore fluid from surrounding material (SCHOLZ *et al.* [1]). Strengthening of the hypocentral region by the drop in pore pressure accompanying the volume change delays the earthquake and the return to normal pore pressure and the consequent decrease in strength triggers the event (NUR [2]). The drop in the speed of compressional waves (reflected in the seismic velocity ratio v_p/v_s) during the interval of undersaturation has received considerable attention and has been reported from widely separated areas (SADOVSKY *et al.* [3], AGGARWAL *et al.* [4], WHITCOMB *et al.* [5]). A typical sequence of events is shown in Fig. 1. At the onset time t_o, v_p/v_s starts to drop as highly compressible cracks form. The maximum decrease in v_p/v_s occurs at t_m and the anomaly disappears at t_s when the dilatant zone is completely resaturated with liquid. Throughout the life of the anomaly, the pore pressure in the center of the dilatant zone is low. It starts to rise again only after resaturation. At t_e, the pore pressure has risen to the point where the strength of the dilatant region again equals the strength at t_o and the earthquake occurs.

In this note I shall point out that the magnitude of the reported v_p/v_s anomalies, the lack of large crustal uplifts before the earthquake and the shortness of the time delay between the disappearance of the anomaly and the earthquake inevitably require that the pore pressure in the hypocentral region at the time of the earthquake is far

---

[1]) Geophysics Program, University of Washington, Seattle, Washington 98195.

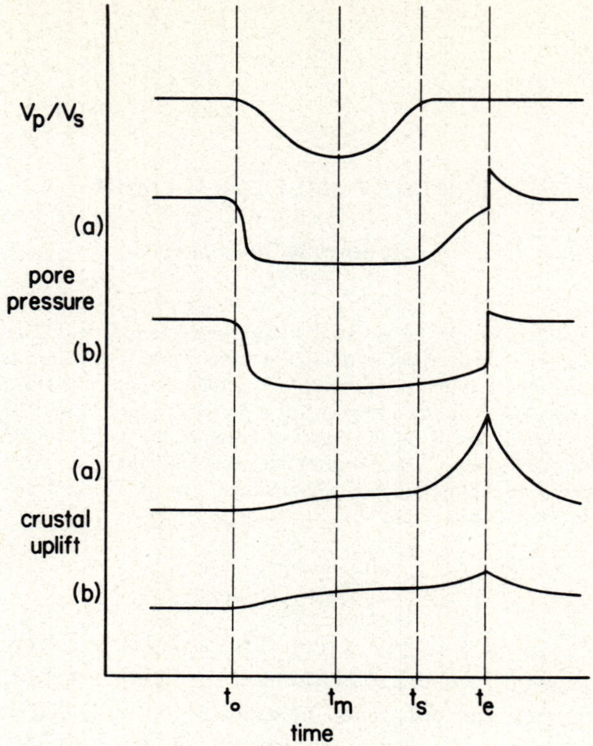

Figure 1

Time variations of velocity ratio v_p/v_s, pore pressure near the center of a dilatant region and crustal uplift over a dilatant region prior to an earthquake at time t_e. Curves (a) are the situation when the pore pressure can recover to nearly its pre-dilatancy value. Curves (b) are the situation when the pore pressure can recover only a small amount before initiating a fracture. As noted in the text, if the pressure prior to t_o is comparable to hydrostatic, curves (a) require $t_e - t_s > t_s - t_o$ which is contrary to observations

below hydrostatic. If, as is generally assumed, the earthquake occurs when the pore pressure rises to a substantial fraction of its pre-dilatancy value, one can conclude that the normal pore pressure in the hypocentral region is also far below hydrostatic.

2. *Crustal Uplift*

The initial dilation of the cracks which produces the undersaturation in a hypocentral region must be accompanied by crustal uplift. Unfortunately, the magnitude of this uplift is indeterminate because the reported drop in v_p/v_s can be achieved with arbitrarily small total volume change if the cracks have sufficiently large aspect ratio (Nur *et al.* [6], Anderson *et al.* [7]) or if the pore pressure prior to dilation is only slightly above that for the liquid–vapor transition (Hanks [8]). Thus, strain measurements during the life of the v_p/v_s anomaly are difficult to interpret without explicit knowledge of the crack geometry.

The rise of pore pressure from the time of resaturation (the disappearance of the v_p/v_s anomaly) to the time of the earthquake should produce further uplift. This additional strain is elastic as opposed to the initial inelastic dilatancy. Since we are concerned with small strains of the order of 10^{-4} or less, the volumetric strain θ is governed by

$$\Delta\theta = \frac{1}{k}\left[\left(1 - \frac{k}{k_s}\right)\Delta p - \Delta\sigma\right] \tag{1}$$

where $\Delta\sigma$ is the change in confining pressure, Δp is the change in pore pressure, k is the bulk modulus of the porous matrix at zero pore pressure (relaxed modulus) and k_s is the bulk modulus of the solid material making up the matrix (NUR and BYERLEE [9]). Positive $\Delta\theta$ implies expansion. I want to argue that k_s, k, and $\Delta\theta$ are usefully bounded by data and that $\Delta\sigma$ can be ignored in most reasonable geometries. This leads to a useful and interesting estimate for Δp.

3. The bulk modulii

Consider a crust which is saturated near the surface but whose cracks are squeezed shut by the lithostatic pressure below some depth. In this model, pore fluid (presumably water) diffuse vertically into the hypocentral region as a result of the permeability of the dilatant rock (see Fig. 2). Horizontal flow occurs only within the dilatant region or

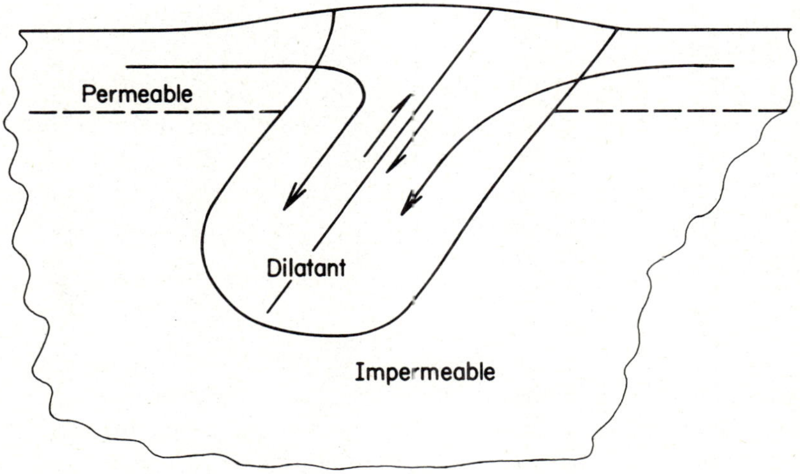

Figure 2
A simple model of a dilatant region which allows calculation of a lower bound to the elastic uplift associated with rise of pore pressure in a dilatant region. The arrows show average fluid flow. Note, however, that within a dilatant zone for a thrust fault, the cracks would be predominantly horizontal. Thus horizontal permeability will be larger than the vertical permeability and horizontal pressure gradients will be small inside the dilatant region

in the shallow, normally permeable, saturated zone. The principal reason for discussing this model is that it gives an easily understood lower bound to the strain in the more commonly envisioned model of a crust which is saturated and permeable to great depth. As an admittedly speculative representation of reality, the model also has the advantages of not requiring permeability or pore fluid in the non-dilatant deep crust, of placing strong horizontal flow in the shallow crust where effects such as well fluctuations, radon emission and electrical conductivity variations are more likely to be detectable and finally of greatly simplifying the geometry of the fluid diffusion problem.

If the lower crust has no cracks, $k_s = \rho(v_p^2 - \frac{4}{3}v_s^2)$, where ρ is the density. Using $v_p = 6.63$ km/s, $v_s = 3.80$ km/s and $\rho = 2.8$ gm/cm^3, $k_s = 6.9 \times 10^{11}$ c.g.s. These values, which are typical for the crust are from WHITCOMB *et al.* [5]. I use their data, not because it is the best, but because we have some information on the pre-earthquake uplift at San Fernando. The conclusions of this paper are insensitive to whose data are used.

At the time of maximum v_p/v_s anomaly, Whitcomb *et al.* find $v_p = 5.4$ km/s and $v_s = 3.5$ km/s which gives an effective bulk modulus $k_e = 3.6 \times 10^{11}$ c.g.s., assuming that the porosity is low enough to neglect the density decrease. Although as GRIGGS *et al.* [10] point out, refraction at the dilatant zone boundary makes it unnecessary that the ray paths lie entirely within the dilatant zone, the minimum observed value of v_p/v_s gives an upper bound to the relaxed bulk modules of the dilatant zone. Thus

$$\Delta\theta \geqslant 2.8 \times 10^{-12}(0.48\Delta p - \Delta\sigma) \tag{2}$$

It is straightforward to extend the above argument to a crust which initially contains saturated cracks. If the pore fluid is water and the porosity is small, the pre-dilatant values of v_p and v_s still give a good estimate of k_s. The effect of additional liquid-filled cracks during the v_p/v_s anomaly is to decrease the matrix bulk modulus relative to the effective bulk modulus deduced from v_p and v_s. This is because the liquid-saturated cracks do not affect the seismic velocities, but can increase the volumetric expansion as the pore pressure slowly rises. Thus (2) is still valid.

4. Strain data

The crustal uplift produced by the rising pore pressure before the February 9, 1971 San Fernando earthquake can be estimated in three distinct ways. The most desirable method would be direct observation between the time of resaturation, t_s, and the earthquake, t_e. Unfortunately, there appears to be no relevant data in the proper time period. The best that one can probably say is that it is unlikely that more than 1 m of uplift could have occurred in such a populous region without detection.

A second approach would be to measure the crustal downwarp after the event. The dilatant volume should collapse after the earthquake on a time scale governed by diffusion of the pore fluid out of the dilatant zone (SCHOLZ [11]). Surveys after the earthquake did detect vertical motions of about 1 m which were attributed to the earthquake itself. (BURFORD *et al.* [12]). Attempts to look for time-dependent motion immediately following the event were restricted to creep on the known fault traces. If a broad downwarp on a time scale of a month or more is retreivable from existing data it has not been done. In any event it is again unlikely that a downwarp of the order magnitude of 1 m would go undetected and the actual motion was probably considerably less.

The final estimation technique is based on the details of the time history of the v_p/v_s anomaly. We know, for instance, that the volume of water which flows into the dilatant region between the onset time t_o and the re-saturation time t_s must be very nearly equal to the initial dilatant increase in volume. We also know that the maximum flow rate into the dilatant volume must occur between t_m and t_s and the flow rate must decrease as the pore pressure begins to rise after t_s. Finally, all experimental data clearly show that the time $t_e - t_s$ is shorter than $t_s - t_o$ or $t_s - t_m$. Thus the volume of water which flows into the dilatant region between t_s and t_e must be less than flows in between t_o and t_s. The conclusion is simply that the crustal uplift in the interval t_s to t_e must be less than the original inelastic dilatant uplift.

Leveling data (CASTLE *et al.* [13]) indicate that anomalous uplift in the San Fernando region did not exceed 10 cm between 1961 and 1968. HANKS [8] therefore suggests 20 cm as a reasonable bound on total dilatant uplift. If we permit an additional 20 cm uplift in the last few months when pore pressure is rising before the event, the total would be 40 cm which is consistent with it being substantially less than 1 m.

5. *Allowable pore pressure rise*

To calculate the crustal uplift from (2) requires knowledge of the overall geometry of the dilatant region and is in general a complicated job. For all reasonable geometries, however, the uplift is approximately

$$h \sim \int_{D-R}^{D+R} \Delta\theta \, dz$$

where D is the average depth of the dilatant zone and $2R$ is the vertical dimension.

The change in confining pressure during the uplift is $\Delta\sigma = \rho gh + S$ where g is the acceleration of gravity and S is the vertical restoring force produced by stretching of the crust above the dilatant region as it expands upwards. For $h < 1$ m, we can neglect ρgh relative to pore pressure changes of a few percent of hydrostatic at any depth below a 100 m. We can also neglect S if the horizontal length scale of the dilatant zone is comparable to or greater than the vertical scale and depth. In any case it is doubtful whether the uplift due to a fixed rise in pore pressure can actually be significantly decreased by making the dilatant zone small. For fixed-ray paths and measured anomalous drop in v_p/v_s, the bulk modulus of the dilatant region decreases faster than its size.

If we let $h = 20$ cm, $R = 5$ km and assume that Δp is constant, (2) gives $\Delta p = 1.5 \times 10^7$ dynes/cm^2 (15 bars). This is equal to the hydrostatic pressure at a depth of 150 m and one can conclude that the pore pressure change at 1.5 km or deeper is at least an order of magnitude below the hydrostatic pressure.

Alternatively one can calculate the uplift one would expect if the pore pressure rises to the hydrostatic value ρgz at every depth z. One obtains $h \gtrsim 26DR$ where h is in cm and D and R are in km. For $D = R = 5$ km, $h \gtrsim 6$ m. To reduce this to $h \gtrsim 20$ cm would require $D \sim R \gtrsim 1$ km. Such a thin, shallow dilatant zone can probably be ruled out as too small to produce the observed v_p/v_s without accepting a lower matrix bulk modulus and consequently larger crustal uplift. We are again led to the conclusion that even the rather weak existing constraints on the crustal uplift require a total pore pressure rise which is at least an order of magnitude below hydrostatic.

6. *Conclusions*

The argument leading to the conclusion that pore pressure at the time of the San Fernando earthquake was very low compared to the hydrostatic value rests on three points: First, the delay of the earthquake after the disappearance of the v_p/v_s anomaly is shorter than the duration of the anomaly. This means the crustal uplift due to the rising pore pressure after saturation must be less than the uplift associated with the initial formation of cracks. Second, the lower bound on the compressibility of the dilatant zone deduced from the magnitude of the v_p/v_s anomaly is large. Raising the pore pressure to hydrostatic in a dilatant zone of 10 km size would result in 6 m of elastic uplift. Finally, uplift before the San Fernando earthquake is at least an order of magnitude less than 6 m. The first two points are equally valid for all reported cases of v_p/v_s anomalies. The uplifts that one predicts assuming a pore pressure rise to hydrostatic are so large that the fact that crustal uplift is not the most obvious forerunner for all earthquakes would seem to imply that pore pressure at the time of the earthquake is always low, or the v_p/v_s anomaly is not related to dilatancy and fluid flow.

It is generally assumed that the earthquake occurs when the pore pressure in the dilatant region returns to its pre-dilatancy value as indicated by curve (a) of Fig. 1. This would imply that the pre-dilatancy pore pressure in the lower crust must be substantially below hydrostatic. There is a viable alternative, however. That is simply that the strength of the crust (measured at constant pore pressure) drops significantly as dilatancy proceeds. Thus catastrophic failure occurs when the pore pressure rises to only a small percentage of its pre-dilatancy value as shown in curve (b) of Fig. 1. In the more general situation one can conceive of a continuum of possibilities, ranging from earthquakes with small uplifts occurring shortly after the v_p/v_s anomaly to earthquakes with large precursory uplifts occurring a long time after the disappearance of the v_p/v_s anomaly. The proper conclusion to draw from the San Fernando data is that either the pre-dilatancy pore pressure was very low or the dilatancy greatly reduced the strength of the hypocentral region or fluid flow is irrelevant. A deep well should answer the question of whether the pore pressure is substantially below hydrostatic and careful laboratory experiments should shed useful light on the strength question.

Acknowledgement

Research sponsored by the US Geological Survey, Department of the Interior, under USGS Grant Number 14-08-001-G-120.

References

[1] C. H. Scholz, L. R. Sykes and Y. P. Aggarwal, *The physical basis for earthquake prediction*, Science *181* (1973), 803–810.

[2] A. Nur, *Dilatancy, pore fluids, and premonitory variations of t_s/t_p travel times*, Bull. Seismol. Soc. Amer. *62* (1972), 1217–1222.

[3] M. A. Sadovsky, I. L. Nersesov, S. K. Nigmatuller, L. A. Latynina, A. A. Lukk, A. N. Semenov, I. G. Simbireva and V. I. Ulomov, *The processes preceding strong earthquakes in some regions of Middle Asia*, Tectonophysics *14* (1972), 295–307.

[4] Y. P. Aggarwal, L. R. Sykes, J. Armbruster and M. L. Sbar, *Premonitory changes in seismic velocities and prediction of earthquakes*, Nature *241* (1973), 101–104.

[5] J. H. Whitcomb, J. D. Garmany and D. L. Anderson, *Earthquake prediction: Variation of seismic velocities before the San Fernando earthquake*, Science *180* (1973), 632–635.

[6] A. Nur, M. L. Bell and P. Talwani, *Fluid flow and faulting, 1, A detailed study of the dilatancy mechanism and premonitory velocity changes*, in *Proceedings of the Conference on Tectonic Problems of the San Andreas Fault*, edited by R. L. Kovach and A. Nur (Stanford University Press, Palo Alto, California 1973), pp. 391–404.

[7] D. L. Anderson, B. Minster and D. Cole, *The effect of oriented cracks on seismic velocities*, J. Geophys. Res. *79* (1974), 4011–4015.

[8] T. C. Hanks, *Constraints on the dilatancy – diffusion model of the earthquake mechanism*, J. Geophys. Res. *79* (1974), 3023–3025.

[9] A. Nur and J. D. Byerlee, *An exact effective stress law for elastic deformation of rock with fluids*, J. Geophys. Res. *76* (1971), 6414–6419.

[10] D. T. Griggs, D. D. Jackson, L. Knopoff and R. L. Shreve, *Earthquake prediction: Modeling the anomalous v_p/v_s source region*, Science *187* (1975), 537–539.

[11] C. H. Scholz, *Postearthquake dilatancy recovery*, unpublished manuscript (1974).

[12] R. O. Burford, R. L. Castle, J. P. Church, W. T. Kinoshita, S. H. Kirby, R. T. Ruthuen and J. C. Savage, *Preliminary measurements of tectonic movement*, in USGS Professional Paper *733* (1971), 80–85.

[13] R. L. Castle, J. N. Alt, J. C. Savage and E. I. Balazs, *Elevation changes preceding the San Fernando earthquake of February 9, 1971*, Geology *2* (1974), 61–66.

(Received 5th December 1974)

Pageoph,
Vol. 113 (1975) Birkhäuser Verlag, Basel

Aseismic Faulting Following the 1973 Nemuro-Oki Earthquake, Hokkaido, Japan (a possibility)

By Keichi Kasahara[1])

Summary – The possibility of delayed strain release by aseismic faulting following a main shock is studied in relation to the recent earthquake off Nemuro peninsula, Hokkaido, Japan, 17 June, 1973. The main purpose of this paper is to obtain a comprehensive view of the land movement in eastern Hokkaido with respect to the seismic cycle there. This region, particularly the Nemuro peninsula area, has been subject to extensive land subsidence in the past several decades. At Hanasaki, for example, it amounted to −60 cm in the past 70 years. The long-term accumulation of subsidence in the same area, as read from remaining marks of the ancient sea level, is only −2 m in the past 5000–6000 years. Therefore, the rate of the recent land movement is twenty times or more as high as the long-term one, which is known from the geomorphological evidence. Such disagreement between the two kinds of rates leads us to expect a future uplift phase which will compensate the previous subsidence.

Delayed strain release by aseismic faulting is suggested in order to account for the former Nemuro-oki earthquake (1894) as inferred from the tidal record at Hanasaki. The model is basically a compound fault system with brittle-ductile transition at a certain depth in the plate boundary. If this were the case, the tide-gauge at Hanasaki would record a significant land uplift following the 1973 earthquake for 40 cm or more in the forthcoming several years. The slip mode in the ductile part is a matter of concern from the viewpoint of earthquake prediction, as a secondary shock may occur if the fault should slip quickly. The recent sea-level change at Hanasaki seems to suggest slow propagation of fracture in the down-dip direction. A brief discussion is added on the rheological properties of the plate boundary to estimate the viscosity coefficient for future reference.

1. *Introduction*

The eastern part of Hokkaido, including the Nemuro peninsula, has been subject to extensive strain accumulation for several recent decades. Levelling and tide-gauge observations (Geographical Survey Institute [1]) discovered large-scale land subsidence which tends to increase south-eastward from midland to the Pacific coast, amounting in the Nemuro area to −60 cm (uplift as positive) in the past 70 years. In sympathy with the vertical movement, there was observed northwest–southeast horizontal contraction over this part of Hokkaido. On the basis of this type of evidence, as well as the vast seismic gap off the coast, several seismologists pointed out the risk of an approaching catastrophe in the present area (e.g. Utsu [2] and Rikitake [3]).

The prediction came true on 17 June, 1973, when an earthquake with $M = 7.4$–7.7 occurred off the Nemuro peninsula. The associated crustal movements, as represented by the land subsidence for about 10 cm in the peninsula area, are interpreted reasonably

[1]) Earthquake Research Institute, University of Tokyo.

by several writers on the basis of the thrust-type fault-origin models (ABE and YOKOYAMA [4], ANDO [5], SHIMAZAKI [6, 7], TADA [8]). We may say that a more or less comprehensive view of the 1973 earthquake has been obtained as far as the coseismic movements are concerned.

It must be remarked, however, that these theoretical models, except SHIMAZAKI [7], do not deal with the tectonic significance of preseismic (or, interseismic) strain accumulation. Geomorphological studies on remaining marks of the ancient sea level discovered that the cumulative land subsidence is small there, only –2 m in the past 5000–6000 years say, i.e. –0.4 mm/yr in the long-term rate (GEOLOGICAL SURVEY OF JAPAN [9]). The interseismic strain rate, as previously stated, is unusually high by this tectonic standard (20 : 1, approximately). It is a reasonable supposition, therefore, that the land subsidence must be compensated by a forthcoming uplift. The mode of uplift process must be a matter of concern from the viewpoint of tectonics as well as of earthquake prediction, because we would have another catastrophe if the uplift occur abruptly.

2. *Interseismic land movements*

As introduced in the foregoing section, we observed extensive strain accumulation both in the horizontal and vertical components. In the present paper, however, the discussion will be developed with respect to the latter component, for which ample data are available.

The distribution of vertical displacement for the period 1900–1955 (GEOGRAPHICAL SURVEY INSTITUTE [1]) is illustrated in Fig. 1, together with the location of the off-coast major earthquakes (JAPAN METEOROLOGICAL AGENCY [10], ANDO [5], UTSU [2]). It is remarkable that the land area is subject to subsidence, in general, which tends to increase south-eastward from midland to the Pacific coast, amounting in the Nemuro area to –35 cm (local subsidence is seen in the Obihiro area, *ca.* 100 km west of Kushiro, but we shall not discuss it in the present paper).

The south-eastward increasing trend of subsidence may be naturally extended to the sea area, leading us to the supposition that the deformation field is generated by underthrusting of the oceanic plate (SHIMAZAKI [11]). The rate of plate motion may be about 3 cm/yr in this case, judging from various sorts of information.

The tide-gauge at Hanasaki seems to have recorded continuously the process of subsidence relative to the sea level (Fig. 2; GEOGRAPHICAL SURVEY INSTITUTE [1]). Originally, the data are discontinuous between 1940 and 1954 due to abandonment of the former station. To draw the figure, therefore, the base line for the new station has been adjusted in such a way that the sea-level difference between 1900 and 1955 harmonizes with the levelling data (–35 cm in the same period). Then the sea-level change for the whole period, except the first several years, can be approximated very well by a straight line indicating a subsidence rate –60 cm per 70 years, i.e. –8 to –9 mm/yr. Off Hokkaido a few major earthquakes occurred in the same period, such as the

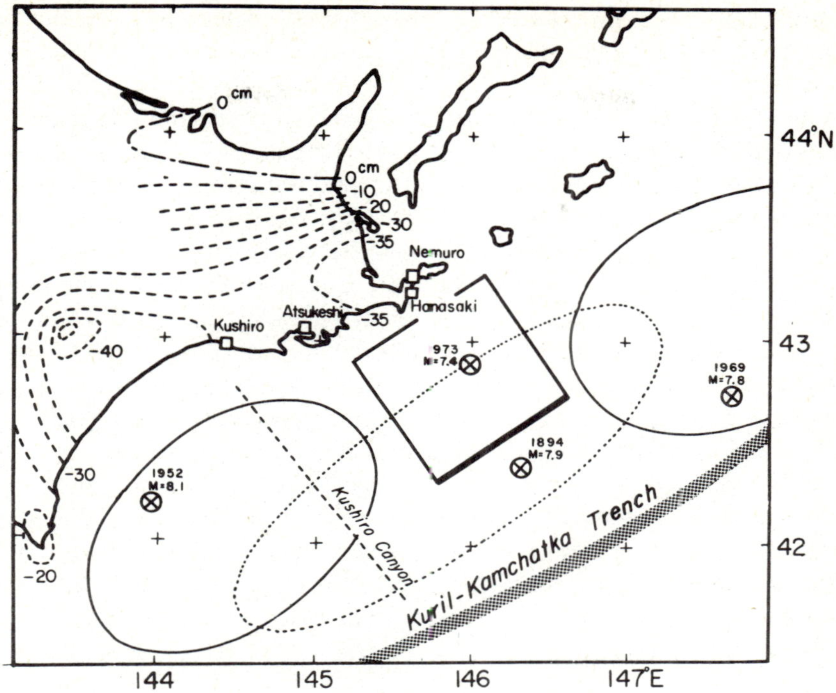

Figure 1
Vertical displacement field in the eastern part of Hokkaido (1900–1955; GEOGRAPHICAL SURVEY INSTITUTE [1]) and the off-coast major earthquakes (JAPAN METEOROLOGICAL AGENCY [10], ANDO [5])

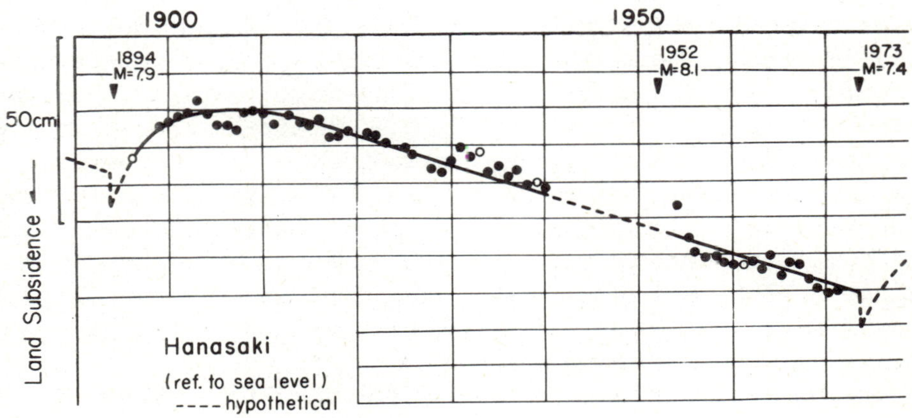

Figure 2
Vertical movement at Hanasaki relative to the sea level (annual mean values; GEOGRAPHICAL SURVEY INSTITUTE [1]). Data contaminated by short interruption of recording are distinguished by open circles

Tokachi-oki (1952, $M = 8.1$) and Kuril (1969, $M = 7.8$) earthquakes. But the data in Fig. 2 indicates no associated anomalies as significant as that noticed in the initial part of the curve.

The mode of sea-level change for a period of several years around 1900 is apparently different from the recent trend of continuous subsidence, and suggests an appreciable land uplift relative to sea level. It is difficult to discriminate the sea-level change due to land movement from the disturbances of oceanographic and meteorological origins. Yet the writer assumes that the present event on the record reflects actual land movement. Firstly because it shows an abnormally large sea-level change following the former Nemuro-oki earthquake and, secondly, a tide-gauge observation, which was initiated at Oshoro (western Hokkaido) in 1905, has recorded a sufficiently stable sea level for many years since then. Little is known about the land movements in the 1894 earthquake. Judging from the tsunami generating source (HATORI [13]), however, the coastal area of eastern Hokkaido seems to have sunk in a manner similar to the recent case (for example, –10 cm at Hanasaki). It is likely, therefore, that the anomalous event in Fig. 2 is the evidence for a postseismic land uplift of the said quake, probably due to delayed strain release, or faulting, at depth. Supposing this mode of movement accompanies every earthquake in this area, they would compensate the interseismic subsidence effectively, resulting in a long-term rate as low as that geomorphologically observed (cf. the previous section).

3. *Coseismic movements and the source models*

The 1973 earthquake caused land subsidence over wide areas of the Kushiro-Nemuro plain, which amounted to –5 to –10 cm at Hanasaki as revealed by levelling and tide-gauge observations (GEOGRAPHICAL SURVEY INSTITUTE [12]). Several source models of the present earthquake have been proposed in order to comprehensively interpret the associated seismic waves, crustal movements and tsunamis (ABE and YOKOYAMA [4]; ANDO [5]; SHIMAZAKI [6, 7]; TADA [8]). They are basically similar to each other, and attribute these seismic effects to the activity of a submarine reverse fault which dips north-westward at a low angle. In these models, the subsidence over the Kushiro-Nemuro plain is well understood as a hinterland deformation associated with the principal uplift in the fault's vicinity (see Fig. 3). If we could explore the crustal deformation off the coast to the south-east, we would certainly observe a significant uplift of the sea bottom adjacent to a submarine fault. Generation of tsunami waves seems to support this prediction (HATORI [13]).

According to ANDO's [5] model, for example, the fault-origin is given by the following parameters:

dimensions: 90 km × 90 km, dip: 20° to N30°W,
slip: 1.5 m (reverse with slight right-lateral),

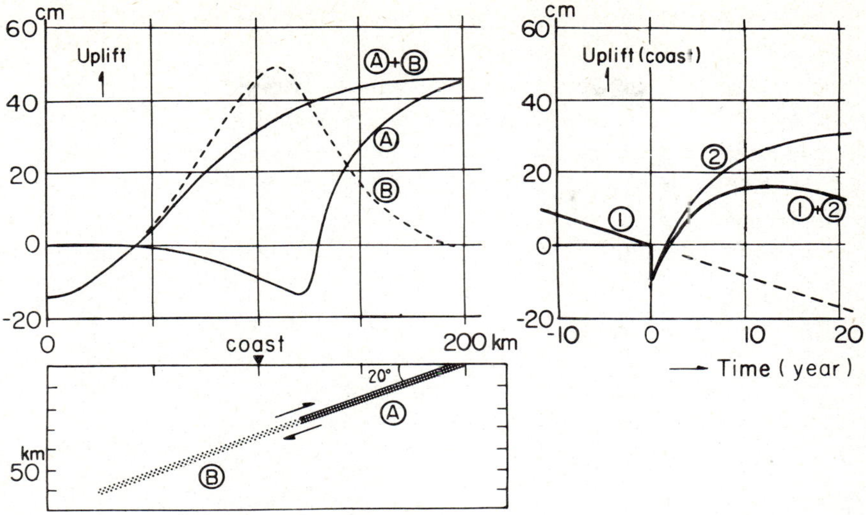

Figure 3

Left. The first model of a retarded faulting (bottom) and the associated vertical displacement (top). The curves Ⓐ and Ⓑ in the top figure represent the contribution from the fault segment Ⓐ (brittle) and Ⓑ (ductile) in the bottom picture, respectively (slip amplitude: 1.5 m; modified from the data by ANDO [5]). Right. Progress of the vertical displacement at a fixed point ('coast' in the left figure). ①: Continuous subsidence due to the plate motion; ②: movements associated with the faultings (brittle and ductile fractures)

where the slip amplitude is taken larger than the original value in relation to other types of data. We shall refer to his model hereafter. Its plan and vertical cross-sectional view are drawn in Figs. 1 and 3 (bottom, left), respectively. Curve A in Fig. 3 (top) represents the vertical movement across the central part of the surface displacement field.

4. *Aseismic faulting at the plate boundary*

The displacement field, as observed in the eastern part of Hokkaido since 1900, is attributed to the motion of an underthrusting sea plate (SHIMAZAKI [11]). In Shimazaki's model, the continental crust is strained by down-dip motion of the sea plate, across its boundary extending from the free surface down to a vertical depth of 100 km (or *ca.* 200 km in the down-dip depth). Although this figure is hypothetical in itself, the wide area of the displacement field (see Fig. 1) is good evidence for the large-scale interaction by the moving sea plate. In comparison with this scale of the field, the 1973 event was only a sectional rebound. In other words, the deeper part of the boundary still conserves the potentiality of slipping.

The speed of such an additional slip is critical for judging the future seismic danger in this area. Should the slip occur instantaneously, it would cause another big earthquake. But, more optimistically, the latent strains might be released aseismically through a rheological process of slipping, as several seismologists consider (e.g. ABE and YOKOYAMA [4]). Let us refer to the latter case. The reason for this choice will be discussed later.

Figure 3 (bottom, left) illustrates the model thus proposed. It is basically a compound fault which undergoes brittle and ductile offsets in this order. Under the critical strain condition, the upper part Ⓐ undergoes brittle fracture over all its area as was the case in the 1973 earthquake. However, the fracture cannot propagate into the Ⓑ region immediately because of the viscous resistance there. Consequently the space adjacent to the Ⓑ boundary is led into an unstable state (interseismic strain accumulation across Ⓑ might not be efficient due to the viscosity in this boundary, yet the space around it could be strained by the plate motion across the Ⓐ boundary). Retarded slip will occur, sooner or later, across the Ⓑ boundary to release the strains remaining in the adjacent space. The time constant of this process is unknown, but it depends essentially on the rheological properties at the Ⓑ boundary.

The system in Fig. 4 (top) is a mechanical analogy of this process. The friction

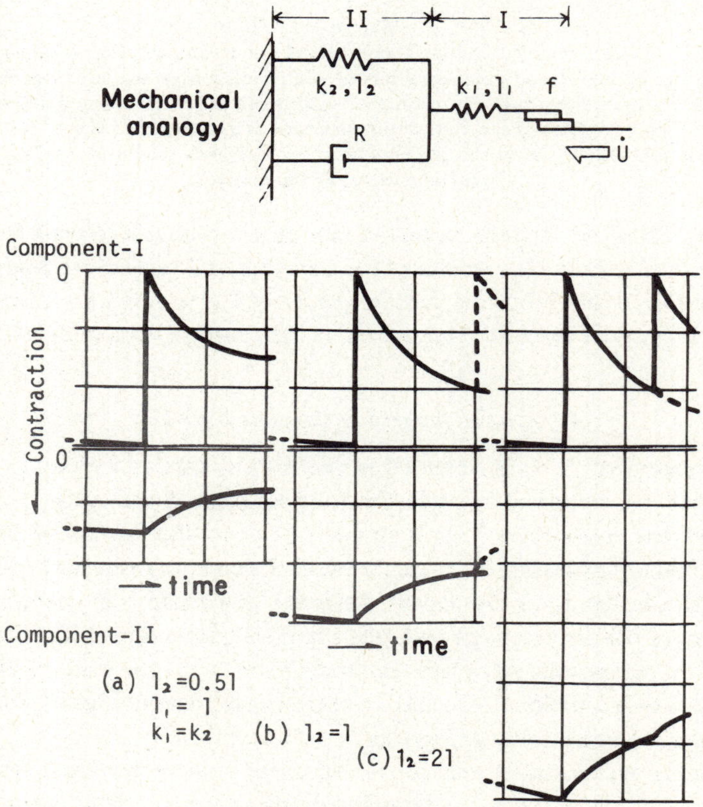

Figure 4
Mechanical analogy of the retarded faulting (top), and its response curves (bottom). Components I and II represent the brittle and ductile parts, respectively

element, characterized as f, represents the brittle fault plane (BENIOFF [14]) and is connected to the continuous plate motion at a rate $\dot{U}$. Between the element and the fixed point on the left side (compared to a remote part of the crust) the system has two springs, 1 and 2 (length: l_1 and l_2; stiffness: k_1 and k_2), which represent the elastic properties of the space adjacent to the boundaries Ⓐ and Ⓑ, respectively. The dash-pot (R) for the viscosity in the Ⓑ region is connected to the fixed point for the sake of simplicity. This expedient may be allowed as the present model is concerned with the short-term process, in which the plate motion may be disregarded. Thus the components I and II represent the brittle and ductile processes, respectively.

Figure 4 (bottom, a, b, c) illustrates the modes of strain release of the system, by assigning three different lengths to the second spring. When the strain across spring 1 (ε_1) is accumulated to the critical level which satisfies $f = k \cdot \varepsilon_1$, the friction element slips to release instantaneously (see extension of component I in the figures). However, component II does not quickly respond to this event due to the viscous resistance by R. Rather, it starts to release the strain across spring 2 gradually with its own time constant. The friction element is locked in this phase, so that the extension of spring 2 and the resultant contraction of spring 1 continue until the two components reach the new equilibrium.

Processes that lead to the final equilibrium depend on the system parameters. If spring 2 is relatively short, as shown in (a), the majority of the energy in the system is released by the initial slip in I. If l_2 is long, on the other hand, the energy release by a single slip in I is a fraction of the total energy. In this case, the energy latent in II would again be great enough to strain spring 1 close to its critical state. Suppose friction level f has been lowered after the initial slip, secondary slip will occur as shown in (c), where lowering of f to two-thirds of its original value is assumed. (b) represents the intermediate case.

5. Vertical movements at Hanasaki compared with models

The broken line in Fig. 3 (top left) illustrates the possible amplitude of the surface displacement due to the retarded slip across Ⓑ. At Hanasaki, which falls on the 'coast', for example, it amounts to +40 cm or more. Progress of the movement in time at an arbitrary point is unknown from the above-stated analogy, because it bears no information about the spatial distribution of retarded displacement.

The simplest assumption for this purpose is that the fault segment Ⓑ undergoes retarded slip uniformly over all its area in such a manner as the analogy predicts. In other words, the surface displacement, having a spatial distribution as shown by the broken line in the figure, grows to its final state without changing its spatial proportion. Curve ② in Fig. 3 (right) is the displacement at the 'coast' thus predicted. Superposing a linear trend of continuous subsidence ① on ②, we finally obtain a curve ① + ②, on the basis of which we may interpret the observations.

The broken line applied to the initial part of the tidal record in Fig. 2 refers to this hypothesis. The parameters chosen for the best fit are as follows:

subsidence rate	0.8 cm/yr
coseismic subsidence	7 cm
retarded uplift	40 cm
time constant	8 yr

where the origin time of the curve is set at 1894. If we apply a similar curve simply to the 1973 event, we obtain a prediction as given in the same figure. From it we learn that land uplift starts immediately after the earthquake and grows to 20–30 cm in the next several years. In this recovery phase, the land returns to its preseismic level in one year following the quake.

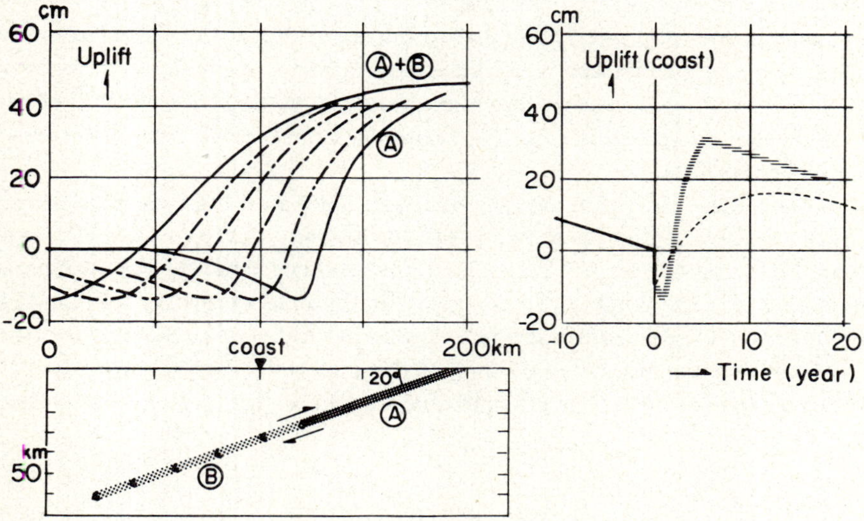

Figure 5

Left. The second model of a retarded faulting (bottom) and the associated vertical movement (top). Chain lines in the top figure represent the transitional displacement field due to aseismic propagation of the fault in Ⓑ. Right. Progress of the vertical displacement at a fixed point ('coast') (lateral hatching). Dashed line represents the previous model for comparison

The model in Fig. 5 (bottom left) hypothesizes an alternative mode of the retarded slipping in the Ⓑ region. In place of the uniform accumulation of slip in the previous model (Fig. 3), the present one assumes slow rupture propagation in the down-dip direction. This process is recognized by the inland progression of the displacement field (see the family of chain lines in the figure (top left)). Consequently, the 'coast' station will observe further subsidence before entering the uplift phase (Fig. 5, right). Difference in uplift between the two models can now be clearly seen by comparing the present curve (lateral hatching) with the broken line for the first model. Besides the postseismic sibsidence phase, the curve for the second model is characterized by an almost linear line of uplift if the fault propagates at a uniform rate. The gradient of the curve depends on the rupture velocity, of course, which is taken provisionally as 20 km/yr in the present case. As is evident from the top figure (left), the maximum subsidence will occur approximately when the tip of the propagating fault is located immediately under the station.

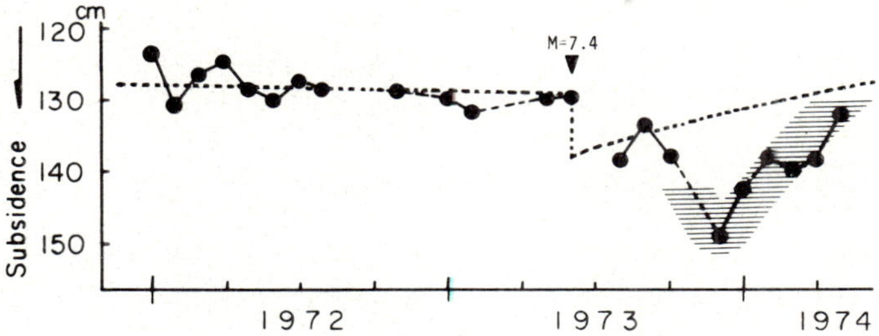

Figure 6
Vertical movement at Hanasaki as observed by the differential tide-gauge method (reference station: Urakawa, southern Hokkaido) (GEOGRAPHICAL SURVEY INSTITUTE [15]). Dashed line and lateral hatching refer to the first and second models, respectively

The two models may be discriminated by reference to the postseismic movements at a key site. For this purpose, Fig. 6 compares the recent movement at Hanasaki (relative to Urakawa, *ca.* 120 km south-west of the present station; GEOGRAPHICAL SURVEY INSTITUTE [15]) with the predictions due to the two models. The broken line reproduces the last portion of the curve in Fig. 2, which predicted a monotonic post-seismic uplift on the basis of the first model. This is not supported by the present data, as is evident from the general trend of the curve in the first several months following the

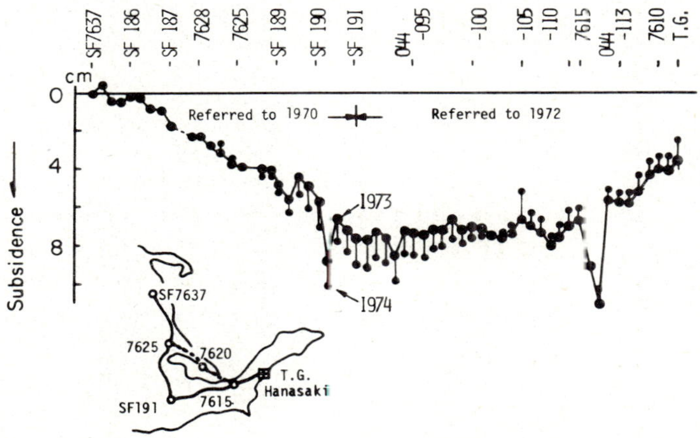

Figure 7
Coseismic and postseismic vertical movements as observed by levellings (GEOGRAPHICAL SURVEY INSTITUTE [15]). The depressed portion of the curve seems to have shifted left during the recent two surveys (1973: large circles; 1974: small circles)

quake. The sea-level change suggests a considerable amount of land subsidence (10 cm, say) prior to the uplift phase. This sense of movement seems to favour the second model, although there was a period of four or five months, during which no marked sea-level change could be seen, immediately before the recorded subsidence. Thus the lateral hatching is applied to the figure on the basis of the second model (see Fig. 5 (right)).

6. *Discussions*

Future movements and long-term rate of tectonic strain

Figure 7 refers to the recent levelling by the GEOGRAPHICAL SURVEY INSTITUTE [15] and illustrates the vertical movement along the route from Shibetsu (SF7637) to Hanasaki (T.G.) in the coseismic and postseismic phases. Strictly, the datum line in the figure is not uniform, the left- and right-half parts referring to the 1970 and 1972 surveys, respectively. However, we may disregard this heterogeneity assuming that the strain accumulation between 1970 and 1972 is negligibly small.

The depressed portion of the curve in Fig. 7 shows a leftward shift during the recent two surveys, suggesting landward propagation of subsidence as predicted by the second model. A more definite answer will be obtained by repeating surveys in future. The evidence for the considerable postseismic subsidence (cf. Fig. 6) is not clear in the present figure, as surveys have missed the period October–December, 1973, in which this particular event seems to have occurred. Yet we notice some qualitative agreement of the levelling data with the tidal record at Hanasaki. As seen in Fig. 7, the momentary level of the bench mark T.G. in 1974 appears between the two former levels, 1972 (datum line) and 1973. This relation seems reasonable when compared with the tidal record in Fig. 6, which suggests that the land has recently returned to an intermediate level between the preseismic and postseismic levels, each corresponding to the first (1972) and second (July, 1973) surveys.

The foregoing discussion proves that the sea level at Hanasaki provides us with critical data for discriminating the models. The sea level regained its preseismic level in one year and continues to record land uplift. If such an uplift should continue, as the models predict, Hanasaki would experience land uplift for 40 cm or more as a result of the retarded faulting. Consequently, the average rate of the land movement there will be low, almost one-tenth of the rate which we observed in the recent several decades prior to the earthquake.

Rheology in the plate boundary

Ductility or anelasticity in the deep plate boundary is the basic assumption for the above-stated models. In spite of its geophysical importance, the writer did not discuss the background of this assumption at all.

KANAMORI [16], who extensively studied source mechanisms of major earthquakes along the Circum-Pacific belt, summarized the seismic activities in the Kuril-Hokkaido region as a Tokachi-oki type plate interaction. In this type, coupling of the continental plate to the oceanic plate is more or less weak in the deep boundary, presumably due to the partial melting of rocks. Seismometrically, this region does not generate such super earthquakes as sometimes occur in the Alaska–Aleutian region. The occurrence of a few deep (or intermediate) earthquakes and the velocity anomaly at depths of 50–100 km (FEDOTOV and KUZIN [17]) also characterizes the present region. YOSHII [18] noticed a trend of subcrustal earthquakes (40–60 km deep) concentrating outside (sea-side) a boundary line which he named the 'aseismic front'. The front extends parallel to the coast and falls on the northern rim of the respective source region of major earthquakes in this area (see Fig. 1). The brittle–ductile transition at a certain depth in the plate boundary may be hypothesized naturally when referring to this type of information.

The progress of the postseismic movement seems to favour the second model as far as the recent sea-level change is concerned. It does not necessarily mean that the retarded faulting will be completed simply in this particular mechanism. Rather, the excellent fit of a hypothetical curve to the sea-level change about 1900 (see Fig. 2) suggests predominance of the first model on that occasion. This experience leads us to suspect that the mode of aseismic slipping following the 1973 quake, now favouring the second type, would be predominated by the first type later. Therefore, for future reference, let us interpret the previously estimated time constant (8 years) in terms of the viscosity and other mechanical parameters.

The basic mode of the present process can be studied, approximately, by taking a model of a pair of parallel plates, coupled viscously. Let us fix the lower plate at its original position and apply a tangential restitutive stress to the upper plate. The movement of the upper plate is then expressed, in general, by an exponential function, with a time constant τ given as follows:

$$\tau = \eta/d \cdot K, \tag{1}$$

where d and η denote the thickness and the viscosity coefficient of the inter-plate medium, respectively. K is a constant which relates the effective restitutive stress by the prestrained crust (S) to u, displacement of the upper plate from its position of the elastic equilibrium, in the following manner:

$$S = K \cdot u. \tag{2}$$

It is not simple to estimate K precisely. The average slip and stress drop in the 1973 earthquake are estimated at 1.6 m and 35 bars, respectively (SHIMAZAKI [7]), suggesting that the crust adjacent to the fault has such apparent stiffness as to effect stress change for 35 bars when the fault's surface moves for 0.8 m, say. Let us assume that the space around the Ⓑ region has an apparent stiffness of this magnitude, we then obtain $K \sim 35$ bars/0.8 m $= 4 \times 10^5$ c.g.s., and

$$\eta/d = \tau \cdot K \sim 10^{14} \text{ c.g.s.},$$

using $\tau = 8$ yr. We have no information about d, so we temporarily assume it as 1 km. The viscosity coefficient is thus estimated at

$$\eta \sim 10^{19} \text{ c.g.s.}$$

Further discussion of the viscosity is not the purpose of the present paper. If a brief remark is to be added, the above estimation of d might be too small as the real thickness of the partial melting layer. Suppose it is 10 km, for example, η becomes ten times the previous estimate. In either case, the average viscosity of this particular layer would be considerably higher in comparison with that of molten rock itself.

7. Conclusions

The writer has hypothesized aseismic faulting of the plate boundary following a main seismic slip. This is for the purpose of obtaining a comprehensive view of the land movements in eastern Hokkaido in relation to the seismic cycle there. If the hypothesis explains the basic mechanisms of the crustal dynamics, we shall observe a significant accumulation of the postseismic land uplift for several years following the 1973 earthquake. At Hanasaki, for example, it will amount to 40 cm or more, compensating most of the land subsidence accumulated in the interseismic period. Thus, the average rate of land movement will be extremely low, as known from geomorphological studies.

This mode of retarded faulting may be attributed to the compound fault system with the brittle–ductile transition at a certain depth along the plate boundary. The slip mode in the ductile part is an interesting problem from the viewpoint of earthquake prediction. The sea-level change at Hanasaki seems to suggest slow fracture propagation in the down-dip direction at present, but further studies are needed before a final conclusion can be reached.

Acknowledgements

Prof. Tsuneji Rikitake of the Cooperative Institute for Research in Environmental Sciences, University of Colorado (on leave from the Earthquake Research Institute, University of Tokyo), and Prof. Lynn R. Sykes of the Lamont-Doherty Geological Observatory, Columbia University, critically reviewed the manuscript and gave helpful suggestions. Also, Dr. Gerald J. Lensen of the Earth Deformation Studies, N.Z. Geological Survey, gave useful advice from the tectonic viewpoint. The writer wishes to acknowledge their assistance.

References

[1] Geographical Survey Institute, *The vertical crustal movements in Hokkaido*[2]), Report, Coordinating Committee for Earthquake Prediction *4* (1970), 1–2 (in Japanese).

[2]) Title translated into English by the present writer.

[2] T. UTSU, *Large earthquakes near Hokkaido and the expectancy of the occurrence of a large earthquake off Nemuro*, Report, Coordinating Committee for Earthquake Prediction *7* (1972), 7–13 (in Japanese).

[3] T. RIKITAKE, *Probability of earthquake occurrence as estimated from crustal strain*, Tectonophysics *23* (1974), 299–312.

[4] K. ABE and I. YOKOYAMA, *An expected major earthquake off the coast of eastern Hokkaido*, Report, Coordinating Committee for Earthquake Prediction, *11* (1974), 45–50 (in Japanese).

[5] M. ANDO, On the Earthquake off Nemuro Peninsula, June 17, 1973[2]), Presented at the 22nd Meeting, Coordinating Committee for Earthquake Prediction (23 August, 1973).

[6] K. SHIMAZKI, *The faulting mechanism of the Nemuro-oki earthquake of June 17, 1973*, Report, Coordinating Committee for Earthquake Prediction *11* (1974), 36–40 (in Japanese).

[7] K. SHIMAZAKI, *Nemuro-Oki earthquake of June 17, 1973: A lithospheric rebound at the upper half of the interface*, Phys. Earth Planet. Interior (in press).

[8] T. TADA, *Fault model and crustal movement of the 1973 Nemuro-Oki earthquake*, *Zisin*, J. Seism. Soc. Japan, Second Ser. *27* (1974), 120–128 (in Japanese).

[9] GEOLOGICAL SURVEY OF JAPAN, *Quaternary tectonic movements in konsen district, eastern Hokkaido*, Report, Coordinating Committee for Earthquake Prediction *11* (1974), 31–35 (in Japanese).

[10] JAPAN METEOROLOGICAL AGENCY, On the Earthquake off Nemuro Peninsula, June 17, 1973[2]), Presented at the 22nd Meeting, Coordinating Committee for Earthquake Prediction (23 August, 1973).

[11] K. SHIMAZAKI, *Preseismic crustal deformation caused by an underthrusting oceanic plate, in eastern Hokkaido, Japan*, Phys. Earth Planet. Interior *8* (1974), 148–157.

[12] GEOGRAPHICAL SURVEY INSTITUTE, *Crustal movements in eastern part of Hokkaido associated with earthquake off Nemuro Peninsula, 1973*, Report, Coordinating Committee for Earthquake Prediction *11* (1974), 7–13 (in Japanese).

[13] T. HATORI, *Tsunami magnitude and source area for the tsunami off the Nemuro Peninsula in June, 1973*, Report, Coordinating Committee for Earthquake Prediction *11* (1974), 41–44 (in Japanese).

[14] H. BENIOFF, *Earthquakes and rock creep, Part I*, Bull. Seism. Soc. Am. *41* (1951), 31–62.

[15] GEOGRAPHICAL SURVEY INSTITUTE, Postseismic Movements in the Eastern Part of Hokkaido[2]), Presented at the 26th Meeting, Coordinating Committee for Earthquake Prediction (26 September, 1974).

[16] H. KANAMORI, *Mode of strain release associated with major earthquakes in Japan*, Ann. Review, Earth Planet. Sci. *1* (1973), 213–239.

[17] S. A. FEDOTOV and L. P. KUZIN, *The velocity profile of the upper mantle in the vicinity of the south Kuril Island*, Bull. Acad. Sci. USSR, Geophys. Ser. (English translation), No. 5 (1963), 415–425.

[18] T. YOSHII, On the Aseismic Front[2]), Presented at the Annual Meeting, Seism. Soc. Japan (14 November 1973).

---

[2]) Title translated into English by the present writer.

(Received 11th November 1974)

Pageoph,
Vol. 113 (1975)

Birkhäuser Verlag, Basel

Dilatancy Model and Empirical Formulas for an Earthquake Area

By T. Rikitake[1])

Summary – A physical interpretation of empirical formulas for earthquake magnitude (M) and spatial extents of land deformation and aftershock area is attempted on the basis of a dilatancy model. It is demonstrated that the precursor time interval (τ) vs. M relation approximately fits in that for the τ vs. typical length of earthquake zone provided the above formulas hold good. A diffusion process in a dilatant region seems also to account for these approximate formulas.

1. Introduction

An anomalous crustal deformation often takes place in association with a large earthquake. Dambara [1] put forward an empirical formula for the relation between the magnitude of an earthquake (M) and the mean radius r_1 of the crustal deformation associated with the earthquake. When r_1 is measured in kilometers, the Dambara formula reads

$$M = 1.96 \log r_1 + 4.45 \tag{1}$$

although its validity for a large M exceeding 7 is somewhat doubtful because of scattered data.

Utsu and Seki [2] obtained an empirical relation between earthquake magnitude and its aftershock area (A). Utsu [3] revised the formula which reads

$$M = \log A + 3.7 \tag{2}$$

in which A is measured in units of km^2.

Tsuboi [4] pointed out that equation (2) reduces to an expression, which is practically the same as equation (1), when we put

$$A = \pi r_1^2 \tag{3}$$

so that we see that the Dambara formula is equivalent to the Utsu one provided we assume that the area of land deformation is the same as that of an aftershock zone.

In an approach to predict the magnitude of a coming earthquake on the basis of a precursory land deformation, Rikitake [5] relied on the Dambara formula as given

---

[1]) Cooperative Institute for Research in Environmental Sciences, University of Colorado/NOAA, Boulder, Colorado 80302, USA.

in (1). It was presumed that almost all the strain energy stored in a region, over which the anomalous land deformation is detected, is released by the earthquake, so that it was assumed that the Dambara formula is also valid for a premonitory crustal deformation although nothing about the physical mechanism of such a deformation was taken into consideration.

In the light of the dilatancy theory developed recently (e.g., SCHOLZ *et al.* [6]), however, it appears that an anomalous crustal movement prior to an earthquake might be explained by a land upheaval caused by rock dilatancy in the earth's crust which is subjected to a stress. It is the aim of this paper to examine to what extent the Dambara–Utsu formula for a precursory land deformation is accounted for by a dilatancy model. As no physical theory has so far been proposed, it is important to see whether the dilatancy theory can afford a basis that accounts for the Dambara–Utsu formula.

2. *Precursor time interval and the Dambara–Utsu formula*

A linear relationship between logarithm of precursor time interval (τ) and earthquake magnitude (M) has been first suggested by TSUBOKAWA [7, 8] entirely on an empirical basis. WHITCOMB *et al.* [9] and SCHOLZ *et al.* [6] later proposed a similar relationship in relation to dilatancy mechanism of earthquake occurrence. If we take the relationship proposed by SCHOLZ *et al.* [6] for a variety of precursory phenomena as an example, we obtain

$$\log \tau = 0.685M - 1.57 \tag{4}$$

where τ is measured in units of days.

If we assume that the area, in which the seismic energy is stored, is given by

$$A = \pi(L/2)^2 \tag{5}$$

in which L may be regarded as an averaged diameter of the earthquake zone. We obtain from (2)

$$M = 2 \log L + 3.6 \tag{6}$$

On introducing (6) into (4), we have

$$\log \tau = 1.37 \log L + 0.88 \tag{7}$$

We now draw $\log \tau$ vs. L relation given by (7) as shown in Fig. 1 in which the data for actual earthquakes as reproduced from SCHOLZ *et al.* [6] are also plotted. Although Scholz *et al.* defined L by the longest dimension of aftershock zone, it would not be unreasonable to compare the $\log \tau$ vs. L relationship to their plots. In Fig. 1 we see that the actual data agree fairly well with the straight line as deduced here.

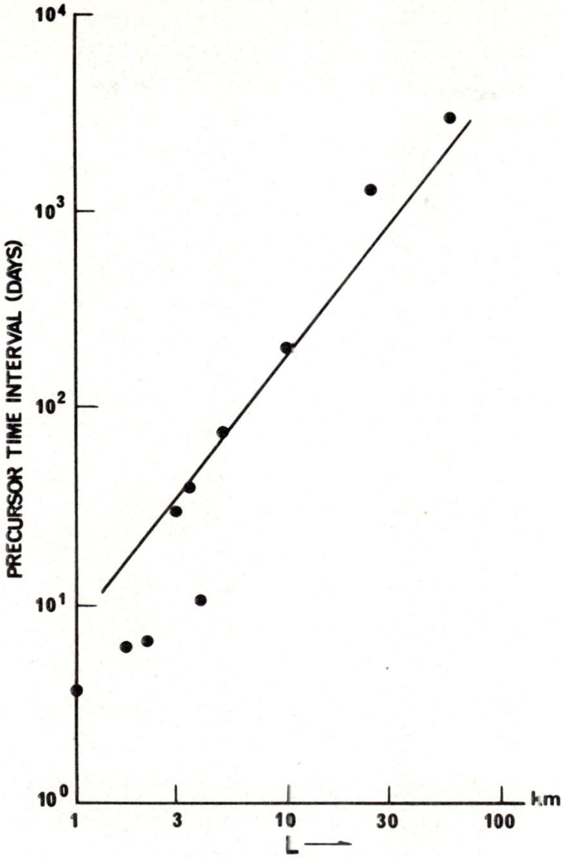

Figure 1
The straight line indicates the precursor time interval vs. typical length L of earthquake zone relationship as obtained from the precursor time interval vs. magnitude relationship due to SCHOLZ *et al.* [6] along with the Dambara–Utsu formula. Solid circles are derived from actual data by Scholz *et al.*

It is therefore seen that the observed $\log \tau$ vs. L relationship is approximately compatible with the empirical $\log \tau$ vs. M one provided the Dambara–Utsu formula can be applied to the $M-L$ relationship. As τ is approximately proportional to L^2, it is indicated that the process is closely correlated to a diffusion one (WHITCOMB *et al.* [9], SCHOLZ *et al.* [6]).

3. *Diffusion process*

Nothing definite is known about diffusion of cracks in a dilatant region. The writer presumes that production of cracks would take place fairly quickly, so that the time needed for that process could be shorter than that of the diffusion of water through the cracks. It would not be unreasonable, therefore, to assume that the rock has zero porosity and that dilatancy occurs suddenly at $t = 0$ in a spherical volume having a radius r_1 for a crude estimate of the diffusion process. In that case fluid pressure P can be obtained by solving a differential equation

$$\partial P/\partial t = c\nabla^2 P \tag{8}$$

in which c is the hydraulic diffusivity and t is the time.

The assumption that the dilatancy occurs at $t = 0$ leads to an initial condition that $P = 0$ within the spherical volume and a boundary condition that $P = P_0$ at $r = r_1$ all the time. To solve (8) under such conditions is a standard problem in thermal conduction analysis, and the solution is given by

$$P/P_0 = 1 + \frac{2r_1}{\pi r}\sum_{s=1}^{\infty}(-1)^s s^{-1}\exp(-cs^2\pi^2 t/r_1^2)\sin(s\pi r/r_1) \tag{9}$$

If we define a mean pressure by

$$\bar{P} = (4/3\pi r_1^3)^{-1}\int_0^{r_1} 4\pi r^2 P\,dr \tag{10}$$

we obtain

$$\bar{P}/P_0 = 1 - 6\pi^{-2}\sum_{s=1}^{\infty} s^{-2}\exp(-cs^2\pi^2 t/r_1^2) \tag{11}$$

Changes in $\bar{P}/P_0$ with dimensionless time $(c\pi^2/r_1^2)t$ are obtained from (11) as shown in Fig. 2. It is then possible to calculate the time needed for completing the diffusion process provided c and r_1 are specified. As for a measure of diffusion time, we may

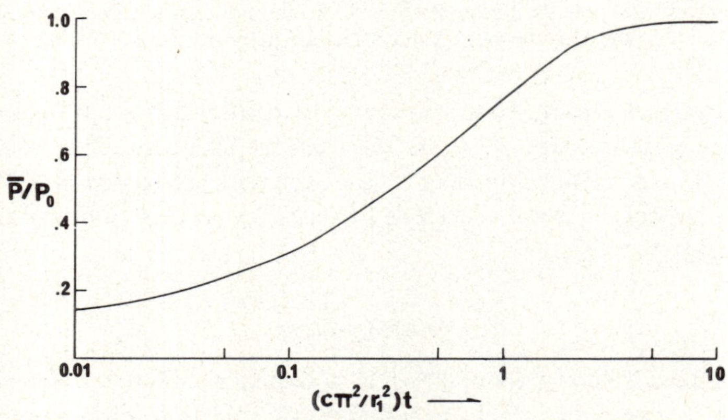

Figure 2
Changes in $\bar{P}/P_0$ (ratio of mean pressure to that in the surrounding medium) with dimensionless time $(c\pi^2/r_1^2)t$

Table 1
Diffusion times and corresponding earthquake magnitudes

r_1	$c = 0.5 \times 10^4$ cm²/s		$c = 1.0 \times 10^4$ cm²/s		$c = 2.0 \times 10^4$ cm²/s	
	τ	M	τ	M	τ	M
2 km	40 day	4.6	20 day	4.2	10 day	3.8
5	250	5.8	125	5.3	63	4.9
10	1040	6.7	522	6.3	261	5.8
20	3990	7.5	1990	7.1	997	6.7
30	9280	8.1	4640	7.6	2320	7.2
50	23600	8.6	11800	8.2	5920	7.8

take τ when $\bar{P}/P_0$ reaches 0.99 as given in Table 1 for a number of combinations of c and r_1.

Going back to (4), which is due to SCHOLZ *et al.* [6], the earthquake magnitudes corresponding to τ's thus obtained are calculated as also given in the table. Figure 3

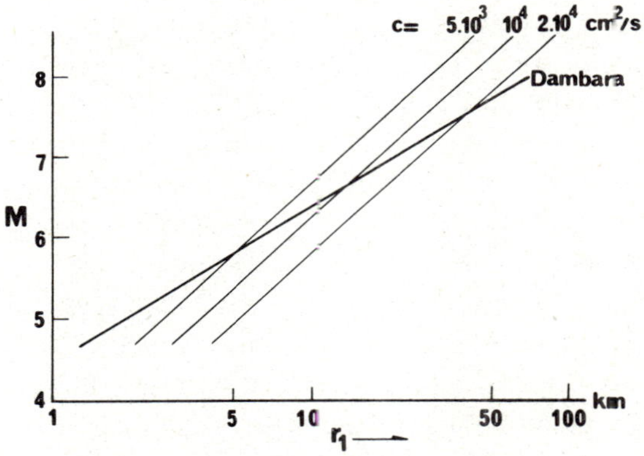

Figure 3
Changes in earthquake magnitude M deduced from the diffusion time as the radius of the spherical volume, in which the diffusion occurs, increases. Three values of hydraulic diffusivity are considered. The thick straight line indicates the $M - r_1$ relationship due to DAMBARA [1]

shows how M increases with r_1 for a few likely values of hydraulic diffusivity. Changes in magnitude with r_1 as calculated from the Dambara formula (1) are also shown in the figure. Although the $M - r_1$ relationship deduced from the dilatancy model does not completely agree with that based on the Dambara formula, it may be said that both results are approximately compatible with one another for a range of c around 10^4 cm²/s. The diffusivity estimated here does not differ much from a value, 5.8×10^4 cm²/s say, as obtained by SCHOLZ *et al.* [6] from a dimensional consideration. It is probably meaningless to discuss the poor agreement for magnitudes larger than 7 because of the low accuracy of the Dambara formula for extremely large earthquakes.

4. Discussion and conclusions

It is first demonstrated in this paper that the relationship between precursor time interval (τ) and typical length of earthquake zone (L) can be approximately deduced from the $\tau - M$ (magnitude) relationship by making use of the Dambara–Utsu formula relevant to the $M - L$ relationship.

A theory of diffusion in a dilatant volume, of which the radius (r_1) and hydraulic diffusivity (c) are specified, enables us to get at the diffusion time which is subsequently converted into earthquake magnitude. Taking c as a parameter, M vs. r_1 graphs can thus be drawn. A similar graph simply based on the Dambara formula approximately fits in the previous ones when c amounts to 10^4 cm^2/s or thereabouts for M's ranging 5–7.

There are arguments (e.g., SCHOLZ *et al.* [6]), however, that the dilated region is considerably wider than the aftershock area or co-seismic crustal deformation area. If it is so, what the writer presents in the above may be in some error because the typical length of earthquake zone is simply presumed to be equal to that of the dilatant region.

DAMBARA [1] showed a few examples for which the spatial extent of pre- and co-seismic deformations of the earth's crust are almost the same. The best examples are the 1930 North Izu ($M = 7.0$) and 1961 Nagaoka ($M = 5.2$) earthquakes. It can certainly be said that the dilatant regions are more or less the same as those for crustal deformation, at least for these two earthquakes which are shallow. The situation is certainly different for earthquakes of which the faulting is predominantly strike-slip. The Dambara formula having been obtained only from the data in Japan, where thrust movements are predominating, may not be applicable to strike-slip earthquakes.

It may be that an underground dilatant region at a certain depth can give rise to an elastic deformation at the earth's surface even though the whole deformed region may not be dilatant. Let us think of a spherical source with hydrostatic pressure P_c placed at a depth z_0 in the crust. The theory of elasticity (e.g., MOGI [10]) implies that the vertical displacement (Δh) at the earth's surface is given by

$$\Delta h = (3a^3 P_c/4\mu)\, z_0(z_0^2 + d^2)^{-3/2} \tag{12}$$

where a, μ, and d denote, respectively, the radius of the sphere, the rigidity, and the radial distance from the z axis which passes the center of the sphere in a vertical direction. (12) is obtained on the assumption that a is much smaller than z_0.

It is apparent from (12) that the vertical displacements at $d = z_0$, $2z_0$ and $3z_0$ amount to some 35, 9 and 3% of that at $d = 0$. If we assume that the depth is a few times larger than the radius of the sphere, a displacement about 0.1 times as large as the maximum one can be observed at a point having a distance several times as large as the radius of the dilatant volume. In the light of the present discussion, it seems possible that all pre-seismic deformation zones may not necessarily be dilatant. Dambara claimed that his formula can only be applied to a precursory crustal deformation having a rate amounting to a value as large as 1 cm/year or more. It is hence supposed that only a land deformation closely associated with a focal region or dilatant region must be taken in the Dambara formula although an extremely sensitive observation may disclose something over an area surrounding that region.

As an alternative one may well think of a highly dilatant volume for the focal region surrounded by a slightly dilatant region of wide extent. Pre-seismic crustal deformation for the latter zone may not be violent enough to apply the Dambara formula to it.

No physical mechanism that accounts for land deformation as well as aftershock areas in relation to earthquake magnitude as obtained by Dambara and Utsu has so far been put forward. It seems likely, however, that a dilatancy model may provide a basis which can possibly be applied to a physical interpretation of the Dambara–Utsu formula.

Acknowledgements

The writer wishes to express his gratitude to Professor CARL KISSLINGER and Dr. MAX WYSS for critically reviewing the manuscript.

REFERENCES

[1] T. DAMBARA, *Vertical movements of the earth's crust in relation to the Matsushiro earthquake*, J. Geod. Soc. Japan *12* (1966), 18–45 (in Japanese).

[2] T. UTSU and A. SEKI, *Relation between the area of aftershock region and the energy of the main shock*, Zisin (J. Seism. Soc. Japan) Ser. 2 *7* (1955), 233–240 (in Japanese).

[3] T. UTSU, *Aftershocks and earthquake statistics (I)*, J. Fac. Sci. Hokkaido Univ.Ser. 7 *3* (1969), 129–195.

[4] C. TSUBOI, *Earthquake energy, earthquake volume, aftershock area, and strength of the earth's crust*, J. Phys. Earth *4* (1956), 63–66.

[5] T. RIKITAKE, *An approach to prediction of magnitude and occurrence time of earthquakes*, Tectonophysics *8* (2) (1969), 81–95.

[6] C. H. SCHOLZ, L. R. SYKES and Y. P. AGGARWAL, *Earthquake prediction: a physical basis*, Science *181* (1973), 803–810.

[7] I. TSUBOKAWA, *On relation between duration of crustal movement and earthquake magnitude*, J. Geod. Soc. Japan *15* (1969), 75–88 (in Japanese).

[8] I. TSUBOKAWA, *On relation between duration of precursory geophysical phenomena and duration of crustal movement before earthquake*, J. Geod. Soc. Japan *19* (1973), 116–119 (in Japanese).

[9] J. H. WHITCOMB, J. D. GARMANY and D. L. ANDERSON, *Earthquake prediction: variation of seismic velocities before the San Fernando earthquake*, Science *180* (1973), 632–635.

[10] K. MOGI, *Relations between the eruptions of various volcanoes and the deformations of the ground surfaces around them*, Bull. Earthquake Res. Inst., Univ. Tokyo *36* (1958), 99–134.

(Received 7th August 1974)

Pageoph,
Vol. 113 (1975) Birkhäuser Verlag, Basel

Theory of Earthquakes

II. Inclusion Theory of Crustal Earthquakes

By B. T. BRADY[1])

Summary – The scale independent inclusion theory of rock failure developed in Part I is applied to the problem of crustal earthquakes and, in particular, to the problem of premonitory phenomena reported to precede such earthquakes. Several well-known premonitory effects such as anomalous variation in the ratio of longitudinal (V_p) and shear (V_s) seismic velocities, V_p/V_s, tilt, regional and local crustal movements and stress axis rotation, to mention a few, are shown to be a natural consequence of the physical processes leading to failure in dry rock. The effects of fluids on failure in the focal region of a potential earthquake are considered in terms of the scale independent inclusion theory.

1. *Introduction*

Observations at Garm, USSR, the New York Adirondacks, and the San Fernando, California, earthquake have shown that prior to these earthquakes the ratio of longitudinal (V_p) and shear (V_s) seismic velocities, V_p/V_s, decreased to anomalously low values and that earthquakes occurred shortly following the return of V_p/V_s to its normal value (SEMENOV, 1969; AGGARWAL *et al.*, 1973; WHITCOMB *et al.*, 1973). A variety of other effects premonitory to earthquakes have also been reported. These include stress axis rotation and anomalous crustal movements and tilt, to mention a few (NERSESOV and SIMBIREVA, 1968; SADOVSKY *et al.*, 1972; SASSA and NICHIMURA, 1956).

These data suggest that a reevaluation of our concepts of what constitutes an earthquake is required. The approximation of an earthquake by a model in which the sudden formation of a shear discontinuity in rocks occurs is an oversimplification in view of these observations. The occurrence of the earthquake following the return of V_p/V_s to its normal value prior to the earthquake further suggests that this and related premonitory phenomena must be related in a causal manner to the processes leading up to and responsible for producing the earthquake. Thus a theory of failure that is applicable to describe failure of rock on the large scale, such as an earthquake, is required. Such a theory must provide not only an explanation of the processes leading up to and including failure but also *all* the precursor phenomena occurring prior to the failure.

---

[1]) U.S. Dept. of the Interior, BuMines, Denver Mining Research Center, Denver, Colo. (present address).

In this paper, the inclusion theory is extended by invoking the scale independent principle to the problem of crustal earthquakes. The paper begins with a recapitulation of the essential concepts of the inclusion theory. This is followed by the application of the theory to earthquake premonitory effects. The subject of earthquake prediction and/or control is considered in terms of the inclusion theory.

2. *Inclusion theory – a recapitulation*

A theory of rock failure, termed the inclusion theory, was formally developed in Part I. The key hypothesis of the inclusion theory is that the intense crack concentrations that develop within localized regions in rock near its ultimate strength (BRADY, 1975) can be mathematically modeled and physically replaced by elastic inclusions whose moduli are lower than the surrounding material. The results of this theory show that there is a three-fold effect of the inclusions. 1) The principal stress axes rotate and change in magnitude both within and outside the inclusion. 2) The least principal stress within the inclusion is tensile and is oriented in a direction normal to the major (long) axis of the inclusion. 3) The magnitude of the least principal stress increases in the focal region of the inclusion; that is, the region into which the inclusion and its associated fault(s) will grow. As the magnitude of the least principal stress in the focal region of the inclusion increases in compression, cracks in the immediate vicinity of the inclusion that opened during the dilatant phase close as the rock begins to store strain energy. As these cracks close, the tensile stress (oriented in a direction normal to the long axis of the inclusion) within the inclusion increases due to the increased elastic contrast between the inclusion and the surrounding rock. Cracks within the inclusion begin to coalesce in response to the increasing tensile stress and a macrocrack(s) forms.

The inclusion theory rests on the postulates that catastrophic failure occurs when all cracks in the focal region of the inclusion that opened during the dilatant phase close and when the aspect ratio of the inclusion, that is, the thickness-to-length ratio of the inclusion, reaches a critical value. When these conditions occur, the length of the macrocrack(s) within the inclusion reaches a critical value. Failure occurs and growth of the fault results from closure of the macrocrack as the inclusion advances into its focal region. In this theory, all faults terminate in an inclusion zone or, simply, a zone of concentrated dilatancy.

The inclusion theory is a scale independent theory. This scale independence in failing rock provides a simple explanation of why the precursor time–magnitude–length relationships hold for both mine failures and a wide range of earthquake magnitudes (see BRADY, 1975, Fig. 11). In the inclusion theory, precursor time ($\Delta\tau$), measured between the initiation of crack closure and the time of failure, depends both on the loading rate in the focal region and on the size of the focal region which in turn is governed by the size of the inclusion. Consequently, the duration of the closure phase is a function of the size of the earthquake that follows it.

3. *Application of inclusion theory to earthquake premonitory effects*

Premonitory effects have been reported to occur prior to earthquakes. These effects include, for example, regional and local crustal movements, anomalous variations in V_p/V_s, tilt, changes in resistivity, changes in microearthquake frequency and stress axis rotation in the focal region of an impending earthquake. Vigorous programs to monitor some of these effects, particularly in Japan and the USSR during the past decade, suggest these effects are real and that they occur together, although not necessarily at the same time (SASSA and NICHIMURA, 1956; NERSESOV and SIMBIREVA, 1968; SADOVSKY *et al.*, 1972; SCHOLZ *et al.*, 1973; WHITCOMB *et al.*, 1973).

Ratio of seismic compressional velocity to seismic shear velocity, V_p/V_s

Figure 1 is a schematic illustration of the variation with time of the ratio of seismic compressional velocity to seismic shear velocity, V_p/V_s, in the focal region of a rock satisfying the constraints of the scale independent inclusion theory. In the following discussion, the behavior of V_p/V_s is considered to represent an average value of V_p/V_s for typical ray paths that pass through the whole focal region. The temporal and spacial variation of V_p/V_s in any small volume of the focal region is considered later.

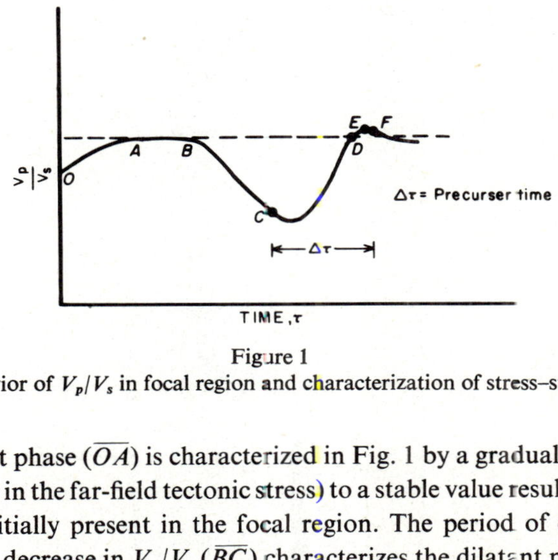

Figure 1
Predicted behavior of V_p/V_s in focal region and characterization of stress–strain behavior

The predilatant phase ($\overline{OA}$) is characterized in Fig. 1 by a gradual increase in V_p/V_s (due to an increase in the far-field tectonic stress) to a stable value resulting from closure of open cracks initially present in the focal region. The period of constant V_p/V_s is denoted by $\overline{AB}$. A decrease in V_p/V_s ($\overline{BC}$) characterizes the dilatant phase. During this phase, new cracks open in the focal region in response to the applied principal stresses. The axes of principal stress begin to rotate both in the focal region and within the region that is to eventually become the inclusion as the inclusion forms. The V_p/V_s value increases during the closure phase ($\overline{CD}$) up to its predilatant value. During this phase, the magnitude of the difference between the maximum and least principal stress (principal stress difference) in the focal region decreases. The least principal stress increases in compression and, as a result, cracks which opened during the dilatant phase ($\overline{BC}$) in response to the applied stresses begin to close or 'heal'. Along pre-existing crack surfaces, the decrease in principal stress difference reverses the shear stress direction and the cracks begin to close in the focal region (Fig. 2). Once all cracks have closed in the focal region (D, Fig. 1), macrocrack growth begins and the stress concentration factor or, equivalently, the principal stress difference in the focal region begins to increase due to transferal of stress into the focal region in response to macrocrack growth within the inclusion. The macrocrack length approaches a critical value and unstable growth occurs (F).

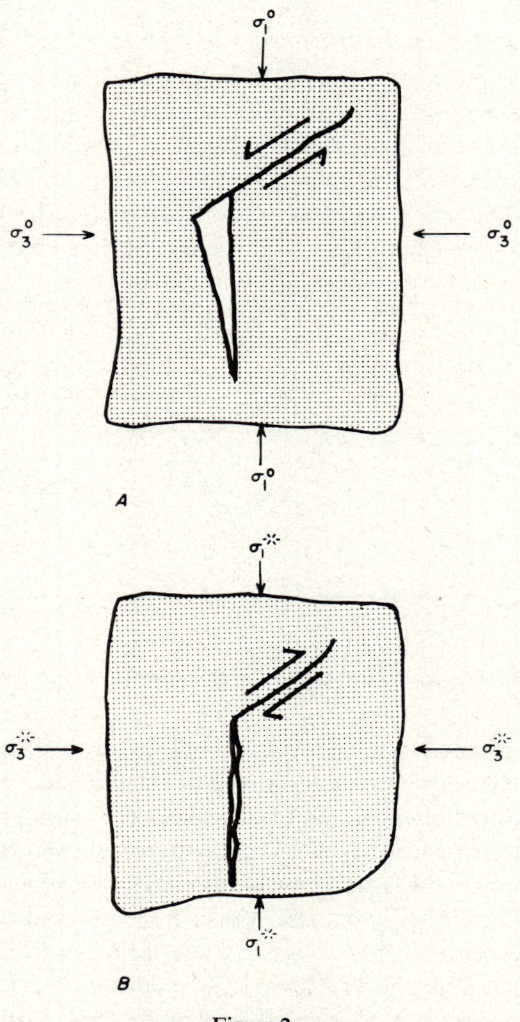

Figure 2
Illustration of crack closure in focal region resulting from a decrease in stress-difference

Since the mean pressure, $\bar{P}$ ($\bar{P} = \frac{1}{3}(\sigma_1 + \sigma_2 + \sigma_3)$, where σ_1, σ_2, and σ_3 denote the values of the local principal stresses), increases in the focal region during the closure phase, the average V_p/V_s value may exceed its predilatant value (D, Fig. 1) before failure if not all the cracks originally present in the rock were closed prior to the initiation of the dilation phase. If failure is preceded by a foreshock sequence (E, Fig. 1); that is, the formation of new cracks in response to the tension stress within the inclusion zone, the average V_p/V_s value may decrease until the elastic contrast between inclusion zone and focal region is a maximum. Failure occurs at this time (F, Fig. 1).

Field observations similar to these were first reported by SEMENOV (1969), who observed unusual variations of the travel–time ratio, t_s/t_p, of shear waves (t_s) to compressional waves (t_p) in the Garm area of the Tadzhik Soviet Socialist Republic. They have since been reported in the United States by AGGARWAL *et al.* (1973) and by WHITCOMB *et al.* (1973).

It is important to note that the anomalous V_p/V_s behavior prior to failure can be explained by the inclusion theory without requiring diffusion of fluids into the focal region since the theory is based entirely on the failure properties of dry rock. The modifying effect of fluids and fluid diffusion (resulting from crack closure in the vicinity of the inclusion) and their effect on the failure process in rock is considered later in the paper.

Stress axis rotation in the focal region

A prediction of the inclusion theory developed in Part I is that the seismically determined stress axes rotate in the focal region prior to an earthquake, and that this rotation occurs during the formation of the inclusion. NERSESOV and SIMBIREVA (1968) reported that stress axis rotation in the focal region occurred prior to an earthquake in the Garm area. The stress axes were determined from fault plane solutions. Their results can be summarized as follows: 1) There is a change in orientation of the seismic stress axes in the foci of weak earthquakes prior to a strong earthquake. 2) A period of approximately 1.5–2.0 months was observed during which the stress axes turned prior to an earthquake of magnitude 4.5 and 2.0 months for a magnitude 5 earthquake The relationship was nearly linear; that is, the larger the magnitude the earlier the stress axes were observed to turn. 3) The intermediate stress axis remained practically in the same position with the compressional and tensional axes turning about it. 4) There was an approximate 90° rotation of the stress axes in the focal region.

Exactly this behavior is predicted from the inclusion theory. For example, the 90° rotation of the seismically determined stress axes in the focal region corresponds to closure of cracks in the closure phase. Here the shear direction is reversed due to an increase in the magnitude of the least principal stress in the focal region (Fig. 2). Note that the orientation of the seismically determined stress axes responsible for producing the main shock will be similar in orientation to the principal stress axes that produce crack closure (CD, Fig. 1).

Microearthquake frequency anomalies prior to major earthquakes

The magnitude-frequency relations of earthquakes in any given place are observed to satisfy an empirical relation of the form

$$\log N = a + bM, \tag{1}$$

where N is the number of earthquakes whose magnitudes are in the range M to $M + \delta M$ and **a** and **b** are constants. Laboratory studies and recent field observations (SCHOLZ *et al.*, 1973; WYSS, 1973) suggest **b** is a function of applied stress.

During the closure phase, the incremental change in seismic activity, dN, and the volumetric inelastic strain associated with this decrease, $d\varepsilon$, can be shown to be proportional to the number of open cracks, N, remaining in the focal region of the inclusion at any time t into the closure phase; that is,

$$dN \approx -AN\,d\varepsilon, \tag{2}$$

where A is a constant of proportionality. Let β denote the average compressibility of the focal region at some time, t, into the closure phase. Let β_h represent the compressibility of the focal region when all cracks which opened during the dilatant phase are closed. The relationship between the volumetric closure strain, $d\varepsilon$, and β and β_h is (BRADY, 1973)

$$d\varepsilon \approx c\Delta\beta\dot{\bar{P}}\,dt, \tag{3}$$

where c is a constant, $\Delta\beta = \beta - \beta_h$ and $\dot{\bar{P}}$ denotes the loading rate averaged throughout the focal region.

The seismicity, $n(t)$, or dN/dt, is found from equations (2) and (3) to be

$$n(t) \approx \bar{B}\,\Delta\beta N_0 \exp\left[-\int_0^t \bar{B}\,\Delta\beta\,dt\right], \tag{4}$$

where $\bar{B} = Ac\dot{\bar{P}}$ and N_0 is the number of open cracks in the focal region prior to initiation of the closure phase. Since $\Delta\beta$ decreases during the closure phase, equation (4) shows that the seismicity must also decrease during the time interval that closure occurs.

The **b**-value for events occurring during the closure phase is found from equations (1), (2) and (3)

$$b \approx b_0 - \frac{1}{M}\int_0^t \bar{B}\,\Delta\beta\,dt, \tag{5}$$

where b_0 is the **b**-value prior to initiation of the closure phase. Equation (5) shows the **b**-value will be positively correlated with the seismicity rate in the focal region; that is, when the seismicity rate decreases the **b**-value also decreases. Similarly, it can also be shown that when the seismicity rate increases the **b**-value increases.

WYSS (1973, Table 1) observed that the **b**-values for the foreshock sequence, b_{fs}, of a major earthquake are lower than the **b**-values of the aftershock sequence, b_{as}, which follows the main earthquake. If the inclusion theory is applied to this problem, it is clear that the compressibility, β_h^{as}, of the fractured focal region (of the main shock) where the aftershocks occur must be greater than the compressibility, β_h^{fs}, of the material where the foreshocks occur. Therefore, equation (5) can be used to give the *b*-value difference of the aftershock and foreshock sequences,

$$b_{as} - b_{fs} \approx \frac{1}{M}\int_0^t \bar{B}(\Delta\beta_{fs} - \Delta\beta_{as})\,dt \approx \frac{1}{M}\int_0^t \bar{B}[\beta_h^{as} - \beta_h^{fs}]\,dt \geqslant 0, \tag{6}$$

where *t* represents the time into the closure phase of either the foreshock or aftershock whose magnitude is *M*. In equation (6), β_h^{as} and β_h^{fs} denote the compressibilities of the focal regions of the aftershock and foreshock sequences, respectively, prior to the initiation of the dilatant phase for these events. Consequently, **b**-values for the aftershock sequence after the main shock will be greater than the **b**-values for the foreshock sequence occurring prior to the main shock. Wyss (1973) has interpreted the **b**-value difference as implying higher tectonic stress for foreshocks than aftershocks. This interpretation is consistent with the inclusion theory, since more strain energy can be stored in the focal region of the main shock than in the focal region of each individual aftershock. The compressibility difference, $\beta_h^{as} - \beta_h^{fs}$, is simply a reflection of the energy difference.

Electrical resistivity

Electrical resistivity anomalies have been observed prior to earthquakes. As a case in point, consider the results of a long baseline (6 km) resistivity profile made at Garm, USSR (Sadovsky *et al.*, 1972). The resistivity data, together with the times of all earthquakes of magnitude greater than 3.0 that occurred within 10 km of the baseline, is shown in Fig. 3. While there may be an obvious correlation between the resistivity profile and the earthquakes that group at the minima of the profile curve, it is difficult

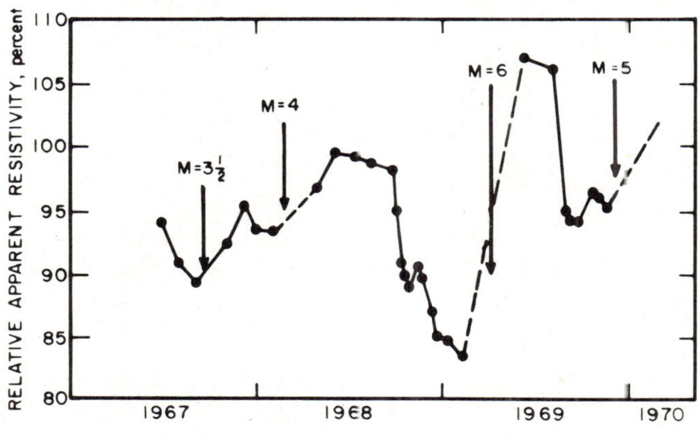

Figure 3
Electrical resistivity anomalies observed prior to earthquakes at Garm, USSR (after Sadovski *et al.*, 1972)

to arrive at a unique interpretation of this data without additional information. For example, the resistivity anomalies in Fig. 3 may be interpreted using the inclusion theory and 'dry' rock by invoking the argument that the apparent resistivity decrease is correlated with crack closure in the focal region. 'Dry' rock in the context used here means that the crack volume is not filled with water. Decrease of resistivity with closure of cracks has been reported to occur in 'dry' rock in the laboratory (PARKHOMENKO and BONDARENKO, 1963). This resistivity behavior is believed to be due to the existence of a thin film of water along the crack surfaces. Under these conditions, the formation of cracks parallel to the direction of maximum principal stress during the dilation phase will produce an increase of resistivity that is greatest in the direction normal to crack growth. Closure of these same cracks during the closure phase will produce a corresponding decrease of resistivity along this direction. However, an alternative explanation of the anomalies in Fig. 3 can be made by postulating that water is present in the dilatant volume that contains both the inclusion and its associated focal region. Crack closure in the focal region prior to the earthquake will produce a migration of water away from the focal region into both the inclusion zone and the diltant volume. This process would produce an increase in pore pressure in these zones. The overall effect that would be produced by this diffusion is a decrease of the apparent resistivity in the dilatant volume.

Regional crustal movements

Anomalous crustal movements are known to precede earthquakes. One well-documented case is the 1964 Niigata earthquake. The magnitude of the main shock was 7.5. Repeated geodetic measurements were made in the vicinity of the main shock. Vertical movements are shown in Fig. 4. The vertical movements had been occurring in the vicinity of the epicenter at a slow, steady rate from 1898 to 1955. Rapid uplift

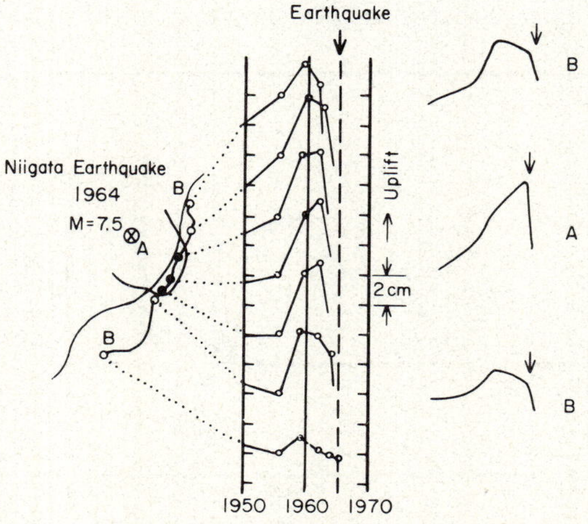

Figure 4
Crustal uplift anomalies in the vicinity of Niigata, Japan, prior to the Niigata 1964 earthquake (after MOGI, personal communication, 1974)

began in 1955. The data in Fig. 4 show that in region A, located near the main shock, the uplift continued unabated until the occurrence of the earthquake. However, in regions *B*, located on either side of *A*, the period of rapid uplift was followed by a decrease in elevation beginning in 1959. If the inclusion theory is applied to this problem, regions *A* and *B* will correspond to the inclusion and its associated focal volume, respectively.

Tilt anomalies

Rapid changes in tilt direction are known to occur shortly before the occurrence of major earthquakes (SASSA and NICHIMIRA, 1956; HOSOYAMA, 1952; WOOD and ALLEN, 1971; LATYNINA and KARMAKYEVA, 1970). These changes place severe constraints on models of the physical processes occurring in the focal region of an impending earthquake and any successful theory of earthquakes must provide a satisfactory explanation of these changes.

A key production of the inclusion theory is that anomalous displacements that may correspond to what has been referred to by some investigators as premonitory 'creep' deformation will occur in the focal region prior to the earthquake. For example, closure of open cracks in the focal region of the inclusion at a time when the V_p/V_s value has attained (or nearly so) its predilatant value will be followed by a period of rapid growth of macrocracks to a critical length within the inclusion. The growth of these cracks produces an increase in the stress concentration factor in the immediate vicinity of the inclusion that results in an elastic volumetric compression and possibly an inelastic volumetric strain component due to closure of any additional open cracks (if they are present) in the focal region during this time. The deformations produced by this stress increase may be detected by both tilt and vertical crustal movements. As an example of this effect, consider the behavior of two tiltmeters, *A* and *B*, located on the surface with *A* and *B* positioned directly above and outside of the focal region of the inclusion, respectively. The reader will note the following discussion applies only to thrust and normal fault earthquakes. Similar behavior will be observed prior to strike-slip failure although the effect will be much less pronounced. The following sequence of events will occur prior to the earthquake. 1) Both *A* and *B* will tilt away from the inclusion or the region which will become the inclusion during the dilatant phase. 2) During the closure phase, *B* will always tilt in a direction where closure is occurring. However, *A* will tilt toward the inclusion zone during the time interval when open cracks are closing between *A* and the inclusion zone. The tilt direction of *A* will reverse during the time when open cracks are closing between *A* and the boundary of the dilatant zone. 3) At a time just prior to the earthquake macrocrack growth occurs within the inclusion zone. At this time both *A* and *B* will tilt away from the inclusion zone. 4) Both *A* and *B* will then tilt toward the inclusion zone due to compaction of rock within the focal region resulting from stress transferral due to growth of the macrocrack. The earthquake will occur shortly following the tilt reversal. This example illustrates two predictions of the inclusion theory for thrust and normal fault earthquakes. 1) Tiltmeters (*B*) located on the surface outside the focal region will exhibit S-bend behavior a short time prior to the earthquake. Tiltmeters (A) located on the surface directly over the focal region will not exhibit S-bend behavior. However, if the tiltmeters are located directly above the inclusion zone, S-bend behavior will be observed. 2) All tiltmeters, with the exception of those located above the inclusion zone, will tilt toward the epicenter (inclusion zone) just before the earthquake.

Tiltmeter behavior similar to the above predictions has been observed. Tiltmeters have recorded several cases of precursor crustal movements. A well-documented case is a magnitude 6 earthquake which occurred at Odaigahara, Japan (SASSA and NICHIMURA, 1956), where six months prior to the earthquake rapid tilting in a direction away from the epicenter was recorded at two locations 40 km and 100 km respectively from the epicenter (closure phase). This tilting continued for three months and then ceased, at which time a station further to the south began tilting. One month before the earthquake, these three stations begin tilting rapidly in a direction toward the epicenter. WOOD and ALLEN (1971) reported anomalous tilting one month prior to the Danville ($M = 4.5$) earthquake. In this earthquake, a short-term precursor tilt occurred toward the epicenter for 10 hours before the main shock. Additional examples of tilt anomalies showing behavior similar to that predicted by the inclusion theory are on record (LATYNINA and KARMAKYEVA, 1970; HOSOYAMA, 1952; SASSA and NICHIMURA, 1956). For instance, tilt measurements in Japan and the Garm region have sometimes shown the 'S-bend' a day, or a few hours, prior to an earthquake. In most cases the tiltmeters tilt in a direction toward the epicenter prior to the earthquake.

Electrical and magnetic precursors

In addition to S-bend tilt behavior, other short-term precursors may result from the extreme compression occurring within the focal region near the inclusion just prior to the earthquake. Precursors of this class may, under certain conditions, include changes in the polarization and/or magnetization vectors in the rock volume being compressed. Stress is known to modify the magnetocrystalline anisotropy of magnetic minerals so that an anisotropy of magnetic susceptibility of the rock in the vicinity of the inclusion would appear just before the earthquake (STACEY, 1969). This effect may be particularly pronounced in basic rocks such as basalt. The stress intensification in the focal region would distort and increase the magnitude of the geomagnetic field in this region. Stress is also recognized to effect the piezoelectric properties of rock. These effects, amounting to 0.1 to 1.0% of the piezoelectric effect produced in single quartz crystals, have been shown to occur in rocks such as granites, gneisses and quartzites (IIDA and KUMAZAWA, 1961).

Anomalous electrical and magnetic phenomena similar to those discussed above have been observed prior to major crustal earthquakes. For example, during the most violent stage of the Matsushiro seismic activity, RIKITAKE (1972) observed the magnetic intensity increased approximately 10 gammas. There are some instances (RIKITAKE, 1972) where no geomagnetic changes exceeding the overall instrument accuracy were observed prior to large earthquakes ($M \geqslant 6.0$). However, the time interval between the readings for this case during which the earthquake occurred was 40 days. Geomagnetic anomalies resulting from a stress-induced mechanism would become detectable a short time (~few hours) before the earthquake and would dissipate rapidly after the earthquake. KATO and UTASHIRO (1949) found a magnetic declination increase of approximately 25 gammas just prior to the 1946 Nankado earthquake. They noted that the field intensity changed dramatically just before the main shock and reverted to normal a few months later. MOORE (1964) observed a geomagnetic field disturbance with an amplitude of approximately 100 gammas at Kodiak, Alaska, an hour before the 1964 earthquake.

Anomalies in the atmospheric–electric potential (on the order of 120 volts/meter) which were characterized by an abrubt onset a few hours prior to an earthquake whose epicenter was 120 km from the observation station have been reported by CHERNYAVSKIY (1963). YOSIMUTSU and NAGATA (SOBOLEV and MOROZOV, 1970) have reported cases of anomalous variations in the telluric current potential several hours before major earthquakes. SOBOLEV and MOROZOV (1970) observed a sharp change in the electric potential gradinet three hours before a magnitude 4.5 earthquake inKamchatka. The electrical disturbance stopped after the earthquake at which time the field resumed its normal behavior.

The above examples suggest a general behavior of electrical and magnetic disturbances just prior to earthquakes that would be expected from the inclusion theory. The inclusion theory requires an increase in the magnitude of the electric polarization vector in the highly stressed rocks of an acidic composition approximately normal to the fault growth direction just prior to the earthquake. It is important to note that the high potential differences produced by the stresses in the focal region would develop only upon an abrupt change in the magnitude and orientation of the local principal stresses, since current leakage would dissipate the polarization effect over a period of time longer than a day, or few days at most. The magnetization effect, on the other hand, if conditions favoring its existence are present, would dissipate more slowly since it is less sensitive to stress changes. The magnetization effect would vanish completely only when all areas of high stress in the fractured focal region are removed.

Precursor phenomena such as electrical resistivity, anomalous crustal movements, tilt and anomalous electrical and magnetic effects are difficult to quantify. At the present time, the best that can be expected is to determine whether these phenomena are qualitatively consistent with any one theory. While these effects are consistent within the frame-work of the inclusion theory the need for additional data which could be obtained from a comprehensive field test program designed to measure all the above precursors is obvious. It is important to note that the inclusion theory predicts that all of the precursor phenomena discussed above bear a fixed relation in time to one another, and further that these phenomena result *only* from *definite physical processes* related to failure and preparation for failure in the focal region and in the inclusion zone of an impending earthquake.

4. Prediction and possible control of crustal earthquakes

The inclusion theory and its associated scale independence principle suggest that the earthquake problem is truly deterministic. Implicit in this concept is the applicability of the theory to the dual problems of earthquake prediction and, possibly, earthquake control. However, the development of possible methods of earthquake prediction and/or control requires an understanding of 1) the effect of fluids on the earthquake process, and 2) the applicability of the inclusion theory to the occurrence of earthquakes within existing fault zones. These subjects are considered before the problem of earthquake prediction and/or control is discussed.

Effect of fluids on inclusion growth

In the inclusion theory, fluids are not required to be present in the focal region to produce most of the precursory effects observed prior to an earthquake. The inclusion theory implies that the effect of fluids in the focal region is that of a modifying agent, an effect discussed at length in rock mechanics literature.

The presence of fluid pressure in the interstices or pores of rock affects the stress state in the solid rock matrix. The fluids within the cracks or pores can be under pressure. The effect of this pressure, P_p, is to cause the cracks to remain open and is conveniently included in the analysis when the total pressure, P, is replaced by the effective pressure $P^* = P - \gamma^* P_p$, where $\gamma^* \approx 1.0$ for most practical problems. Failure of rock containing fluids with pore pressure is then determined in the inclusion theory only by the effective stresses ($\sigma_1 - P_p$, $\sigma_2 - P_p$, $\sigma_3 - P_p$, where σ_1, σ_2, σ_3 denote the applied principal stresses) acting on the rock.

Assume fluids are present in both the inclusion and dilatant volumes of an impending earthquake. During the closure phase, fluids in the focal region must diffuse out from that part of the focal region where closure is occurring. Fluid will also diffuse into the inclusion zone. Thus, pore pressure in the inclusion zone will increase during this phase. As a result of the increased pore pressure in the inclusion zone, the tensile stress, σ_t, within the inclusion will increase to a value $\sigma_t + P_p$. This increase in tension in the inclusion zone produces a concurrent increase in stress concentration in the focal region. When the scale independent principle is applied to this situation, it is clear that diffusion and increased pore pressure in the inclusion zone have a much larger influence on focal region modulus than on inclusion modulus. Consequently, inclusion thickness must decrease due to elastic deformation resulting from increased tension in the inclusion and increased compression in the focal region. Therefore, the effect of fluids in the potential earthquake region and fluid diffusion produced by closure is merely to enhance failure; that is, the system, primary inclusion – focal region, tends toward a condition of instability earlier than it would if the rock were completely dry. Once the main earthquake has occurred, the fluid will tend to diffuse into the fractured secondary inclusion zone in response to the stress transferral resulting from the earthquake. The effect of fluid diffusion in the secondary inclusion zone on aftershock sequences will be similar to its effect on the main shock.

BOOKER (1972) showed that the effect of fluid diffusion on aftershock sequences is to result in a $1/t$ type decay law for the aftershocks. This result suggests some aftershock activity may be caused by transient pore pressure change induced by the main shock. However, it is possible to show that a $1/t$ aftershock decay law can also occur without fluid diffusion. This result strengthens the hypothesis that the effect of fluids and fluid diffusion on the earthquake process is only one of a modifying agent.

Earthquakes within existing fault zones

Figure 5 shows precursor time ($\Delta\tau$) versus failure length (L) for several earthquakes, including a rock burst from northern Idaho and a coal mine roof fall in Pennsylvania (BRADY, 1974). Both the rock burst and roof fall were fresh failures that occurred in

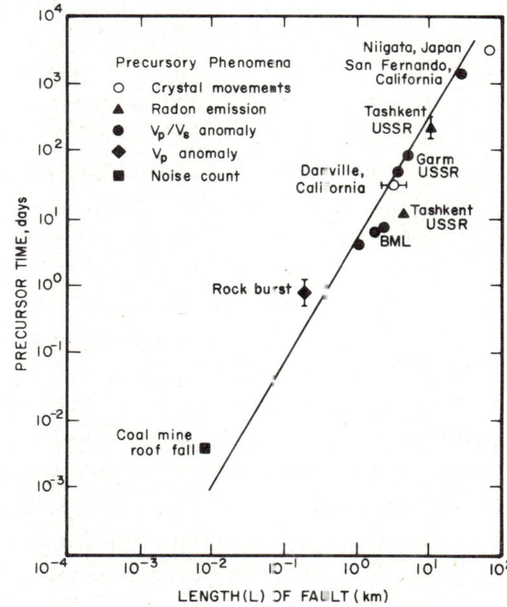

Figure 5
Precursor time-failure length relationships for earthquakes and mine failures

dry unjointed rock. Most major earthquakes in the world, including the earthquakes listed in Fig. 5, are known to occur within pre-existing fault zones. However, the data in Fig. 5 suggest a similar mechanism of failure must be operative both for failures along pre-existing fault zones and failures in fresh unfaulted rock masses. This observation, and the fact that earthquakes re-occur in basically the same locations time and time again, suggest that old faults may 'heal' with time resulting from crack closure due to either a reorientation of the applied principal stresses or to crack cementation resulting from high temperature fluids depositing minerals within these cracks after an earthquake.

The inclusion theory is directly applicable to describe earthquakes for the case where effective healing of any kind has occurred. In addition, the inclusion theory obviously has application to earthquakes generated along 'old' fault surfaces where 'lock' points or asperities are present. Effects similar to this are known to occur in laboratory-size rock specimens (SCHOLZ *et al.*, 1973; BYERLEE, 1966). These studies have shown that sliding along pre-existing surfaces in rock can be characterized as either stable or unstable (stick-slip). BYERLEE (1966) has proposed that whenever two surfaces of a brittle material are brought together, the asperities on the surfaces in contact become locked together. If the stiffness of the system normal to the surfaces is large enough, the asperities must be sheared off before stable sliding can occur. The application of this type of failure mechanism to an existing fault surface implies that during sliding some asperities become locked along the fault and must be broken before sliding can resume. In terms of the inclusion theory, the asperities or 'lock' points become nuclei for inclusion formation and growth. Failure in the vicinity of these inclusions follows the same process discussed in Part I.

There are some consequences of failure of 'lock' points within pre-existing fault zones. For example, the method of estimating what has been called 'effective fault length' in the literature (Fig. 5) from the aftershock region and relating this to the precursor time for the main shock is, in general, consistent with the inclusion theory. However, there may be situations where the calculated 'fault length' is too large. For example, consider the problem where a major 'lock' zone exists along a pre-existing fault. This zone must be destroyed before slippage along the fault can occur. Therefore this zone becomes the nucleus for the inclusion and its associated focal region. Thus, the focal region or, alternatively, the aftershock region, will be contained within the 'lock' zone. When the earthquake occurs, there will be slippage along the fault in the vicinity of the 'lock' zone. Failure of additional smaller 'lock' zones in the vicinity of the major 'lock' may occur due to the rapid stress transferral from the earthquake. As a consequence, the aftershock area will include the aftershock areas of the major 'lock' zone and the additional smaller zones along the fault and appear somewhat larger than it really is. Consequently, care must be exercised in interpreting the numerical relationship between $\Delta\tau$ and the 'effective fault length' L.

Precursor time as an earthquake predictor

Precursor time, $\Delta\tau$, should be used with caution as a possible prediction of earthquake magnitude, because in addition to being dependent on focal region volume it is also functionally related to boundary conditions applied to the focal region of an impending earthquake. Thus, if loading rate suddenly increases or decreases in the focal region, precursor time will be shortened or lengthened in proportion to the excess rate of load increase or decrease, respectively, in the focal region. As a practical example, recent studies by the author in northern Idaho by the U.S. Bureau of Mines have demonstrated that a rock burst in one mine pillar can result in rapid stress transferral to neighboring portions of the pillar or adjacent pillars, resulting in secondary rock bursts.

Precursor times for the secondary bursts were thereby shortened, because loading rate increased over and above that which existed prior to the secondary bursts. A large scale example of stress transferral may be the occurrence of multiple earthquake sequences that appear to be causally connected. Precursor times for the secondary shock may be shortened in proportion to the amount of stress transferral to the second shock region resulting from the main event.

Caution must also be exercised in using velocity anomalies such as V_p/V_s or V_p-residuals (WYSS and HOLCOMB, 1974) to predict the time of occurrence and the magnitude of an impending earthquake. *P*-wave velocity decreases prior to earthquakes have been observed by monitoring the average V_p-residual at seismic stations near the focal region of an impending earthquake. The V_p-residual is the difference between the observed travel time from an epicenter to a station and that computed theoretically based on some earth model. Thus, if the *P*-velocity in the crustal material underneath a seismic station decreases temporarily before an earthquake, each seismic wave reaching this station will arrive somewhat later during the premonitory period. WYSS (personal communication, 1974) observed that prior to the Sitka earthquake the V_p-residual increased for a time and then decreased back to zero well before the occurrence of the earthquake. Consequently, both the 'precursor time' as well as the magnitude of the earthquake to follow would be underestimated if it were assumed that the earthquake occurs when the V_p-residual returns to zero. Figure 6 is a schematic illustration of the processes that would be expected to occur in an earthquake region based on the scale independent inclusion theory. The boundary between the closure and the dilatant zones at various times, $t_1, t_2, \ldots, t_5$ ($t_1 > t_2 \ldots > t_5$), into the closure phase are shown in both cross-section (A) and plan view (B). If a seismic station is located on the surface directly above the dilatant zone at point *C* in Fig. 6B, *an observer* would first detect an increase in V_p-residual due to the formation of the dilatant zone. This would be followed at time t_2 by a decrease to its normal value due to crack closure beneath the station. However, the earthquake would not occur until some later time, say t_5, well after closure had occurred at point *C*. To accurately determine the precursor time and magnitude, additional velocity measurements throughout the dilatant zone are required.

Methods of earthquake control

Water injection and withdrawal in high stress zones has been considered a method of earthquake control (HEALY *et al.*, 1972). *One procedure consists of drilling three wells parallel to and within an active fault zone where natural water is present.* Water is withdrawn from the two outlying wells, increasing normal stress and thus theoretically locking the fault at these points. Water is pumped into the central well and the stored energy is released sequentially by increasing the pore pressure in the vicinity of the central well. It is of interest to consider this technique in terms of the inclusion theory.

Consider first the case where no water is present in either the focal region or in the inclusion. Assume crack closure has occurred or is occurring within the focal region. Two possibilities arise. This first case involves 1): Direct water injection into the focal

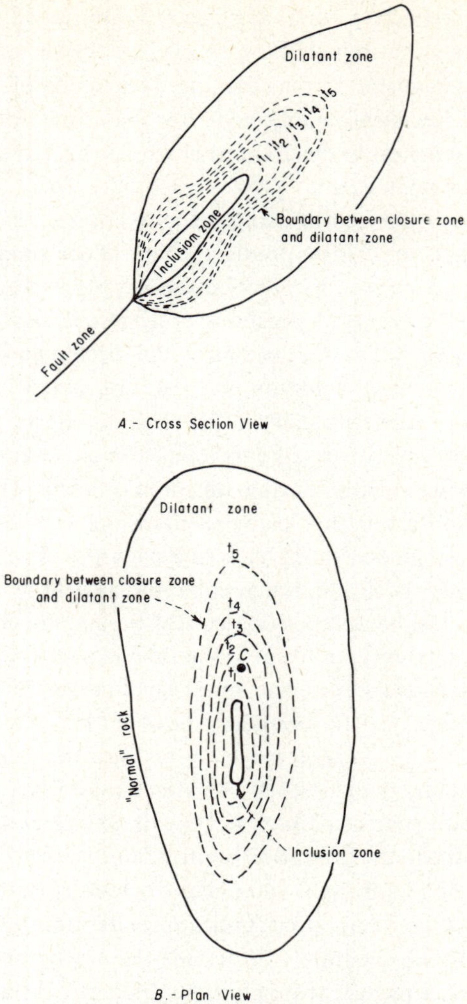

Figure 6
Spatial and temporal variation of crack closure in focal region of the primary inclusion

region. The effective pressure decreases in the focal region and tensional stress in the inclusion decreases. This procedure locally destresses the system and prevents failure until possibly some later time. The method should probably be rejected on the grounds that if conditions responsible for producing instability prevail, namely high stress, this method could only serve to cause a larger shock in the focal region at some later time. 2) The second case is direct injection of water into the inclusion. This produces an increase in tensional stress in the inclusion and an increase of the effective pressure in the focal region. This technique tends to produce instability, and there is a distinct possibility of undesirable side-effects, such as triggering a large earthquake where only a series of small quakes was desired. Since the fault or focal region is not 'locked', this method of control should also be tentatively rejected. Consequently, when natural waters are not present, the injection and withdrawal method of control is unsatisfactory.

Now consider the case where water is present in both the focal region and the inclusion. Assume for simplicity the system is in the dilatant phase. Four possibilities arise, as shown in Fig. 7. 1) *All three wells in the focal region* (Fig. 7A): This procedure enhances instability as withdrawal increases the effective pressure in that part of the focal region nearest to the inclusion. The tension stress increases in the inclusion.

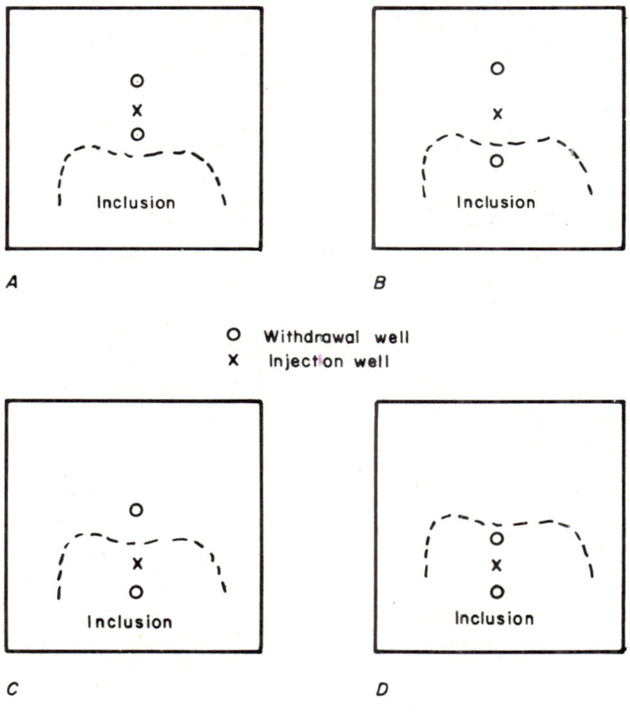

Figure 7
Plan view of possible injection-withdrawal sequences in the vicinity of the primary inclusion

2) *Two wells in the focal region, one withdrawal well in the inclusion* (Fig. 7B): The effective pressure decreases in the focal region and tension stress in the inclusion decrease. The system tends towards stability. 3) *One withdrawal well in the focal region, withdrawal and injection wells in the inclusion* (Fig. 7C): The effective pressure increases in the focal region and tension stress in the inclusion increases. The system tends toward a state of instability. 4) *All three wells in the inclusion* (Fig. 7D): Tension stress in the inclusion at outlying wells decreases and the tension stress increases in the central region. Instability develops between the two withdrawal wells. Methods 1 and 3 should be rejected as dangerous because of the possibility of triggering a large earthquake. A modification

method 2 is discussed in the following paragraph. Method 4 produces the desired effect of destressing the inclusion. If the inclusion represents a 'lock' point along an active fault, this method could be successful provided water is initially present and the location of the 'lock' point is known beforehand.

A third control possibility not considered is a modification of method 2 above. Assume the inclusion is saturated with water of pore pressure P_p and that the system is in the dilatant phase. If wells are drilled into the inclusion zone and water withdrawn P_p will decrease. The inclusion will begin to 'tighten' up or close as the water is withdrawn. The effective inclusion modulus increases and the stress difference decreases in the focal region as tension stress in the inclusion decreases. The system behaves as if the inclusion is vanishing. If the tectonic stresses are large enough by themselves to close all the cracks in the inclusion when the water is completely withdrawn, this method of control could prove to be '100%' effective in destressing both the inclusion and focal region. If the tectonic stresses in the absence of the inclusion are not large enough to initiate further crack formation and/or growth, this method of control could permanently eliminate an earthquake in this region provided the boundary stresses remain fixed. However, if the tectonic stresses continue to increase, a condition will again eventually be reached where a larger magnitude earthquake may occur where only a smaller one would have taken place had the zone been left alone.

Destressing by using high explosives in high stressed zones is a proven method in many instances of rock burst control (BLAKE, 1968). If the scale independence principle is applied to this problem, then methods used to control rock bursts may apply in earthquake control problems. Very simply, destressing removes a high stress on a small area and redistributes it over a larger area. The inclusion theory provides an explanation of why destressing is a valuable control tool for earthquakes. Destressing of the focal region results in a decrease in focal region modulus. Since inclusion modulus is less affected by destressing, the inclusion length must decrease due to crack closure produced by the decrease in tensile stress within the inclusion. Thus destressing has the effect of decreasing both the tensile stress within the inclusion and the stress concentration factor at any point in the focal region. Aside from numerous practical problems involved in destressing focal regions of potential earthquakes, this approach may prove of some value in either preventing or controlling low magnitude earthquakes.

5. *Conclusions*

The scale independent inclusion failure theory has been applied in this paper to the problem of crustal earthquakes and, in particular, to the precursory phenomena that precede these earthquakes. In this theory, fluids are not required to be present in the focal region of an impending earthquake to produce either the precursor effects or the earthquake. Short-term precursors, such as S-bend tilt behavior and changes in the electric polarization and magnetization vectors in the focal region occurring shortly before some crustal earthquakes, were also shown to be qualitatively consistent with the inclusion theory.

Acknowledgements

My friend and colleague, G. FITZPATRICK, has offered valuable insight and constructive criticism throughout this study. K. MOGI, M. S. PATTERSON, D. ANDERSON, C. KISSLINGER, J. DEWEY, M. WYSS, W. SPENCE, AND W. SILL read early drafts of this work and offered valuable criticism.

REFERENCES

AGGARWAL, Y. P., SYKES, LYNN R., ARMBRUSTER, J., and SBAR, M. L. (1973), *Premonitory changes in seismic velocities and prediction of earthquakes*, Nature *241*.

BLAKE, W., Rock Burst Research at the Galena Mine, Wallace, Idaho, USA, BuMines Tech. Progress Report, TPR 39, 1971.

BOOKER, J. R. (1974), *Time dependent strain following faulting of a porous medium*, Geophys. Res. *79*, (14), 2037–2044.

BRADY, B. T. (1973), *A mechanical equation of state for brittle rock, Part II*, Int. J. Rock Mech. Min. Sci. *10*, 291–309.

BRADY, B. T. (1974), *Seismic precursors prior to rock failures in underground mines*, Nature *252*, 549–552.

BRADY, B. T. (1974) *Theory of earthquakes – Part I – A scale independent theory of rock failure*, Pure appl. Geophys. *112*, 701–725.

BYERLEE, J. D., The Fractional Characteristics of Westerly Granite, Ph.D Thesis, Mass. Institute of Technology, 1966.

CHERMYAVSKY, A. Atmospheric-Electric and Electrotelluric Phenomena in Earthquakes, Cots Nauka I Telonika, Tashkent, 1963.

HEALY, J. H., LEE, W. H. K., PAKISER, L. C., RAYLEIGH, C. B., and WOOD, M. D. (1972), *Prospects for earthquake prediction and control*, Tectonophysics 319–332.

HOSOYAMA, K. (1952), *Characteristic tilt of the ground that preceded the occurrence of the strong earthquake of March 7, 1952*, J. Phys. Earth *1*, 75–81.

IIDA, K., and KUMAZAWA, M. (1961), *Measurements of elastic wave velocities in volcanic rocks at high temperatures by means of ultrasonic impulse transmission*, J. Earth Sci. *9*.

KATO, Y., and UTASHIRO, S. (1949), *On the changes of terrestrial magnetic field accompanying the great Nankaido earthquake of 1946*, Sci. Rep. Tôhoku Univ., Serces 5, Geophysics *1*.

LATYNINA, L. A., and KARMALEYEVA, R. M. (1970), *On certain anomalies in the variations of crustal strains before strong earthquakes*, Tectonophysics *9*, 239–247.

MOORE, G. W. (1964), *Magnetic disturbances preceding the 1964 Alaska earthquake*, Nature *203*.

NERSESOV, I. L., and SIMBIREVA, I. G., *Regularities in the Distribution of Stresses in the Foci of Weak Earthquakes in the Garm Area and Their Relationship to Seismicity*, Third All-Union Symposium of the Seismic Regime, June 3–7, 1968, Part I, pp. 147–181.

PARKHOMENKO, E. I., and BANDARENKO, A. T. (1963), *Investigation of the electric resistivity of rocks at pressures up to 40,000 kg/m² and temperatures up to 400°C*, Izv. Akad, Nauk SSSR, Geofiz. Ser. *12*, 1106–1111.

RIKITAKE, T. (1972), *Earthquake prediction studies in Japan*, Geophys. Surv. *1*, 4–26.

SADOVSKY, M. A., NERSESOV, I. L., NIGMATULLAEV, S. K., LATYNINA, L. A., LUKK, A. A., SEMENOV, A. N., SIMBIREVA, I. G., and ULOMOV, R. I. (1972), *The processes preceding strong earthquakes in some regions of Middle Asia*, Tectonophysics *14*, 296–307.

SASSA, K., and NICHIMURA, E. (1956), *On phenomena forerunning earthquakes*, Bull. Disast. Prev. Res. Inst., Kyoto Univ. *13*, (1), 1–8.

SCHOLZ, C. H., SYKES, L. R., and AGGARWAL, Y. P. (1973), *The physical basis for earthquake prediction*, Science *181* (4102), 803–810.

SEMENOV, A. M. (1969), *Variations in the travel time of transverse and Longitudinal waves before violent earthquakes*, Isv. Earth Physics *4*, 245–248.

SOBOLEV, G. A., and MOROZOV, V. N. *Local disturbances of the electrical field on Kamachtka and their relation to earthquakes*, in *Physical Bases of Seeking Methods of Predicting Earthquakes*, M. Sadovskii, Ed. 1970.

STACEY, F. D., *Physics of the Earth* (John Wiley and Sons, Inc., New York 1969), 324 pp.

WHITCOMB, J., GARMANY, J., and ANDERSON, D. (1973), *Earthquake prediction: Variation of seismic velocities before the San Fernando earthquake*, Science *180*, 632–635.

WOOD, M. D., and ALLEN, R. V. (1971), *Anomalous Microtilt preceding a local earthquake*, Bull. Seism. Soc. Am. *61*, 1801–1811.

WYSS, MAX (1973), *Towards a physical understanding of the earthquake frequency distribution*, J. Geophys. Res. *31*, 341–359.

WYSS, M., and HOLCOMB, D. J. (1973), *Earthquake prediction based on station residuals*, Nature *245*.

(Received 14th August 1974)

Pageoph,
Vol. 113 (1975) Birkhäuser Verlag, Basel

Two Models for Earthquake Forerunners

By V. I. MJACHKIN[1]), W. F. BRACE[2]), G. A. SOBOLEV[1]) and J. H. DIETERICH[3])

Summary – Similar precursory phenomena have been observed before earthquakes in the United States, the Soviet Union, Japan, and China. Two quite different physical models are used to explain these phenomena. According to a model developed by US seismologists, the so-called dilatancy diffusion model, the earthquake occurs near maximum stress, following a period of dilatant crack expansion. Diffusion of water in and out of the dilatant volume is required to explain the recovery of seismic velocity before the earthquake. According to a model developed by Soviet scientists growth of cracks is also involved but diffusion of water in and out of the focal region is not required. With this model, the earthquake is assumed to occur during a period of falling stress and recovery of velocity here is due to crack closure as stress relaxes. In general, the dilatancy diffusion model gives a peaked precursor form, whereas the dry model gives a bay form, in which recovery is well under way before the earthquake. A number of field observations should help to distinguish between the two models: study of post-earthquake recovery, time variation of stress and pore pressure in the focal region, the occurrence of pre-existing faults, and any changes in direction of precursory phenomena during the anomalous period.

1. Introduction

In 1969 seismologists in the USSR reported premonitory changes in the seismic velocities prior to a series of earthquakes near Garm [1]. Later, electrical resistivity precursors and changes in radon content of wells were also reported prior to several earthquakes [2, 3]. US seismologists learned a great deal more of these developments during the 1971 IUGG Congress in Moscow. Following this it was discovered that similar changes in velocity ratio had occurred prior to a number of earthquakes in the United States. The first report came from the Lamont group [4], who detected changes in velocity ratio very similar to those reported at Garm. Shortly after, similar changes were reported prior to the San Fernando earthquake [5] and, very recently, prior to small earthquakes in central California on the San Andreas fault [6].

There are currently two quite different ways of explaining these and other precursors. One has been put forward by Soviet scientists at the Institute of Physics of the Earth, Moscow, the other by geophysicists in the US. Actually the Russian scheme resembles in many respects models recently proposed by BRADY [39], MOGI [40], and by STUART [41]. However, because the Russian model (abbreviated IPE) has precedence,

[1]) Institute of Earth Physics, Moscow, USSR.
[2]) MIT Earth Sciences Department, Cambridge, Massachusetts.
[3]) US Geological Survey, Menlo Park, California.

is more highly developed, and is poorly understood outside the USSR, we restrict attention here to comparison of it with the US model involving dilatancy and diffusion (abbreviated DD). In this note we compare the two particularly as to the physical basis and, following STUART [41], note some diagnostic field criteria. We avoid a direct judgement at this time of which model is more nearly correct; perhaps both will be discarded in favor of an as yet unknown third scheme.

2. The IPE model

The physics of the earthquake source is, in essence, the physics of the fracture of material due to tectonic stresses. Here we focus attention on the least-developed part of the physics of the earthquake source, namely on the processes of preparation [7]. The main ideas of this paper were first reported at the IUGG Assembly in Moscow, 1971 [8]. Any quantitative description of the earthquake preparation process is very difficult and theoretical studies have only just begun. However, the concepts and experimental data of fracture mechanics allow us to create a qualitative picture of the earthquake preparation process and to explain the character of the precursors, both discovered and anticipated. These concepts may be summarized as follows:

1) Fracture of statistically heterogeneous materials is caused by the increase of number and size of crack-like defects.
2) The defects may develop in time under approximately constant stress and the rate of their formation increases with the increase of stress.
3) The total deformation consists of intrinsic elastic deformation and deformation caused by mutual displacement of crack edges.
4) Macrofracture (development of the main fault) is the result of the avalanche-like growth and resulting instability which occurs on reaching a certain density of cracks.
5) Formation of the main fault results in the decrease of stress level in the surrounding volume; as a result, the growth of new defects stops and the number of active cracks is decreased.
6) The fracture process does not depend strongly on scale.

On the basis of these concepts, the temporal–spatial course of the earthquake preparation process can be visualized as follows. In real rocks there always exist randomly distributed defects (microcracks). Under the influence of tectonic shear stress the number and size of favorably oriented defects slowly increases and new ones are formed. In statistically homogeneous media, such cracking takes place throughout the whole volume. This quasi-homogeneous cracking is well known in engineering materials and corresponds to a precritical stage of failure. One of the causes of uniformity of cracking is the formation of stable tension cracks at the end of shear cracks [9, 10]. Already at this stage there should be changes in medium properties such as effective modulus of elasticity and anisotropy.

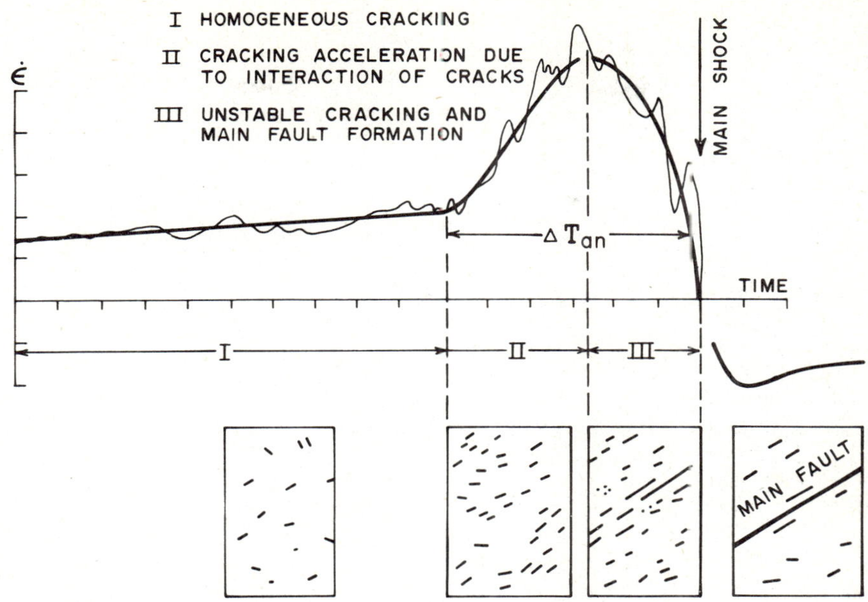

Figure 1
Change of average deformation velocity during the seismic cycle

Figure 1 shows the temporal course of the velocity of average deformation in the earthquake preparation zone. I indicates the stage of uniform cracking in which precursors do not yet develop. Transition to stage II occurs when a certain critical average density of cracks is reached in most of the volume.

Due to the interaction of cracks an accelerating, or avalanche, stage (II in Fig. 1) takes place [11]. For engineering materials the beginning of acceleration occurs at a critical density of microcracks which does not depend on the loading rate. The rapid increase in number and size of cracks in II results in a sharp increase in the rate of total deformation and in sudden changes in physical characteristics of the medium.

If this process leads to the earthquake then it is, by definition, unstable. It means that the further increase of deformation is accompanied by the decrease of stress. Due to heterogeneity of the medium, unstable deformation is restricted to a narrow zone where several relatively large cracks are formed. This stage has often been observed in modeling experiments and geology [12–14]. The material actually divides at this moment into two parts (Fig. 2). In narrow zone A the process of unstable deformation occurs, while in the surrounding zone B the load drops. Because of the general decrease of stresses in III the cracks in B stop developing. Rate of deformation of the whole zone is reduced during this stage of instability and restoration of many original characteristics of the rock takes place. The narrow zone of unstable deformation, A, is characterized by increased concentration of small faults parallel to the future main fault. This main fault, assumed to form during the earthquake, is formed by collapse of the solid

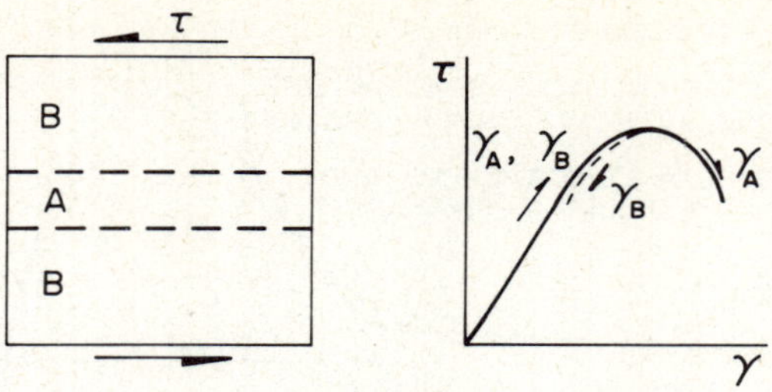

Figure 2
An idealized scheme of formation of the zone of unstable deformation. A – zone of unstable deformation, B – zone of unloading

material between the small faults. Qualitatively, formation of the main fault is similar to the growth or extension of small cracks in II and III. Therefore those small events, too, should be preceded by a short-term, small-amplitude change of deformation velocity. There may be several such short-term changes of deformation velocity; these fluctuations might appear as foreshocks to the main event. In Fig. 1, the thin line shows such possible short-term precursors.

After the earthquake the zone adjacent to the surface of the fault appears to be unloaded. Stresses are transferred to the edges of the new fault. Deformation velocity

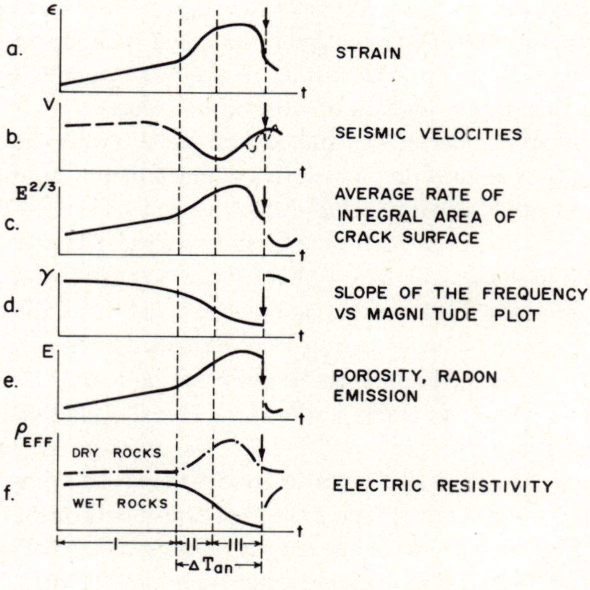

Figure 3
General view of temporal course of earthquake precursors which corresponds to the diagram in Fig. 1: a – strain; b – velocities of seismic waves V; c – earthquake energy E; d – slope of repetition graph; e – porosity radon concentration; f – electrical resistivity and telluric potentials

after the earthquake may be negative because many small ruptures, being freed of loads, can slip in a reverse sense.

To sum up, the above model of the faulting process is based on the generalization of theoretical and experimental results from fracture mechanics. Two features are essential in this model: 1) sharp acceleration of total deformation due to avalanche-like development of interacting cracks (stage II); and 2) deceleration of both deformation velocity and stress in a zone near fault as it develops locally (stage III).

The form of precursors which correspond to the process under discussion are shown in Fig. 3. All precursors, such as changes of distance and altitude between points of the earth's surface or tilts, are connected with the amount of total deformation obtained by simple integration of the deformation velocity.

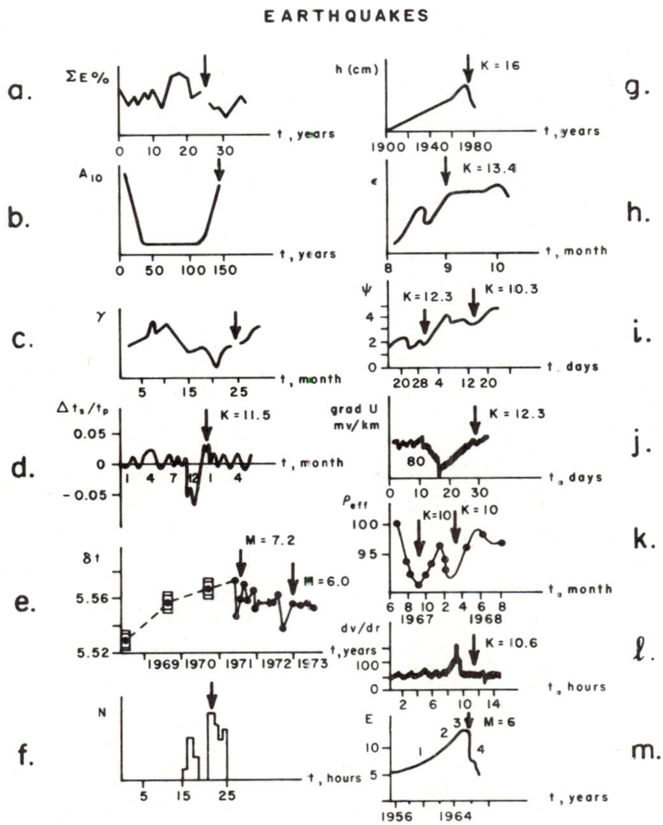

Figure 4

Examples of precursors before earthquakes: a – energy of earthquakes [15]; b – seismic activity [10]; c – slope of repetition graph [17]; d – ratio of travel time of V_p and V_s from weak earthquakes [18]; e – travel time of longitudinal waves from explosion [19]; f – number of acoustic impulses [20]; g – change of altitude [21]; h – change of strain [22]; i – tilts [23]; j – electrotelluric potentials [24]; k – electroresistivity [2]; l – electric potential of air [42]; m – radon concentration in hydrothermal wells [3]

Velocities of elastic waves depend principally on the effective modulus of elasticity of the medium. Since the effective modulus during microcracking drops during stage II velocities of elastic waves sharply decrease. At stage III, with closing of most of the smaller cracks, the effective modulus recovers since the number of big cracks has little effect. The V_p/V_s ratio will change in the same way.

The decrease in the slope of the frequency-magnitude diagram, which results from an increase of the number of strong foreshocks, should be large in stage II and probably somewhat less in stage III as weak connections become successively eliminated.

The increase of radon content and other products of radioactive decay, and the increased flow of springs, is connected with the amount of cracking of rocks and, therefore, is sharply increased at stage II and flattens out at stage III. Decrease can be expected before the earthquake when many small cracks become closed.

The electrical resistivity of dry rocks should increase in stage II and decrease in stage III. In water-saturated rocks, on the contrary, resistivity should sharply decrease in stage II and continue to drop, though slowly, in stage III. Electrotelluric currents evidently should change in the same way.

The above behavior of precursors applies to the region where the processes of avalanche-like increase of cracking and their further localization to the fault surface takes place. When crustal deformation and physical characteristics are measured outside this region the effects will be connected, not with cracking, but, with relatively sharp change in the stress field.

The above model of the earthquake source preparation is based on concepts of fracture mechanics alone; no special mechanism is needed for application to the earth. In particular, there is no requirement for water diffusion as the main factor provocating earthquakes and for the explanation of the observed form of precursors. In Fig. 4 there are examples of precursors before earthquakes. In most cases, their temporal course has a peaked form and it agrees well with the suggested model. Also, since their form is the same as that observed in mines and in the laboratory, the IPE model evidently is not strongly dependent on scale.

3. The dilatancy–diffusion model

In the USA much effort since 1971 has been devoted to explaining the earthquake precursors on the basis of dilatancy and fluid flow. Dilatancy is the non-elastic increase in volume caused by stress and it has been well known in the field of soil mechanics for some thirty years, first being noted for rocks by BRIDGMAN [25] in 1949. In soils, the dilatant volume change is in the form of void space between the grains. In rocks, on the other hand, the dilatant expansion takes the form of new cracks which open up between and through the grains. Most effort has gone into the study of microcrack dilatancy, that is, the form of dilatant cracking that one sees in small laboratory samples under compressive stress. At about half the fracture strength of rock cracks begin to form whose plane is approximately parallel to the plane of the least compressive strength. As stress is increased the cracks lengthen and multiply. As stress is further increased the cracks coalesce by a mechanism, as yet poorly understood, into one or more faults. In most hard crystalline rocks, like granite, diabase or quartzite, the fault grows almost instantaneously once the peak stress is reached. There is some indication from studies made in very stiff testing machines that there is a narrow range of stress over which dilatant cracking becomes concentrated along the eventual direction of the fault [26]. This is also shown indirectly in study of the acoustic emission in rocks close to fracture [27].

The above refers to microcracks and some thought has been given to the large-scale form which these cracks might take. In rock masses one might expect dilatant cracks in two forms – first, as simply a large-scale (on the scale of meters) version of the microcrack. Examples are the tension gashes and oriented veins found in some ore deposits [28]. They have the same relation to the principal stresses as dilatant microcracks. The second large-scale form of dilatant crack might be along those joints, faults, or other natural discontinuities which happen to be oriented in the appropriate direction relative to maximum compression. The opening of such features is clearly shown in model experiments on artificially cracked media [29].

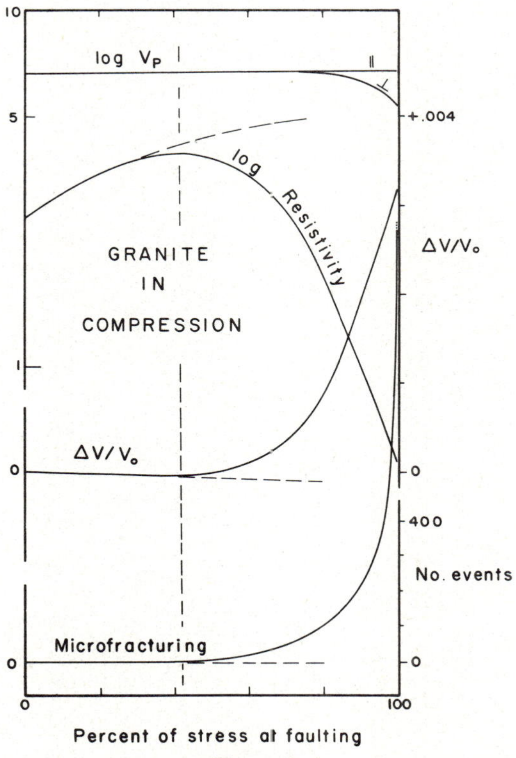

Figure 5
Changes in properties of granite with compressive stress [35]

The physical effects which result from dilatancy have been well known for some time although in the original work the term dilatancy did not appear. As early as 1957, TOCHER [30] showed that compressional velocity decreased in certain directions at high stress in rock. MATSUSHIMA [31] duplicated this at pressures of up to several kilobars. The velocity changes in parallel and perpendicular directions were related to the volume changes in the specimens and it was shown that velocity decrease began just when the volume of the specimen began to increase. The electrical resistivity changes in saturated rocks were related to dilatancy [32] where it was found that resistivity measured parallel with maximum compression could decrease by as much as an order of magnitude (Fig. 5). It was suggested that such changes might be used to predict earthquakes.

With the report by Soviet seismologists of the dramatic velocity changes at Garm

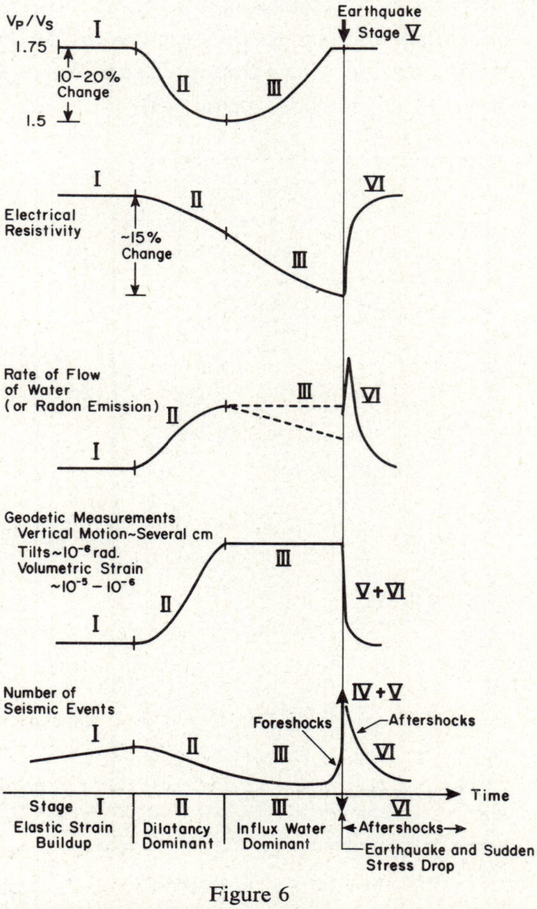

Figure 6
Predicted changes in various physical parameters as a function of time during the earthquake cycle for the dilatancy model [34]. Roman numerals indicate various stages in the cycle

further thought was given to their explanation using the dilatancy model. It was necessary to find some mechanism to cause recovery – that is, to bring the velocity back to the value before the earthquake. The key suggestion was that of NUR [33] who showed that the effect of dilatancy on velocity might depend strongly on whether the dilatant cracks contained water or mixtures of water and air or vapor. The changes in V_p/V_s ratio should be the reverse in these two cases.

The different steps in the complete model are as follows [34] (Fig. 6). During stage I stress is increasing but dilatant cracks have not yet begun to open or form. In stage II dilatant cracks form and become undersaturated. The various properties like velocity and resistivity respond appropriately to this effect [34]. In stage III it is assumed that water re-enters the dilatant rock causing a recovery of the various physical properties back to their initial level. This re-entering of water and subsequent increase in the pore pressure lowers the effective confining pressure and hastens the fracture or frictional sliding which causes the earthquake at the end of stage III. The influx of water in stage

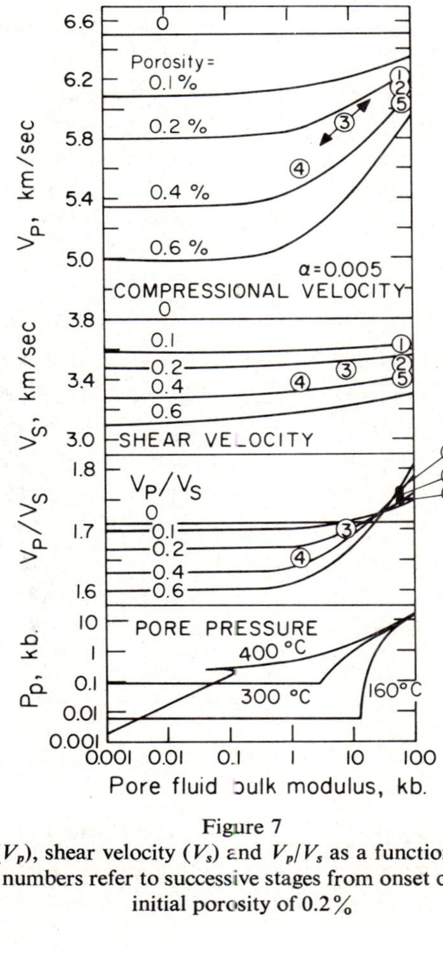

Figure 7
Compressional velocity (V_p), shear velocity (V_s) and V_p/V_s as a function of pore fluid bulk modulus and porosity [5]. Circled numbers refer to successive stages from onset of dilatancy to fracture 5 with initial porosity of 0.2%

II comes about because permeability also increases extremely rapidly in dilatant rock. This is suggested by studies of permeability under high pressure.

At mid-crustal depths one has to consider the effects of high temperature. Water undergoes a transition, depending on pressure and temperature, from liquid to vapor and the effect of transforming pore water to a mixture of water and vapor has been studied in detail by ANDERSON and WHITCOMB [36]. Some results are shown in Fig. 7. As pore water approaches the liquid-to-vapor transition the bulk modulus undergoes very dramatic changes. This is, in turn, reflected in changes in compressional and shear velocity and in their ratio. Consider in Fig. 7 the changes which might take place in dilating rock prior to an earthquake where the different stages are represented by the numbers 1 through 5. Dilatancy and falling pore pressure at first produce only minor changes, namely from 1 to 2. However, as the transition is approached, the properties begin to change more dramatically from 3 down to 4. The changes in compressional velocity are seen to be considerably greater than those in shear velocity and as a result, the ratio V_p/V_s drops. The recovery phase is between 4 and 5 as water, owing to increased permeability, re-enters the cracks and moves the pore fluids away from the liquid–vapor transition. The velocities – particularly compressional velocity – recover and the ratio returns to something close to the normal value.

A number of the changes caused by dilatancy and diffusion are shown in relation to the well-known velocity anomaly in Fig. 6. These include electrical resistivity, radon emission, geodetic uplift or tilt, and the number of small foreshocks. Changes in all of these features are anticipated qualitatively by the dilatancy–diffusion model. Only in the case of velocity changes have the observed effects been explained quantitatively assuming crack aspect ratios, density of cracks, and so on.

In a general way the DD model has been directed principally at earthquakes of the shallow, crustal type common in California. Earthquakes there typically occur along well-established faults so that the stresses probably remain well below the level necessary to fracture intact rock. Thus the amount of dilatancy is probably a lot less than would occur were fresh rock being fractured and during dilatancy one has to consider the re-opening of old cracks as stress is cyclically raised and lowered [37]. Many of the features observed in fault formation may not occur prior to frictional sliding. Nonetheless there is probably a sufficient amount of dilatant volume change prior to frictional sliding to produce the electrical and elastic changes typically observed in the laboratory [38].

4. Comparison and Tests of the Models

With the above brief sketch of the IPE and DD models we now emphasize the essential differences in the models, and suggest field tests which might be diagnostic.

1) According to the IPE model the earthquake occurs as a fault is formed; it thus involves fracture of intact rock on a scale comparable with the dimensions of the

main rupture. This might involve either new faulting, the extension of an old fault into new rock, or the re-fracture of a healed fault. According to the DD model an earthquake can result from motion on a pre-existing fault; it does not require large-scale fracture.

2) In the IPE model, the earthquake occurs after a stress drop having a duration of approximately half the anomaly time. The main earthquake occurs at significantly less than the maximum stress prior to the earthquake. By contrast, the earthquake in the DD model occurs near the peak stress.
3) In both models cracks develop uniformly at first over some volume of rock adjacent to the future rupture surface. In the IPE model cracks of a different orientation form in a zone near the future fault (zone A in Fig. 2) just prior to the earthquake, even in the central part of a long future fault. No such zone is required according to the dilatancy–diffusion hypothesis; cracks may enlarge prior to the earthquake, but orientation should remain the same throughout the anomalous period.
4) Pore fluids play a central role in the DD model; they are not required in the IPE model.
5) In the IPE model, the cracks which form in response to stress have an orientation parallel to the main fault prior to the earthquake. In the DD model cracks are parallel to the plane of the least compression and, thus, inclined to the main fault.

The two models could be distinguished if stress near the focal region could be measured. For example, the decrease prior to the main event in the IPE model should be evident. Also the principal stress directions should probably change with time prior to the earthquake in zone A (Fig. 2) if this feature of the IPE model is indeed significant.

Measurement of pore pressure in the focal region may also be diagnostic. The DD model requires significant changes with a time duration which scales with the earthquake magnitude and with a magnitude great enough to give the observed precursor anomalies. No such changes are *required* in the IPE model, although some changes could accompany the changes in crack geometry inherent in that model.

Development of zone A (Fig. 2) should be reflected in differences in the seismic velocities, resistivity and other phenomena relative to the values outside the zone; therefore comparison of values normal to the eventual ruptures during the development of an anomaly should serve as a test of the IPE model.

The different orientation of cracks in the two models ought to be reflected in the anisotrophy of various properties. For example, maximum conductivity increase during an anomalous period should be in a direction parallel with the eventual fault in the IPE model but inclined to it for the DD model. There will be similar differences for velocity. Finally, any changes in direction of properties during the anomalous period would favor the IPE model.

Acknowledgements

This manuscript was prepared while V. I. Mjachkin and G. A. Sobolev were guests of the US Geological Survey at Menlo Park, and portions of the experimental work reported by W. F. Brace was supported by the US Geological Survey. Nevertheless, the authors assume responsibility for the conclusions reached here. Comments on the manuscript by P. Molnar, W. Stuart, and A. Nur were particularly helpful.

References

[1] I. L. Nersesov, A. N. Semionov and I. G. Simbireva, *Space time distribution of ratios of travel times of transverse and longitudinal waves in Garm region*, Exp. Seismology (1971), Sbornik, 334.

[2] O. M. Barsukov, *Relationship between the electrical resistivity of rocks and tectonic processes*, Izv. Earth Physics No. 1 (1970), 84.

[3] V. I. Ulomov and B. Z. Malashev, *O predvestike silnogo tecktonicheskogo zemletria senia*, Doke. A.N. SSSR, *176* No. 2 (1967).

[4] Y. P. Aggarwal, L. R. Sykes, J. Armbruster and M. L. Sbar, *Premonitory changes in seismic velocity and earthquake prediction*, Nature *241* (1973), 101.

[5] J. H. Whitcomb, J. D. Garmany and D. L. Anderson, *Earthquake prediction: Variation of seismic velocities before the San Fernando earthquake*, Science *180* (1973), 632.

[6] R. Robinson, R. L. Wesson and W. L. Ellsworth, *Variation of P-wave velocity before the Bear Valley, California, earthquake of 24 February 1972*, Science *184* (1974), 1281.

[7] H. A. Sadovsky, *Otvetstennij redactor*. Sbornik. Physica ochaga zemletriasenia, Nauka, Moscow (in press).

[8] V. I. Mjachkin, G. A. Sobolev, N. A. Dolbilkina, V. N. Morosov and V. B. Preobrazensky, *The study of variations in geophysical fields near focal zones of Kamchatka*, Tectonophysics *14* (1972), 287.

[9] W. F. Brace and E. G. Bombolakis, *A note on brittle crack growth in compression*, J. Geophys. Res. *68* (1963), 3709.

[10] B. V. Kostrov. *Teoria Ochaga zemletriasenia* Nauka, Moscow (in press).

[11] V. M. Finkel, *Physica razruschenia*, Mettalurgia, Moscow (1970).

[12] K. Mogi, *Source locations of elastic shocks in the fracturing process in rocks*, Bull. Seism. Soc. Japan *46* (1968), 5.

[13] M. V. Raz and S. N. Chernischov, *Treschinovatost i svoistra treschinovatosty gornick porod*, Nedra, Moscow (1970).

[14] G. A. Sobolev, *The study of precursors of the failure under biaxial compression*, this issue Pure and Appl. Geophys. (in press).

[15] V. I. Keilis-Borok and L. N. Molinovskaja, *Ob odnoi zakonomernosty v vozniknovenii selnikh zemletriasenii*. Sbornik Seismikheskie metody issledovania, Nauka, Moscow (1966).

[16] S. A. Fedotov, *O seismicheskom ziele, vozmozhnosty kolichestvenogo seismicheskogo raionirovania i dolgosrochnom seismikhiskom prognoze*. Seismicheskie raionirovanie SSSR, Nauka, Moscow (1968).

[17] U. A. Mamadaliev, *Ob issledovanii parametrov seismicheskogo regima vo vremeny i prostranstve Voprosy regionalnoi seismichnosty Szednei Azii*, Ilim, Phrunze (1962).

[18] A. N. Semionov, *Izmenenie otnoschenia vremioi probega poperechnich i prodolnich voln pered silnimy zemletriaseniamji*, Izvestia AN SSSR, Physika Zemly, No. 4 (1969).

[19] V. I. Mjachkin and N. A. Dolbilkina, *Seismicheskoe prosvechivanie ochagorikh zon*, VINITI, Moscow (1973).

[20] M. S. Anziferov, *O primenenii geoakusticheskide metodov k rescheniu problemy prognoza zemletziasenij*, Trudy Geophysicheskogo Instituta, No. 25 (1954).

[21] I. Tsubokava, V. Ogava and T. Hayashi, *Crustal movements before and after the Niigata earthquake*, J. Geol. Soc. Japan *10* (1964).

[22] L. A. LATININA and R. M. KARMALEEVA, *Izmerenie medlennikh dvizhenii v zemnoi core kak metod poiska predvestnikov zemletziosenii*. Sbornik. Physicheskie osnovania poiskov metodov prognoza zemletriasenii, Nauka, Moscow (1970).

[23] A. E. OSTROVSKII, *Ob izmeneniakh naklonov zemnoi poverkhnosty pered silnimji blizkimi zemletriaseniami*. Sbornik. Physicheskie osnovania poiskov metodov prognoza zemletriasenii, Nauke, Moscow (1970).

[24] G. A. SOBOLEV and V. N. MOROSOV, *Localnie vozmuschenia electricheskogo polja na kamchatke i ikh sviazi s zemletriaseniamy*. Sbornik. Physicheskie osnovania poiskov metodov prognoza zemletriasenii, Nauka, Moscow (1970).

[25] P. BRIDGMAN, *Volume changes in the plastic stages of simple compression*, J. Appl. Phys. *20* (1949), 1241.

[26] B. W. PAULDING, JR., Crack growth during brittle fracture in compression, PhD thesis, Mass. Inst. of Technology (1965).

[27] C. H. SCHOLZ, *Experimental study of the fracturing process in brittle rock*, J. Geophys. Res. *73* (1968), 1447.

[28] H. E. MCKINSTRY, *Structural control of ore deposition in fissure veins*, Amer. Inst. Min. Met. Engr. Tech. Pub. No. 1267, Mining Technology (Jan., 1941), 25 pp.

[29] E. T. BROWN and J. A. HUDSON, *Progressive collapse of simple block-jointed systems*, Australian Geomechanics Jour. *22* (1972), 49.

[30] D. TOCHER, *Anisotropy in rocks under simple compression*, Trans. Amer. Geophys. Un. *38* (1957), 89.

[31] S. MATSUCHIMA, *Variation of the elastic wave velocities of rocks in the process of deformation and fracture under high pressure*, J. Physics of the Earth *8*, Kyoto Univ. Bull. No. 32 (1960), 1.

[32] W. F. BRACE and A. S. ORANGE, *Electrical resistivity changes in saturated rocks during fracture and frictional sliding*, J. Geophys. Res. *73* (1968), 1433.

[33] A. NUR, *Dilatancy, pore fluids, and premonitory variations of t_s/t_p travel times*, Bull. Seismol. Soc. Amer. *62* (1972), 1217.

[34] C. H. SCHOLZ, L. R. SYKES and Y. P. AGGARWAL, *Earthquake prediction: A physical basis*, Science *181* (1973), 803.

[35] W. F. BRACE, *Current laboratory studies pertaining to earthquake prediction*, Tectonophysics *6* (1968), 75.

[36] D. L. ANDERSON and J. H. WHITCOMB, *The dilatancy–diffusion model of earthquake prediction*, Proc. Conf. on Tectonic Problems of the San Andreas Fault System, edited by R. L. KOVACH and A. NUR, Stanford Univ. Publ. XIII (1973), 417.

[37] K. HADLEY, *Laboratory investigation of dilatancy and motion on fault surfaces at low confining pressures*, Proc. Conf. on Tectonic Problems of the San Andreas Fault System, edited by R. L. KOVACH and A. NUR, Stanford Univ. Publ. XIII (1973), 427.

[38] W. F. BRACE, *Dilatancy-related electrical resistivity changes in rocks*, this issue Pure and Appl. Geophys. *112* (1974), 701.

[39] B. BRADY, *Theory of earthquakes, Part I, A scale independent theory of rock failure*, Pure and Appl. Geophys. (in press).

[40] K. MOGI, *Regularities in the spatial and temporal distribution of large earthquakes and earthquake prediction*, in *Symposium on Earthquake Forerunners Searching Tashkent*, USSR (1974).

[41] W. S. STUART, *Diffusionless dilatancy model for earthquake precursors*, Geophys. Res. Letters, Vol. *1* (1974), 261.

[42] V. F. BOUCHKOVSKIJ, *Izmenenio gradienta electricheskogo potenziala atmospheri kak odin iz vos mogenich predvestnikov zemletria senij*, *Problemi prognoza zemletriasenij*, Moskva, Uzdatelstvo A.N. SSSR (1954).

(Received 9th December 1974)

Pageoph,
Vol. 113 (1975) Birkhäuser Verlag, Basel

A Theoretical Model of the Dilatant Behavior of a Brittle Rock

By J. THEODORE CHERRY,[1]) ROBERT N. SCHOCK[2]) and JOEL SWEET[1])

Summary – A dilatancy model is presented which seems capable of simulating the results of laboratory tests on rock samples. The fundamental assumption incorporated in the model is that dilatancy is caused by the opening of cracks, where the cracks open in the least compressive stress direction (BRACE *et al.*, 1966). Its simplicity and compatibility with numerical techniques, involving the simulation of both static and dynamic stress fields, permits prediction of dilatancy induced effects for stress states and loading conditions which are not experimentally attainable.

1. Introduction

During the past several years, beginning with the classic work of BRACE *et al.* (1966), the volume expansion (dilation) of brittle rock, prior to the onset of unstable fracture propagation, has been consistently observed during laboratory tests on rock specimens. The experimental results of BRACE *et al.* (1966), SCHOCK *et al.* (1973), THILL (1973) and GUPTA (1973a) may be summarized as follows:

1) Dilation begins at approximately one-half to two-thirds of the fracture strength of the rock.
2) Dilation increases as the induced stress state approaches the fracture strength of the material.
3) Dilation decreases as the confining pressure on the test specimen is increased.
4) Dilation causes not only a severe departure from a linear stress–strain behavior but induces a pronounced anisotropic behavior in both the strain field and body wave velocities of the dilating rock specimen.

These experimental results suggest that prior to an earthquake the rock mass surrounding the fault should enter the dilatant phase of material behavior.

Direct field evidence for dilation in the regional strains field across the San Andreas fault, near Parkfield, California, has been reported by CHERRY and SAVAGE (1972). Indirect field evidence for dilation, involving a reduction in the ratio of compressional to shear wave velocity (V_p/V_s) prior to an earthquake, has been reported by Semenov (1969), AGGARWAL *et al.* (1973), WHITCOMB *et al.* (1973), WYSS and HOLCOMB (1973), ROBINSON *et al.* (1974) and BUFE *et al.* (1974).

[1]) Systems, Science and Software, Inc., La Jolla, California 92037.
[2]) Lawrence Livermore Laboratory, Livermore, California 94550.

An earthquake prediction capability will result when field measurements of these types are used to obtain a quantitative estimate of the stress state in the dilatant region and the proximity of this stress state to the fracture strength. The first step toward achieving this predictive capability is to develop a dilatancy model that is not only capable of reproducing laboratory measurements of rock deformation and body wave velocity during dilatant behavior but is also able to predict the effects of dilatancy for stress states and loading conditions that are not experimentally attainable. Therefore, the dilatancy model should be normalized to laboratory data and then used to predict dilatancy related effects in a realistic earth environment.

Interpretation of field measurements, especially those which give an indirect measure of the dilatant state of a fault zone, will depend on utilization of the model to match the field measurements. Therefore, the final requirement for the model is that it should be capable of being merged with appropriate analytical or numerical techniques in order to guide the field measurements program and interpret the field data.

In this paper a dilatancy model is presented that appears to satisfy these requirements. It is certainly not the only dilatancy model possible. However, it has been used to match laboratory measurements of the stress–strain behavior of a brittle rock as it enters the dilatant state (CHERRY and SCHOCK (1971)). It has also predicted the pre-earthquake anisotropic strain distribution around thrust, transcurrent and normal faults (GUPTA (1973a)) and has been used to explain the regional strain field across the San Andreas fault near Parkfield, California (CHERRY and SAVAGE (1972)). We show here that the model also predicts anisotropy in the *P* wave velocity. Furthermore, its simplicity and compatibility with numerical techniques involving the simulation of static and dynamic stress fields makes it an obvious starting point in establishing if the physical properties of a dilatant rock mass may be used to predict earthquakes.

2. Tension failure and dilatancy

The fundamental assumption in non-linear continuum mechanics is that the total strain rate, $\dot{\varepsilon}_{rs}$, is the sum of elastic, $\dot{\eta}_{rs}$, and inelastic, $\dot{\pi}_{rs}$, components. Thus

$$\dot{\varepsilon}_{rs} = \dot{\eta}_{rs} + \dot{\pi}_{rs}. \tag{1}$$

The stress rate, $\dot{\sigma}_{rs}$, is determined using only the elastic component of strain rate. Hooke's law gives

$$\dot{\sigma}_{rs} = \lambda \dot{\eta}_{kk} \delta_{rs} + 2\mu \dot{\eta}_{rs} = \lambda \dot{\theta} \delta_{rs} + 2\mu \dot{\varepsilon}_{rs} - (\lambda \dot{\pi} \delta_{rs} + 2\mu \dot{\pi}_{rs}) \tag{2}$$

where $\dot{\theta}$ is the total volumetric strain rate, $\dot{\pi}$ is the inelastic volume strain rate, λ and μ are the intrinsic Lamé constants of the material.

Since $\dot{\varepsilon}_{rs}$ is obtained from the symmetric part of the velocity gradient, then equation (2) may be used to calculate $\dot{\sigma}_{rs}$ if $\dot{\pi}_{rs}$ is known. In plasticity theory a flow rule of the type

$$\dot{\pi}_{rs} = M_{rsmn} \frac{\partial Y}{\partial \sigma_{mn}} \qquad \text{when } Y(\sigma_{mn}) = 0$$

$$\dot{\pi}_{rs} = 0 \qquad \text{when } Y(\sigma_{mn}) < 0 \tag{3}$$

is often assumed when M_{rsmn} is the flow rule multiplier, and Y is the yield function such that

$$Y(\sigma_{mn}) \leqslant 0. \tag{4}$$

An excellent critique of this type of flow rule and the computational procedure required to determine the flow rule multiplier has been given by CAMERON and SCORGIE (1968). The cap model, presented by NELSON *et al.* (1971), extended this procedure to include the nonlinear behavior of a geologic material at stress states for which $Y(\sigma_{mn}) < 0$. This model's basic limitation seems to be that, for the mathematics to be tractable, the yield strength should depend only on the first invariant of the stress tensor. HANDIN *et al.* (1967) and MOGI (1967) have shown that a yield function dependent only on the first invariant of the stress tensor will not uniquely describe the results of laboratory experiments in which the rock specimens were loaded to failure under a variety of test conditions. Also, HILL (1950) has shown that this type of flow rule may be derived by assuming that the rate of plastic work is maximized. This assumption is open to question during the dilatant phase of material behavior.

The basic mechanism that produces dilatant behavior in brittle rocks has been identified by BRACE *et al.* (1966). They concluded that dilatancy is caused by the opening of cracks in the specimen where the cracks open in the least compressive stress direction. Any 'flow rule' that specifies the inelastic strain rate during dilatant behavior must bias this strain rate in the direction of the least compressive stress.

If $\dot{\pi}$ is the inelastic volumetric strain rate due to the crack, where $\dot{\pi}$ is positive if the crack opens and negative if the crack closes, and if v_r are the direction cosines of the unit vector normal to the crack, then the inelastic strain rate $\dot{\pi}_{rs}$ may be written (CAMERON and SCORGIE (1970))

$$\dot{\pi}_{rs} = \dot{\pi} v_r v_s. \tag{5}$$

Equation (2) becomes

$$\dot{\sigma}_{rs} = \dot{\hat{\sigma}}_{rs} - \dot{\pi}(\lambda \delta_{rs} + 2\mu v_r v_s) \tag{6}$$

where $\dot{\hat{\sigma}}_{rs}$ is obtained by assuming that the total strain rate is elastic, i.e.,

$$\dot{\hat{\sigma}}_{rs} = \lambda \dot{\theta} \delta_{rs} + 2\mu \dot{\varepsilon}_{rs} \tag{7}$$

This model for tension failure was first proposed by MAENCHEN and SACK (1964) and used by CHERRY *et al.* (1973) and HASTINGS (1974) in order to predict the tension fractures induced in a material by an impulsive source. Figure 1 is taken from HASTINGS

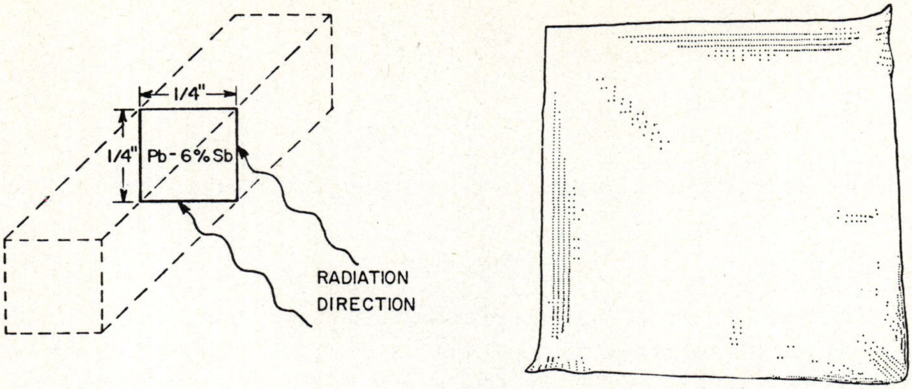

Pretest calculation results at 3.4×10^{-6} sec. The tension failure in the solid material is indicated by the dots. The size of the Pb–6% Sb specimen was initially $\frac{1}{4}$ inch by $\frac{1}{4}$ inch.

Post-test photograph of a Pb–6% Sb specimen which was $\frac{1}{2}$ inch by $\frac{1}{2}$ inch. The number, location and orientation of the cracks are as predicted.

Figure 1
Comparison of predicted and observed tension failure in a specimen subjected to radiation deposition (HASTINGS (1974))

(1974) and compares predicted tension failure against that observed in a post-test specimen. The material has dilated because of tension failure, i.e., because one of the principal stresses has exceeded the tensile strength of the material. The inelastic strain component, $\dot{\pi}$, represents the rate at which void space is developed in the material.

This tension failure model may be extended (CHERRY and SCHOCK (1971) in order to permit crack formation at stress states where all the principal stresses are compressive, the rationale being that stress concentrations at inclusion boundaries produce local tensile stresses even though the macroscopic stress state is compressive. This extension requires knowledge of the fracture strength of the rock, the onset of dilatancy and a 'flow rule' which specifies the rate at which void space is developed during the dilatant phase of material behavior.

As noted earlier, dilation is observed as being at one-half to two-thirds the fracture strength of the rock. Figure 2 shows the fracture strength, Y_s, of Climax Stock grano-

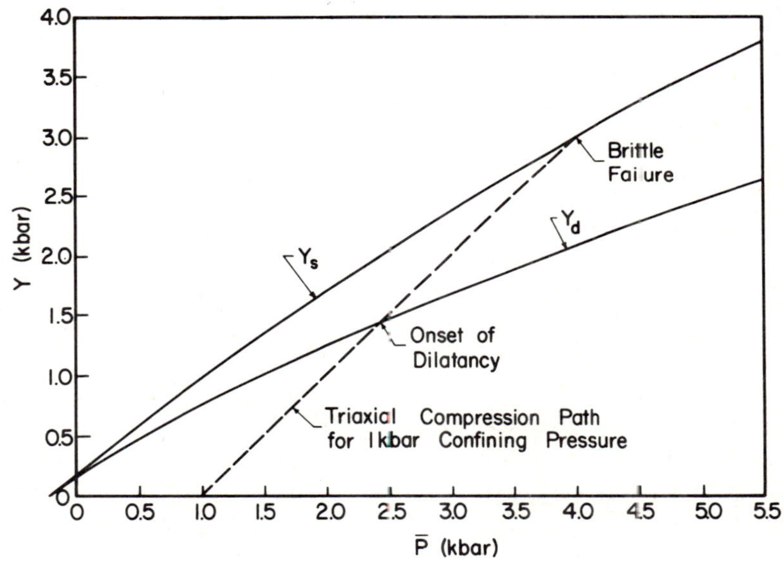

Figure 2
Strength (Y_s) and dilatancy onset (Y_d) of Climax Granidiorite. Also shown is the triaxial stress loading path at a confining pressure of 1.0 kbar. The strength data are from HEARD (1970)

diorite as a function of $\bar{P}$, determined from the triaxial compression tests of HEARD (1970). Both Y and $\bar{P}$ are obtained from stress invariants according to CHERRY and PETERSEN (1970) where

$$Y = (\tfrac{3}{4} J_2')^{1/2}, \tag{8}$$

$$\bar{P} = P - \frac{1}{2}\left(\frac{J_3'}{2}\right)^{1/3}, \tag{9}$$

$$P = -\frac{J_1}{3}, \tag{10}$$

J_1, J_2' and J_3' are the first, second deviatoric and third deviatoric stress invariants, i.e., if σ_{11}, σ_{22}, σ_{33} are principal stresses then

$$J_1 = \sigma_{11} + \sigma_{22} + \sigma_{33}, \tag{11}$$

$$J_2' = \frac{(\sigma_{11} + P)^2 + (\sigma_{22} + P)^2 + (\sigma_{33} + P)^2}{2}, \tag{12}$$

$$J_3' = (\sigma_{11} + P)(\sigma_{22} + P)(\sigma_{33} + P). \tag{13}$$

Note that when the intermediate principal stress, σ_{22}, is equal to either the maximum, σ_{11}, or minimum, σ_{33}, principal stress then

$$Y = \left| \frac{\sigma_{11} - \sigma_{33}}{2} \right|, \tag{14}$$

$$\bar{P} = -\left(\frac{\sigma_{11} + \sigma_{33}}{2} \right). \tag{15}$$

Figure 3 shows the improvement in fracture strength definition when $Y_s(\bar{P})$ replaces $Y_s(P)$.

In Fig. 2 the onset of dilatancy is given by the curve labeled Y_d. This curve was somewhat arbitrarily established by assuming that dilatancy begins at half the maximum strength obtained during the triaxial compression test. A triaxial compression loading path is also shown by the line having unit slope, with dilatancy beginning at half the fracture strength. This assumption is not necessary when triaxial data are used to establish the onset of dilatancy, i.e., when the pressure versus specific volume curve obtained from triaxial compression deviate from the hydrostat, where pressure, P, is determined according to equation (10). Therefore, the Y_d curve should be regarded as a measurable material property (Cherry and Schock (1971), Schock *et al.* (1973)) since it can be obtained from a standard laboratory test.

Given the ability to use laboratory data in order to establish a unique fracture strength, $Y_s(\bar{P})$, a unique dilatancy onset surface, $Y_d(\bar{P})$, then one is ready to address the problem of determining $\hat{\pi}$ in equation (6) when the stress state falls between Y_d and Y_s. As noted by Cherry and Schock (1971), if a relation can be found between the rate of change of the second deviatoric invariants of the unknown stress state (σ_{rs}) and the tilda stress state ($\hat{\sigma}_{rs}$) then $\hat{\pi}$ is uniquely determined. The relation that they assumed may be written as follows:

$$\begin{aligned} \dot{Y} &= \dot{\hat{Y}} - (\dot{\hat{Y}} - \dot{Y}_n)\beta \qquad && Y_d < \hat{Y} < Y_s, \\ \dot{Y} &= \dot{\hat{Y}} && \hat{Y} < Y_d. \end{aligned} \tag{16}$$

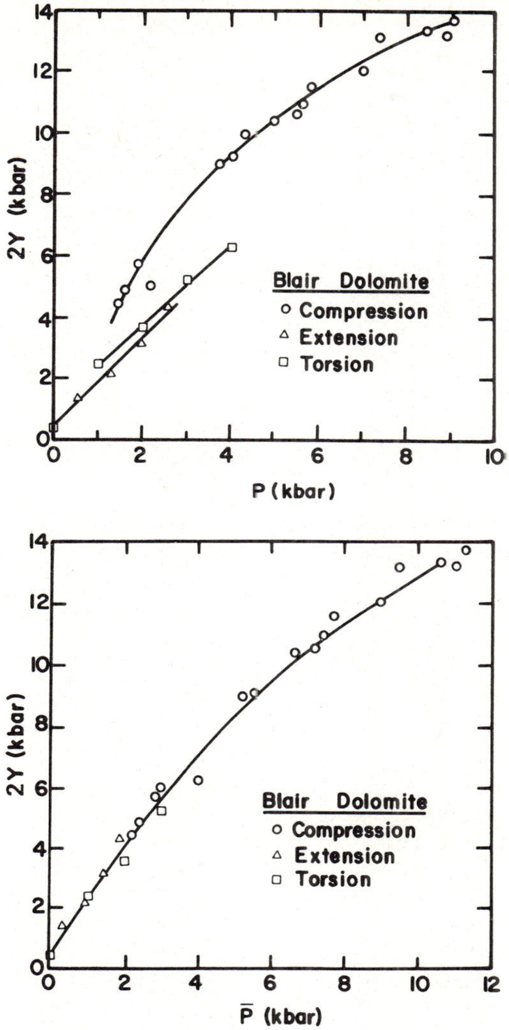

Figure 3
These figures are from CHERRY and PETERSEN (1969) and show the improvement in fracture strength definition when P is replaced by $\bar{P}$. The strength data are from HANDIN *et al.* (1967)

Here β depends on the stress state and its proximity to the fracture strength $(0 < \beta < 1)$ and $\dot{Y}_n$ is a specified neutral path in the dilatancy region. (If $\dot{\hat{Y}} > \dot{Y}_n$, then $\dot{\pi} > 0$ and the cracks open. If $\dot{\hat{Y}} < \dot{Y}_n$, then $\dot{\pi} < 0$ and the cracks close).

If $\hat{\sigma}_{11}$ is the least compressive (maximum) principal stress then equations (6), (8), (12), and (16) give

$$\dot{\pi} = (\dot{\hat{Y}} - \dot{Y}_n)\alpha \tag{17}$$

where

$$\alpha = \frac{4}{3}\frac{\beta}{\mu}\frac{\hat{Y}}{\hat{\sigma}_{11} + \hat{P}}. \tag{18}$$

The choice of a neutral path, Y_n, is the remaining problem to consider. Experimental data obtained by BRACE and RILEY (1972) and SCHOCK *et al.* (1973) indicate that the dilatant region is not entered during uniaxial strain loading. In the Y–$\bar{P}$ plane this loading path intercepts and then follows the Y_d curve. Therefore, if σ_{11} is the maximum principal stress and if $\dot{\varepsilon}_{11} = 0$, so that there is no total strain rate permitted in the maximum stress direction, then cracks will neither open nor close in that direction.

If we assume that this statement remains valid for stress states within the dilatant region then,

$$\dot{Y}_n = \frac{3}{2}\frac{(1-2\nu)}{1+\nu}\dot{P} = \mu(\dot{\pi} - \dot{\theta}), \tag{19}$$

where ν is Poisson's ratio. This assumption can be checked experimentally by loading a specimen into the dilatant region and then changing the loading to a uniaxial strain condition. Experimental data that define the neutral path in the dilatant region should be obtained.

Substituting equation (19) into equation (17) and solving for $\dot{\pi}$ gives

$$\dot{\pi} = \alpha\left(\frac{\hat{Y} + \mu\dot{\theta}}{1 + \mu\alpha}\right). \tag{20}$$

Finally, if σ_{11} is the maximum principal stress direction, then equation (6) may be written

$$\dot{\sigma}_{11} = \dot{\hat{\sigma}}_{11} - (\lambda + 2\mu)\dot{\pi},$$

$$\dot{\sigma}_{22} = \dot{\hat{\sigma}}_{22} - \lambda\dot{\pi},$$

$$\dot{\sigma}_{33} = \dot{\hat{\sigma}}_{33} - \lambda\dot{\pi}. \tag{21}$$

Equations (20) and (21), along with a specification of the fracture strength, Y_s, and dilatancy onset, Y_d, are apparently all that is required in order to explain the main results of both static and dynamic laboratory observations on the dilatant behavior of brittle geologic materials. The model has been formulated in terms of stress invariants with the basic assumption that dilatancy is caused by the opening of cracks in the material and the cracks open in the direction of the least compressive stress.

Any change in the effective elastic properties of the material, as the material is loaded toward the fracture strength, will not be due to changes in λ and μ. These intrinsic Lamé constants for the rock constituent are not very sensitive to fairly large changes in the stress state. For example, a change in pressure of the order of 5 kbar produces about 10% change in the intrinsic Lamé constants of granite (SCHOCK (1970).

Effective elastic constants are changed by the removal of the inelastic strain rate from the total strain rate. Application of first-order perturbation theory (DAHLEN (1972)) to the problem of superposition of stress waves on a previously existing stress field, with the material response governed by equation (20), gives identical results to those obtained using an effective elastic constant approach.

Consider the effective bulk modulus $\bar{k}$. From equations (10) and (21)

$$\bar{k} = k(1 - \dot{\pi}/\dot{\theta}). \tag{22}$$

Suppose we now allow a plane compression wave to propagate in the direction of the maximum principal stress, σ_{11}, so that the only applied strain rate is $\dot{\varepsilon}_{11}$. If $\sigma_{22} = \sigma_{11}$ then

$$\alpha = \frac{2\beta}{\mu}, \qquad \dot{Y} = \frac{\mu\varepsilon_{11}}{2}.$$

Since $\dot{\theta} = \dot{\varepsilon}_{11}$, the

$$\dot{\pi} = \frac{3\beta\dot{\varepsilon}_{11}}{1 + 2\beta},$$

and

$$\bar{k} = k\left(\frac{1 - \beta}{1 + 2\beta}\right). \tag{23}$$

Equation (23) shows that the effective bulk modulus is reduced when the wave propagates in the maximum principal stress direction. A similar analysis of the effective modulus, $\bar{\lambda} + 2\bar{\mu}$, that determines compression wave velocity $\bar{V}_p$, gives

$$\bar{V}_{p(1,0,0)} = \sqrt{\frac{\lambda + 2\mu}{\rho}\,\frac{1 - \beta}{1 + 2\beta}} = V_p\sqrt{\frac{1 - \beta}{1 + 2\beta}}. \tag{24}$$

If the plane wave propagates in the minimum principal stress direction, then

$$\alpha = \frac{2\beta}{\mu}, \qquad \dot{Y} = \mu\dot{\varepsilon}_{33}.$$

Since $\dot{\theta} = \dot{\varepsilon}_{33}$ then $\dot{\pi} = 0$. The effective moduli equal the intrinsic moduli and the wave speed is unaltered, i.e., $\bar{V}_{p(0,0,1)} = V_p$. The ratio of the compression wave velocities in the two directions is

$$\frac{\bar{V}_{p(1,0,0)}}{\bar{V}_{p(0,0,1)}} = \sqrt{\frac{1 - \beta}{1 + 2\beta}}. \tag{25}$$

The analysis of a shear wave propagating in a principal stress direction requires that the off diagonal stresses be included in the expression for J_2', i.e.,

$$J_2' = \frac{S_{ij} S_{ji}}{2}, \tag{26}$$

where S_{ij} is a deviatoric component of the stress tensor. If $\dot{\varepsilon}_{13}$ is the only nonzero component of strain rate and $\dot{\theta} = 0$. Then

$$\alpha = \frac{2\beta}{\mu} \hat{Y} = \frac{3}{2} \frac{\mu^2}{\hat{Y}} \frac{d(\varepsilon_{13})^2}{dt}, \qquad Y_n = 2\beta \frac{\dot{\hat{Y}}}{1 + 2\beta}.$$

Substitution into equation (16) gives the effective shear modulus, $\bar{\mu}$, where

$$\bar{\mu} = \mu \sqrt{\frac{1+\beta}{1+2\beta}}. \tag{27}$$

Therefore, the shear wave velocity, $\bar{V}_s$, does not exhibit anisotropic behavior in the (1, 0, 0) or (0, 0, 1) directions. However, this velocity is affected by the dilatant behavior of the material since

$$\bar{V}_s = V_s \left(\frac{1+\beta}{1+2\beta} \right)^{1/4}. \tag{28}$$

Equations (25) and (28) show the relation between body wave velocities in the principal stress directions and the proximity of the rock to the failure surface. Equation (25) is consistent with the experimental data of THILL (1973) and shows that the factor $(1-\beta)/(1+2\beta)$ is obtained from the ratio of P wave velocities in the (1, 0, 0) and (0, 0, 1) directions. If equation (28) is experimentally verified then the factor $(1+\beta/1+2\beta)$ may be obtained from $\bar{V}_s/V_s$. This implies that the intrinsic shear wave velocity, V_s, is known.

Measurement of β and the interpretation of this parameter in terms of the stress state in the rock and the proximity of this stress state to the fracture strength should permit an assessment of earthquake probability. Regarding β, CHERRY and SCHOCK (1971) assumed that

$$\sqrt{\beta} = \left[Y_d' + \left(\frac{Y - Y_d}{Y_s - Y_d} \right) (Y_s' - Y_d') \right] \left(\frac{\hat{Y} - Y_d}{Y_s - Y_d} \right), \tag{29}$$

where Y_s' and Y_d' are the slopes of the strength and dilatancy onset curves at $\bar{P}$ (Fig. 2). They used this model (equations (18), (20) and (21)) to simulate the rock deformation experienced under triaxial loading conditions.

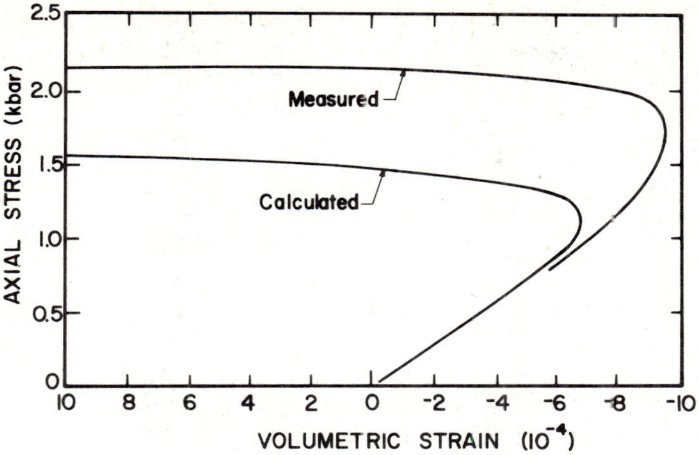

Figure 4
Axial stress versus volumetric strain in granite for unconfined compression loading. The assumed maximum load for the calculation was 1.6 kbar. The sample failed in the experiment at 2.3 kbar (CHERRY and SCHOCK (1971))

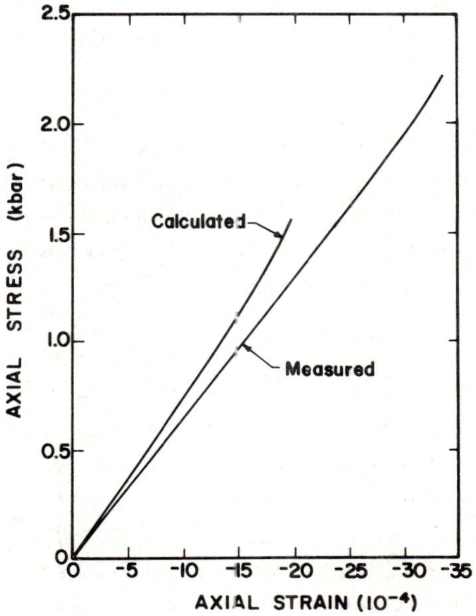

Figure 5
Axial stress versus axial strain in granite for unconfined compression loading (CHERRY and SCHOCK (1971))

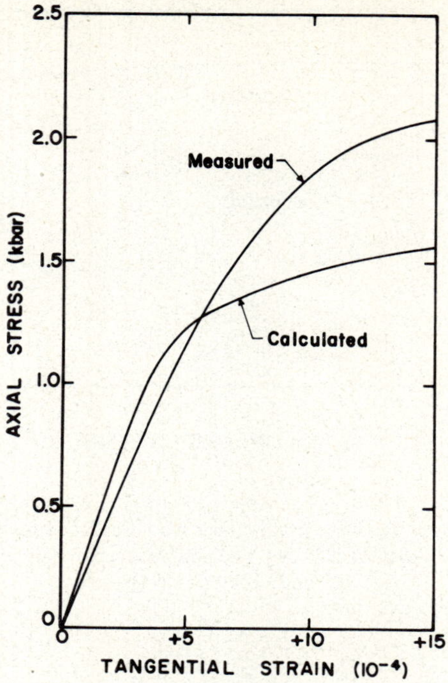

Figure 6
Axial stress versus tangential strain in granite for unconfined compression loading (Cherry and Schock (1971))

Figures 4, 5 and 6 compare their results with data from an unconfined compression test on granodiorite. The main features of the laboratory data are reproduced by the model. Note especially the severe effect that the dilatancy model induces in the tangential strain component shown in Fig. 6. This strain component is responsible for the dilatant behavior of the material shown in Fig. 4.

3. Discussion

The dilatancy model presented here should be regarded as a first step toward obtaining a consistent approach to the inversion of body wave velocity measurements for earthquake prediction purposes. Additional improvements will certainly be required in order to model the dilatancy–diffusion phenomenon proposed by Nur (1972) and shear wave birefringence (Gupta (1973b); Bonner (1974)). However, our basic philosophy should not change, i.e., to use laboratory test data in order to place constraints on the parameters in the model and then to quantitatively interpret field data in terms of earthquake probability.

Measurement of body wave anisotropy may prove a rapid means of identifying specific regions on a fault system which are earthquake prone and which should be subjected to intensive temporal monitoring. An artificial seismic source will certainly be required in order to perform these measurements.

Acknowledgements

The authors are indebted to Mr. CARL HASTINGS, staff physicist at Systems, Science and Software for permission to publish his tension fracture prediction shown in Fig. 1. J. T. CHERRY and J. SWEET received considerable support from Systems, Science and Software during the time required to investigate the body wave anisotropy aspects of the dilatancy model.

REFERENCES

Y. P. AGGARWAL, L. R. SYKES, J. ARMBRUSTER and M. L. SBAR, *Premonitory changes in seismic velocities and prediction of earthquakes*, Nature *241* (1973), 101–104.

B. P. BONNER, *Shear wave birefringence in dilating granite*, Geophys. Res. Ltrs. *1* (1974), 217–220.

W. F. BRACE, B. W. PAULDING, Jr. and C. SCHOLZ, *Dilatancy in the fracture of crystalline rocks*, J. Geophys. Res. *71* (1966), 3939–3953.

W. F. BRACE and D. K. RILEY, *Static uniaxial deformation of 15 rocks to 30 kbar*, Int. J. Rock Mech. Min. Sci. *9* (1972), 271–288.

C. G. BUFE, J. H. PFLUKE and R. L. WESSON, *Premonitory vertical migration of microearthquakes in central California – Evidence of dilatancy biasing*", Geophys. Res. Ltrs. *1* (1974), 221–224.

I. G. CAMERON, *Theoretical model of the early phases of an underground explosion*, Proceedings on Engineering with Nuclear Explosives *1* (1970), 221–229.

I. G. CAMERON and G. C. SCORGIE, *Dynamics of intense underground explosions*, J. Inst. Maths. Applics. *4* (1968), 194–222.

J. T. CHERRY and F. L. PETERSEN, *Numerical simulation of stress wave propagation from underground nuclear explosions*, Proceedings on Engineering with Nuclear Explosions *1* (1970), 142–220.

J. T. CHERRY and T. C. SAVAGE, *Rock dilatancy and strain accumulation near Parkfield, California*, Bull. Seism. Soc. Am. *62* (1972), 1343–1347

J. T. CHERRY and R. N. SCHOCK, *A dilatancy model for Granodiorite*, Lawrence Livermore Laboratory Report UCRL-72953 (1971).

J. T. CHERRY, J. SWEET and E. J. HALDA, *Prediction of surface vertical and horizontal displacements in non-homogeneous media resulting from subsurface nuclear explosions*, Systems, Science and Software Report, SSS-R-73-1517 (1973).

F. A. DAHLEN, *Elastic velocity anisotropy in the presence of an anisotropic initial stress*, Bull. Seism. Soc. Am. *62* (1972), 1183–1193.

I. N. GUPTA, *Dilatancy and premonitory variations of P, S travel times*, Bull. Seism. Soc. Am. *63* (1973), 1157–1161.

I. N. GUPTA, *Premonitory variations in S-Wave velocity anisotropy before earthquakes in Nevada*, Science *182* (1973), 1129–1132.

J. HANDIN, H. C. HEARD and J. N. MAGOUIRK, *Effects of the intermediate principal stress on the failure of limestone, dolomite and glass at different temperatures and strain rates*, J. Geophys. Res. *72* (1967), 611–640.

C. R. HASTINGS, *Underground Test No. 4 – Experiments to characterize 2-D damage for validation of Minuteman components*, Project Officers Report, Systems, Science and Software Report SSS-CR-74-2318 (Draft). (S/RD) (1974).

H. C. HEARD, *The influence of environment on the inelastic behavior of rocks*, Proceedings on Engineering With Nuclear Explosives *1* (1970), 127–141.

R. HILL, *The Mathematical Theory of Plasticity* (Oxford 1950).

G. Maenchen and S. Sack, *The TENSOR code*, *Methods in Computational Physics 3* (Academic Press, New York 1964).

K. Mogi, *Effects of the intermediate principal stress on rock failure*, J. Geophys. Res. *72* (1967), 5117.

I. Nelson, M. L. Baron and I. Sandler, *Mathematical models for geological materials for wave-propagation studies*, in *Shock Waves and the Mechanical Properties of Solids*, edited by J. J. Burke and Volker Weiss (Syracuse University Press, Syracuse, New York 1971).

A. Nur, *Dilatancy, pore fluids and premonitory variations of t_s/t_p travel times*, Bull. Seism. Soc. Am. *62* (1972), 1217–1222.

R. Robinson, R. L. Wesson and W. L. Ellsworth, *Variation of P-wave velocity before the Bear Valley, California earthquake of 24 February 1972*, Science *184* (1974), 1281–1283.

R. N. Schock, *Dynamic elastic moduli of rocks under pressure*, Proceedings on Engineering with Nuclear Explosions *1* (1970), 110–126.

R. N. Schock, H. C. Heard and D. R. Stephens, *Stress–strain behavior of a granodiorite and two graywackes on compression to 20 kilobars*, J. Geophys. Res. *78* (1973), 5922–5941.

A. N. Semenov, *Variations in the travel time of transverse and longitudinal waves before violent earthquakes*, Izv., Earth Phys. No. 4 (1969), 72–77.

R. E. Thill, *Acoustical methods of monitoring failure in rock*, Proceedings of Fourteenth Symposium on Rock Mechanics, (1973), 649–688.

J. H. Whitcomb, J. D. Garmany and D. L. Anderson, *Earthquake prediction: Variation of seismic velocities before the San Fernando earthquake*, Science *180* (1973), 632–635.

M. Wyss and D. J. Holcomb, *Earthquake prediction based on station residuals*, Nature *249* (1973), 139–140.

(Received 23rd December 1974)

Pageoph,
Vol. 113 (1975)

Birkhäuser Verlag, Basel

A Note on the Constitutive Law for Dilatancy

By Amos Nur[1])

Summary – The spatial extent of the dilatant zone and its time variations associated with preseismic fault creep are very sensitive to the dilatant constitutive law. For a power law of the form $\gamma = \delta \cdot I_2^{n/2}$ (where γ is dilatant strain, δ is a constant and I_2 is the deviatoric stress invariant) the ratio of dilatant scale to depth of faulting is significantly less than unity for $n = 2$ ('microcrack dilatancy'), about unity for $n = 1$ ('joint dilatancy'), and much larger than unity for $n = 1/2$ ('sand pile dilatancy'). Consequently it is necessary to establish the *in situ* nature of the dilatant process in order to understand the onset, distribution and development of crustal dilatant regions.

1. Introduction

The hypothesis that dilatancy is responsible for *in situ* premonitory changes of velocities and other rock properties (Nur, 1972; Whitcomb *et al.*, 1973; Scholz *et al.*, 1973) was based on dilatancy observed in the laboratory (Brace *et al.*, 1966), in low porosity competent and brittle rocks. Thus not only the phenomenon of dilatancy itself, but also the details of how it depends on shear stress, at what stress level it begins and how large it is at failure were inferred for the crust directly from laboratory data. Several serious problems immediately arose: the bulk of microcrack dilatancy, as obtained in the laboratory, develops just prior to fracture. Only limited dilatant volume expansion has been observed, however, prior to frictional sliding on preexisting cuts in laboratory specimens (Hadley, 1973). As most earthquakes occur on preexisting faults, it appears that *in situ* dilatancy inferred from results on intact specimen, will be very limited in magnitude. Furthermore, as we show later, the spatial extent of the dilatant volume inferred from intact rock data is very small. And yet, field observations suggest that the size of the dilatant region is appreciable relative to the largest dimension of the active fault segment (e.g., Aggarwal *et al.*, 1975).

Theoretical considerations (Byerlee, 1970; Nur and Schultz, 1973) and recent observations (Stuart and Johnston, 1974) suggest that earthquakes are preceded by aseismic fault creep in a region surrounding the impending rupture. In this case, the state of shear stress in the focal region changes with time, and could markedly influence the time-dependent spatial distribution of dilatant strain. Since this kind of creep associated with frictional inhomogeneities (Nur and Schultz, 1973) is not amenable to laboratory studies, its relevance for dilatancy cannot be evaluated from laboratory microcrack dilatancy.

[1]) Department of Geophysics, Stanford University, Stanford, Ca. 94305.

It is quite possible that the dilatant process *in situ* is quite different from that in competent rock with microcracks. Dilatancy – the increase of longitudinal strain with shear stress – is also observed in arrays of blocks, jointed rocks and in loose sand. In these diverse materials the micromechanical processes vary greatly, but the macromechanical process – dilatant strain – is similar. Furthermore the details of stress-dilatant strain relation and the onset of dilatancy depend greatly on the micromechanical behavior of the material.

In this note I explore the importance of the character of constitutive laws for dilatancy in the earth's shallow crust. This is done by first exploring the general character of the one-dimensional stress-dilatant strain relation for microcrack dilatancy, sand dilatancy and fracture dilatancy. This is then generalized into a two-dimensional law and incorporated into an inhomogeneous deformation model of crustal faulting. The problem of spatial extent and onset is explored by models of a strike slip fault and a thrust fault with dilatancy. And, finally, the problem of time history associated with preseismic, aseismic fault creep is explored by simple dimensional analysis.

2. *One-dimensional behavior*

If dilatancy is responsible for premonitory velocity changes, then dilatant strain must occupy a volume surrounding the impending focal zone. It is necessary therefore to understand the dilatant process *off* the slip zone itself, in regions which experience relatively high stress before fault slip, and a stress decline during slip.

Microcrack dilatancy

BRACE *et al.* (1966) suggested that axial cracks may open up due to frictional sliding at grain boundaries or preexisting cracks, as shown schematically in Fig. 1. As the axial stress in a uniaxial test increases, friction is overcome along an increasing number of microcracks, more and more sliding is initiated, and consequently the dilatant volume increases more rapidly with stress at high stress, as seen in laboratory tests on rocks containing microcracks. For simplicity the microcrack dilatancy stress relation can be crudely approximated by a power law

$$\gamma \approx \delta \tau^n \tag{1}$$

where γ is dilatant strain, τ is shear stress, and δ is a constant. A value of $n \geqslant 2$ leads to behavior similar to crack dilatancy: the dilatant strain develops significantly only at large stress – suggesting a high stress onset, and its magnitude accelerates with stress. In this work the (arbitrary) value of $n = 2$ is used.

Sand dilatancy

The origin of the concept of dilatancy is in the observation of the deformation of wet sand (REYNOLDS, 1885). Like microcrack dilatancy, volumetric strain can be induced in sand aggregates by shearing. The mechanism of the dilatation however is quite

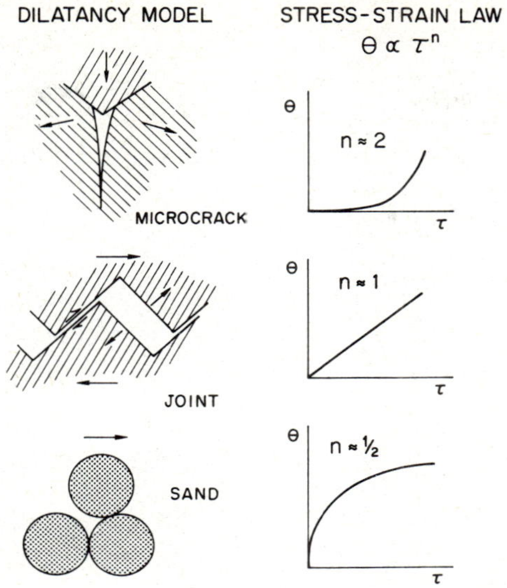

Figure 1
Schematic models for dilatancy. On the left are three types of dilatant deformation processes, with their corresponding strain-shear stress relations on the right

different. As shown by FRANK (1965) shear deformation is accommodated by relative rotation between grains, as illustrated in Fig. 1. As rolling begins, pore space increases with displacement, and when shear decreases, pore space closes again. The pore volume first increases rapidly with shear, but less so at higher shear. This relation between shear stress τ and dilatant strain γ, in a uniaxial test, can be roughly approximated by a square root law $\gamma = \delta\tau^n$ with $n = 1/2$. Dilatancy in this case develops significantly at low stress levels, in contrast with microcrack dilatancy.

Intermediate cases

In all likelihood, the crust cannot be adequately characterized as either granular, or as competent material with only microcracks. It is quite probable that crustal behavior is influenced and perhaps controlled by joints and macro-, not micro-, fractures. Although there is sizeable evidence that fracture and joint porosity decreases rapidly with depth (BRACE, this issue), the coherent patterns of joints and fractures extending over hundreds of kilometers (e.g. WISE, 1968) suggest that they are relatively deep. If true, dilatancy in the crust may be controlled by joint dilatation (e.g. GOODMAN, 1974), in which the pore space along preexisting fractures changes with shear. The character of the constitutive law for dilatancy in jointed material depends on the geometry of joints, their roughness, spacing and distribution of directions. A simple possible model is illustrated in Fig. 1. Here it is assumed that the fracture surface consists of short plane segments which undergo relative shear slip. As slip occurs, spaces open up along joints, producing overall dilatation. For this model the dilatant strain is proportional to the shear stress, leading to $\gamma = \delta\tau^n$ with $n = 1$.

Obviously other models, equally reasonable, could lead to other constitutive laws. But for the present it suffices to consider the range of $n = 1/2$, 1, 2 to simply evaluate the importance of the law in determining the distribution of dilatancy *in situ*.

3. Two-dimensional continuum theory

If the local inhomogeneities in rock responsible for dilatancy are small compared to the scale of spatial variation of strain in the crust, it appears reasonable to describe the elastic and dilatant solid as a continuum. In two dimensions the three strain components are

$$\begin{aligned} \varepsilon_{11} &= \alpha\sigma_{11} + \beta\sigma_{22} + \gamma_{11} \\ \varepsilon_{22} &= \beta\sigma_{11} + \alpha\sigma_{22} + \gamma_{22} \\ \varepsilon_{12} &= \eta\sigma_{12} + \gamma_{12} \end{aligned} \tag{2}$$

where σ_{11}, σ_{22}, σ_{12} are stresses, α and β are elastic compliances, and γ_{11}, γ_{22}, γ_{12} are the dilatant strain components. These strain components have to fulfil two conditions. They have to undergo a transformation like a second-rank tensor, and be a (rather arbitrary) function of the stress invariants (PRAGER, 1961).

$$\begin{aligned} I_1 &= \sigma_{11} + \sigma_{22} \\ I_2 &= \tfrac{1}{4}\{(\sigma_{11} - \sigma_{22})^2 + 4\sigma_{12}^2\}. \end{aligned} \tag{3}$$

The first invariant I_1 is the hydrostatic pressure, which does not induce dilatancy. For simplicity, this term is ignored, so that only I_2 is assumed to influence the dilatant strain. In this study the strains γ_{ji} were chosen to be

$$\begin{aligned} \gamma_{11} &= f(I_2)\ (\cos^2\alpha + k\sin^2\alpha) \\ \gamma_{22} &= f(I_2)\ (\sin^2\alpha + k\cos^2\alpha) \\ \gamma_{12} &= f(I_2)(2\sin\alpha\cos\alpha)(1 - k) \end{aligned} \tag{4}$$

where $f(I_2)$ is a continuous function of I_2, k is a constant and the angle α is given by

$$\begin{aligned} \cos^2\alpha &= \tfrac{1}{2}\left\{1 + \left(\frac{1}{4R^2+1}\right)^{1/2}\right\} \\ \sin^2\alpha &= \tfrac{1}{2}\left\{1 - \left(\frac{1}{4R^2+1}\right)^{1/2}\right\} \end{aligned} \tag{5}$$

where

$$R = \frac{\sigma_{12}}{\sigma_{11} - \sigma_{22}}.$$

The strains γ_{ij} transform as a reduced second-rank tensor. The definition of $\cos\alpha$ and $\sin\alpha$ in equations (5) implies that the symmetry axes of the dilatant strain ellipse coincide with the axes of the stress, and consequently elastic strain ellipse. In other words, the principal directions of stresses and strains, both elastic and dilatant, are identical. The parameter k in equations (4) permits a wide flexibility in describing dilatant strain: when $k = 1$, dilatant strain is isotropic, without any dependence on direction. When $k = 0$, strain is uniaxial in the direction of least compression. Finally the expression for volumetric dilatant strain

$$\gamma = \gamma_{11} + \gamma_{22} = (1 + k) f(I_2) \tag{6}$$

shows that $k = -1$ describes pure dilatant shear strain. Because volumetric expansion is an essential feature of dilatancy, the proper range is $0 \leqslant k \leqslant 1$.

The most important unknown in equations (4) is the function $f(I_2)$, which relates the dilatant strain to the applied stress. At present there is no adequate experimental data, nor theoretical micromechanical models, which could provide the behavior of a dilatant material subject to a general, not just uniaxial, state of stress. Instead, one has to make a sensible guess of the form of $f(I_2)$. One constraint on such a guess is that it must reduce to known experimental or theoretical uniaxial behavior. One of the simplest forms is given by the power law

$$f(I_2) = \delta I_2^{n/2} \tag{7}$$

where δ is a constant. Equations (3) indicates that under uniaxial stress $\tau = \sigma_{11}/2$, $\sigma_{22} = \sigma_{21} = 0$, equation (7) reduces to $f = \delta\tau^n$. This is identical to the one-dimensional expression of the previous section.

Combining equations (3), (4), (5) and (7) yields the expressions for two-dimensional dilatant strains in terms of stresses at each point in the dilatant solid

$$\gamma_{11} = \tfrac{1}{2}\delta\{\sqrt{(\sigma_{11} - \sigma_{22})^2 + 4\sigma_{12}^2}\}^n \left\{(1 + k) + (1 - k)\sqrt{\frac{(\sigma_{11} - \sigma_{22})^2}{(\sigma_{11} - \sigma_{22})^2 + 4\sigma_{12}^2}}\right\}$$

$$\gamma_{22} = \tfrac{1}{2}\delta\{\sqrt{(\sigma_{11} - \sigma_{22})^2 + 4\sigma_{12}^2}\}^n \left\{(1 + k) - (1 - k)\sqrt{\frac{(\sigma_{11} - \sigma_{22})^2}{(\sigma_{11} - \sigma_{22})^2 + 4\sigma_{12}^2}}\right\}$$

$$\gamma_{12} = \tfrac{1}{2}\delta\{\sqrt{(\sigma_{11} - \sigma_{22})^2 + 4\sigma_{12}}\}^n \left\{(1 - k)\sqrt{\frac{4\sigma_{12}^2}{(\sigma_{11} - \sigma_{22})^2 + 4\sigma_{12}^2}}\right\}. \tag{8}$$

It is now relatively simple to compute the distribution of elastic and dilatant strains, stresses, and displacements in dilatant elastic solids subject to inhomogeneous deformation. This can best be done by iteration and numerical integration of equations (2), using the expressions of equations (8). However, for exploring the role of the form of the dilatant law $f(I_2)$, it probably suffices to consider only the volumetric term of γ_{ij}

$$\gamma(n) = \delta I_2^{n/2} = \delta\{\sqrt{(\sigma_{11} - \sigma_{22})^2 + 4\sigma_{12}^2}\}^n \tag{9}$$

where $k = 1$.

4. *Results*

Spatial distribution of dilatancy

It is not possible yet to accurately predict strain accumulation, stress drop vs. total stress, and the exact distribution of shear stress before faulting. Instead, in this study I assume that an adequate approximation for the preseismic state of stress is given by an antidislocation which is identical with the actual seismic dislocation, except for a reversed sign.

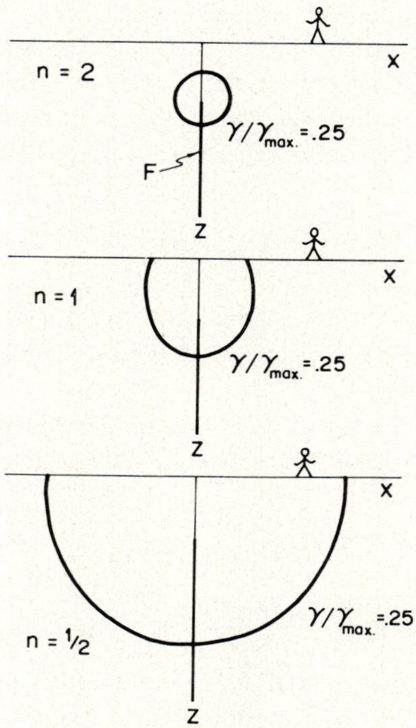

Figure 2
Dilatant volume strain distribution for a screw antidislocation. The region $0 \leqslant z \leqslant 1$ has not yet slipped, whereas the region $x = 0$, $z \geqslant 1$ has slipped. The strain is normalized with respect to strain at point $P(0, 1)$. Figures show normalized dilatant strains for 'microcrack dilatancy' ($n = 2$), 'joint dilatancy' ($n = 1$) and 'sand dilatancy' ($n = 1/2$)

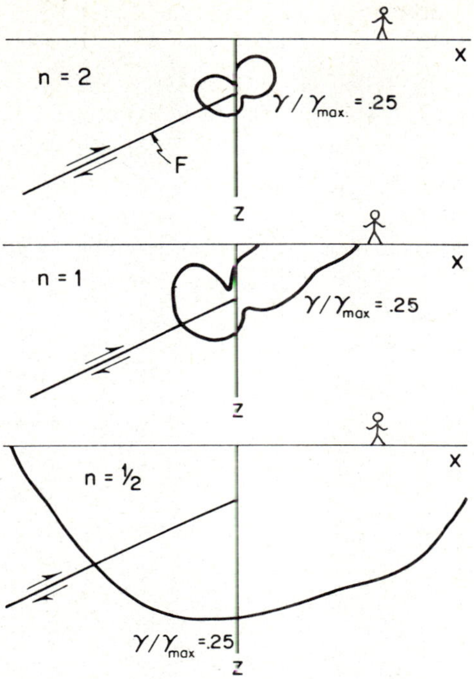

Figure 3

Dilatant volume strain distribution for an edge antidislocation. The region $x = 0$, $0 \leqslant z \leqslant 1$ has not yet slipped. Strain is normalized with respect to strain at point $P(0, 1)$

Consider now two cases: a screw antidislocation representing the state of elastic strain prior to strike slip type earthquake, and an edge antidislocation preceding thrust-type faulting. These models correspond, as shown in Figs. 2 and 3, to 'stick' planes (vertical for strike slip, inclined for thrust) near the free surface which are going to slip in a future earthquake. The slip zone at greater depth has already yielded, in accordance with models by SAVAGE (1970) and TURCOTTE and SPENCE (1974). The stresses associated with these two antidislocations in an elastic solid are known (COTTRELL, 1953; HEAD, 1953). The second stress invariants I_2 at each point $P(x, z)$ were then derived from these stresses, for both the screw antidislocation and the edge antidislocation. In order to evaluate now the effect of the power n in the power law of equation (9), the value of $\gamma(x, z; n)$ – the approximate volumetric dilatant strain component – was computed for two kinds of dislocations and for the values of $n = 2$ ('microcrack dilatancy'), $n = 1/2$ ('sand dilatancy') and $n = 1$ ('joint dilatancy'). Although this is only a first estimate of the strain, which should be used in equations (2) and (8) to reevaluate the stresses, and then recalculate the strain etc., it is a good indicator of the spatial distribution of dilatancy.

The strain results are shown in Figs. 2 and 3. In both vertical screw dislocation and inclined edge dislocation (with dip of 60°), the contour of $\gamma = 0.25$, relative to $\gamma_0 = 1.0$

at the tip of the antidislocation, indicates that the dilatant zone is extremely localized for $n = 2$ (similar to microcrack dilatancy results by STUART and DIETRICH, 1974). The dilatant region is, in contrast, very broad for $n = 1/2$, and the case of $n = 1$ is of course intermediate between the two. The ratio R between the width of the dilatant zone and a typical dimension of the earthquake fault – length of rupture zone or, more appropriately, the depth – strongly depends on the type of dilatancy. Figures 2 and 3 show $R \ll 1$ for $n = 2$, $R \approx 1$ for $n = 1$, and $R > 4$ for $n = 1/2$. The dependence of size of the dilatant region of the power n is further exhibited by the dilatant surface deformation, as shown in Fig. 4 for the screw, obtained by numerical integration of the volume

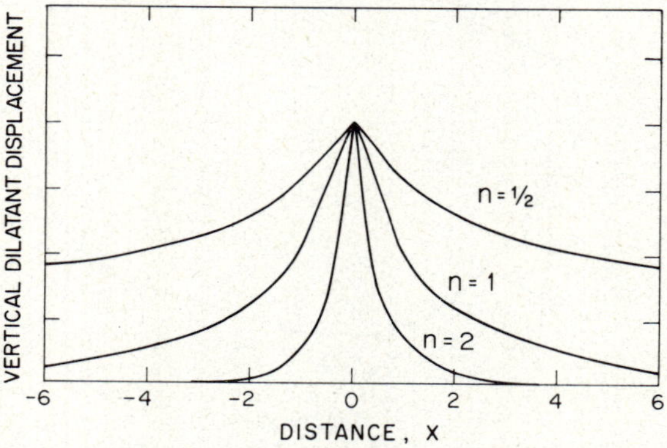

Figure 4
Estimated vertical dilatant displacement of the ground surface ($z = 0$) for the screw antidislocation. All three cases ($n = 2, 1, 1/2$) are normalized to the same displacement at $x = 0$

strain fields. The vertical displacement is very localized ($R \ll 1$) for $n = 2$, and very broad ($R \gg 4$) for $n = 1/2$. For the sake of comparison the three displacement fields were equalized at $x = 0$.

Thus, in spite of the gross approximations made in deriving the governing equations, and the rough estimate of the constitutive law in the form of a simple power, as well as the approximate computation of dilatant strain, the results clearly indicate that *in situ* dilatant volumes cannot be predicted directly from laboratory-derived microcrack dilatancy. Instead, it is essential to determine the type of dilatancy – microcrack, joint or sand – before making reasonable estimates of the validity and extent of the dilatant process *in situ*.

Preseismic fault creep and time variations of dilatancy

If fault creep precedes failure, then the distribution of shear stress in the focal region is varying with time. Because the dilatant strain is induced by shear stress, its distribution may also be time dependent. Consider a fault segment of area A_0, across which an effective shear force increment ΔF is added. The average effective shear stress (failure stress minus actual stress) increment is then $\Delta\tau_0 = \Delta F/A_0$. Suppose now that an area $(A_0 - A)$ of the segment creeps; stress remains approximately unchanged on the creeping part, but it increases on the sticking part A. The increased stress $\Delta\tau$ is roughly given by

$$\Delta\tau \approx \Delta\tau_0 \left(\frac{A_0}{A}\right)$$

where A is time dependent.

The volume increase ΔV due to dilatancy is then given by

$$\Delta V(n) \propto A^{3/2} \Delta\gamma \propto A^{3/2} (\Delta\tau)^n$$

and

$$\Delta V(2) \propto \frac{1}{\sqrt{A}} \qquad \text{microcrack dilatancy}$$

$$\Delta V(1) \propto \sqrt{A} \qquad \text{joint dilatancy}$$

$$\Delta V(\tfrac{1}{2}) \propto A \qquad \text{sand dilatancy.}$$

These relations indicate that the changes in the dilatant volume due to preseismic fault slip greatly depend on the dilatancy constitutive law. As creep develops with time, A – the 'stick' area – diminishes with time. In microcrack dilatancy, the dilatant volume, which is proportional to $1/\sqrt{A}$ will increase with creep, and time. In sand and joint dilatancy, creep causes the reversed effect. The dilatant volume actually decreases with time.

5. Conclusions

The continuum theory for dilatancy in this study is based on several (reasonable) assumptions, and the numerical results for dilatancy distribution are only rough approximations. Furthermore, the analysis of dilatancy associated with fault creep is very crude. Nevertheless, several general conclusions emerge.

Of all the unknown features of dilatancy, the most important beside the occurrence of dilatancy itself, is the general character of the relation between dilatation and shear stress. Microcrack-type behavior leads to relatively localized dilatancy, concentrated in *small* volumes. If preseismic fault creep takes place, it may cause dilatancy *increase* with slip, and consequently time. Dilatancy typical of granular material leads to a *large* dilatant volume, and it *decreases* when preseismic fault slip begins. Intermediate behavior leads to intermediate size dilatant zones and no dependence on preseismic creep. These results suggest also that preseismic fault creep could perhaps be responsible for dilatancy recovery, without fluid flow. However, unlike the nonlinear models of BRADY (1975), STUART (1975), or MJACHKIN *et al.* (1975), the reason for the nonlinear process for dilatancy recovery is fault slip itself. This is compatible with the notion that earthquakes occur in preexisting faults, and therefore do not involve the higher stresses required to break unfractured rock.

Acknowledgement

This study was supported by a grant from Chevron Oil Field Research Co., and by the USGS, contract No. USDI 14-08-0001-14566. The views and conclusions contained in this document are those of the author, and do not represent official policies of the US Government.

REFERENCES

AGGARWAL, N. P., SYKES, L. R., SIMPSON, D. W. and RICHARDS, P. G. (1975), *Spatial and temporal variations in t_s/t_p and in P wave residuals at Blue Mountain Lake, New York*, J. Geophys. Res. *80*, 718–732.

BRACE, W. F., PAULDING, B. W., JR, and SCHOLZ, C. (1966), *Dilatancy in the fracture of Crystalline rocks*, J. Geophys. Res. *71*, 3939–3956.

BRADY, B. (1975), *Theory of earthquakes I*, Pure appl. Geophys., this issue.

BYERLEE, J. D. (1970), *The mechanics of stick-slip*, Tectonophys. *9*, 475–586.

COTTRELL, A. H., *Dislocations and Plastic Flow in Crystals* (Clarendon Press, Oxford 1953).

FRANK, F. C. (1965), *On dilatancy in relation to seismic sources*, Rev. of Geophys. *3*, 485–503.

GOODMAN, R. (1974), *The mechanical properties of joints*, in *Advances in Rock Mechanics*, Vol. IA, pp. 127–140.

HEAD, A. D. (1953), *Edge dislocations in inhomogeneous media*, Proc. Phys. Soc., London *B66*, 793–801.

MJACHKIN, V. I., BRACE, W. F., SOBOLEV, G. A., and DIETRICH, J. H. (1975), *Two models for earthquake forerunners*, Pure appl. Geophys., this issue.

NUR, A. (1972), *Dilatancy, pore fluids, and premonitory variations of t_s/t_p travel times*, Bull. Seism. Soc. Am. *62*, 1217.

NUR, A., and SCHULTZ, P., *Fluid flow and faulting, 2. Proc. Conf. on Tectonic Problems of the San Andreas Fault System* (Stanford Univ. Press 1973).

PRAGER, W., *Introduction to Mechanics of Continua* (Ginn & Co., Boston 1961).

REYNOLDS, O. (1885), *On the dilatancy of media composed of rigid particles in contact*, Phil. Mag. Ser. 5, *20*, 469–481.

SAVAGE, J. C., and BURFORD, R. O. (1970), *Accumulation of tectonic strain in California*, Bull. Seism. Soc. Am. *60*, 1877–1896.

SCHOLZ, C. H., SYKES, L. R., and AGGARWAL, Y. P. (1973), *Earthquake prediction: a physical basis*, Science *181*, 803–810.

STUART, W. D. (1975), *Diffusionless dilatancy*, Geophys. Res. Lett., in press.

STUART, W. D., and DIETRICH, J. D. (1974), *Continuum theory of rock dilatancy*, in *Advances in Rock Mechanics*, Vol. IIA, p. 530.

STUART, W. D., and JOHNSTON, M. J. S. (1974), *Tectonic implications of anomalous tilt before central California earthquakes*, EOS, *56*, 1196.

TURCOTTE, D. L., and SPENCE, D. A. (1974), *An analysis of strain accumulation on a strike slip fault*, J. Geophys. Res. *79*, 4407–4412.

WISE, D. U. (1968), *Regional and subcontinental sized fracture system detectable by topographic shadow techniques*, in *Proc. Conf. on Research in Tectonics*, Geol. Survey Canada, GSC 68–52, pp. 175–199.

WHITCOMB, J. H., GARMANY, J. D., and ANDERSON, D. L. (1973), *Earthquake prediction: variation of seismic velocities before the San Fernando earthquake*, Science *180*, 632.

(Received 23rd December 1974)

Pageoph,
Vol. 113 (1975) Birkhäuser Verlag, Basel

Dilatancy-Related Electrical Resistivity Changes in Rocks

By W. F. BRACE

Summary – Dilatancy may lower pore pressure at mid-crustal depths to near the liquid–vapor transition of water. However, the resistivity of typical pore fluids changes very little down the transition so that changes in rock resistivity will be almost solely due to dilatant volume change. Experiments with partly saturated rocks at room temperature suggest that rocks containing mixtures of fluid and vapor, along the transition, will generally behave electrically during dilatancy as though they were fluid saturated. Observed resistivity changes before earthquakes give the dilatant volume change directly and thus may be used to further constrain the dilatancy–diffusion model, independent of observed uplift or velocity changes.

1. Introduction

Dilatancy is widely assumed to be responsible for certain effects observed just before earthquakes as discussed most recently by ANDERSON and WHITCOMB [1], HANKS [2], and NUR [3]. Synchronous changes in seismic velocities, electrical resistivity, crustal strains and other phenomena have been explained in a general way by a model which combines dilatancy with flow of pore fluids into the seismic region. Of these various precursory phenomena, most attention has been focused on changes in the seismic velocities; such changes are frequently observed. However, resistivity changes have now been observed both in the USSR [4] and in the USA [5]. ANDERSON and WHITCOMB [1] showed that the observed seismic velocity variations could be explained by a change in the properties of pore water, assuming that rocks were initially saturated. Water undergoes a liquid-to-vapor transition (abbreviated as the LVT) and many characteristics of water change dramatically near this transition. ANDERSON and WHITCOMB [1] calculated the change in seismic velocities near the LVT for comparison with seismic observations. To our knowledge this has not yet been done for electrical resistivity. In this paper we apply the same methods used before for purely hydrostatic pressure [6] to stress and dilatant rock in which the pore solutions are assumed to approach the LVT. At the LVT some resistive vapor may replace conductive fluid; as a guide to behavior under these conditions we present the results of experiments on stressed rock containing a mixture of air and water in the pore spaces [7]. Finally, we suggest how resistivity observations, particularly from seismic areas, may constrain the general dilatancy–diffusion model of the earthquake mechanism [2].

2. *Resistivity of aqueous solutions near the LVT*

The resistivity of most rocks up to modest temperatures is controlled by the resistivity of the pore fluids. In an estimate of crustal resistivity to 40 km depth [6] we assumed that rocks were saturated with typical ground water at a pressure somewhere between hydrostatic and lithostatic. The resistivity of a wide range of geologically realistic aqueous solutions is fortunately known to about 10 kbar and 800° from the efforts of QUIST *et al.* [8], for example. Here, we simply extend the results given in [6] to the new set of conditions implied by the dilatancy model. Again, the published studies of Quist will be used as a guide to the all-important properties of the pore fluids.

According to the current dilatancy model, as discussed for example, by HANKS [2], dilatancy caused by the build-up of stress prior to an earthquake may reduce pore pressure more or less isothermally. For pore fluids which are dilute aqueous solutions, the properties of pure water will control the detailed changes in specific volume which accompany the drop in pressure. These changes are easily visualized in the *P–T* diagram for water based on data of KENNEDY and HOLSER [9] as shown in Fig. 1. Point C is the *critical point* below which there is a discontinuous change in specific volume. Along the heavy line below C the liquid–vapor transition is therefore very abrupt, whereas above C it is more gradual. By the LVT here we will refer both to the regions below and above C, with the latter assumed to be near the 3 cm^3/gm line.

Two geotherms are also given in Fig. 1 for somewhat extreme conditions [6]; EUS LITH refers to the eastern United States, assuming pore fluids at lithostatic pressure, and BR HYD to the Basin and Range tectonic province with pore fluids at hydrostatic pressure. As noted previously, water specific volume remains fairly close to 1 cm^3/gm along these geotherms to depths of 15 to 20 km.

As a result of dilatancy, pore volume increases and we might imagine pore pressure falling from a point on a geotherm such as BR HYD as the fluids expand. If the volume increase is sufficient the LVT is reached; with continued pore expansion relatively dense water is converted to a less dense mixture of liquid and vapor.

ANDERSON and WHITCOMB [1] calculated the change in bulk modulus of water as pressure dropped isothermally to the LVT; this led to an estimation of seismic velocity change in an initially water-saturated rock as a result of dilatancy. Elastic properties were shown to be highly sensitive to the first appearance of vapor, as the LVT was reached.

For estimation of the effect of dilatancy on resistivity, we need the electrical characteristics of typical pore water at least down to the LVT. The chemical nature of ground water at mid-crustal depths is not known, but we will assume, following the arguments in [6], that its resistivity is 1 to 30 ohm m. Typical aqueous solutions with resistivity in this range (0.001- to 0.1-molar NaCl solutions, for example) differ but little in the way their resistivity depends on pressure and temperature [6]. Accordingly, the effect of dilatancy can be seen by selecting a single solution in this range and examining the way its resistivity varies down to the LVT.

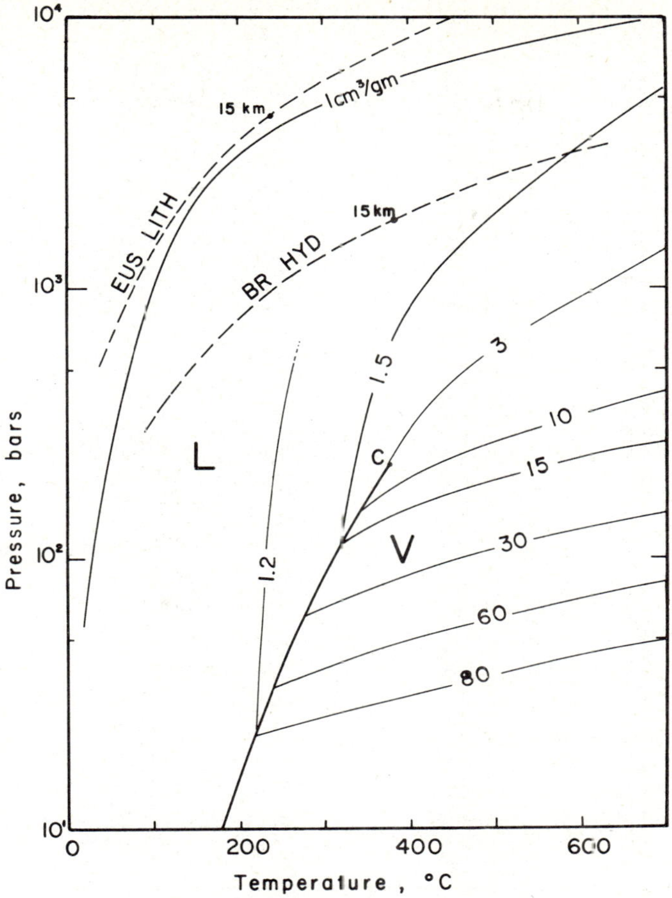

Figure 1
The pressure–temperature–volume relationships of H_2O, from data in [9]. C is the critical point. The two dotted lines are geotherms as explained in text

Resistivity of 0.01 molar KCl is shown in Fig. 2 from data in QUIST *et al.* [10] and QUIST and MARSHALL [11]. For the supercritical region at specific volumes greater than 3 cm^3/gm, relative changes of pure water were used (FOGO *et al.* [12]). Some of the features of Fig. 1 are repeated in Fig. 2 including the two geotherms and several lines of constant specific volume. The crustal profiles in [6] were taken from values such as shown here, along the geotherms. Several new features are evident in this more complete picture.

First, below the critical temperature (373°C), solution resistivity changes little along an isotherm, down to the LVT. Near C, resistivity may increase by a factor of up to two, close to the LVT. At higher temperatures an increase of an order of magnitude may occur down to the LVT. On the vapor side of the LVT resistivity rapidly increases.

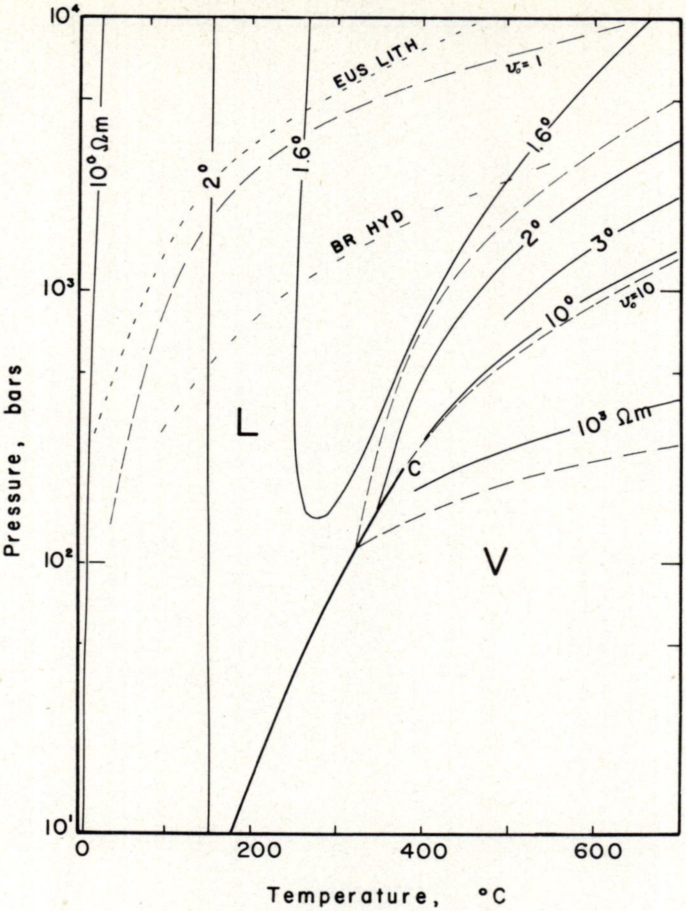

Figure 2
Same as Fig. 1 with values of electrical resistivity of 0.01 molar KCl solution. 3° means 3 × 10° ohm meters

Using changes in the dielectric properties of water as a guide [8] resistivity increases three orders of magnitude per tenfold increase in specific volume.

What may we conclude about saturated dilatant rock? For isothermal changes below C, we may expect, in contrast to seismic velocity [1], very little change in electrical properties of the pore fluids. Rock resistivity, then, will be determined almost solely by the changes in porosity, just as at room temperature [13]. Above C, the situation is more complicated. Pore fluid resistivity increases toward the LVT at a rate, perhaps comparable with the changes arising due to increase in pore volume; the two effects would tend to cancel one another. Still a third effect may also need to be considered. At temperatures in excess of about 500°C mineral conduction in typical rocks exceeds condution through pore fluids [6]. In this case dilatancy should have almost no effect on resistivity of saturated rocks.

Still a final factor must be considered. If dilatancy reaches the stage that some vapor, which will be highly resistive, becomes mixed with the pore solutions, will rock resistivity suddenly change, as do the elastic properties? This matter is treated in the next section.

3. The effect of liquid and vapor mixtures on resistivity

The electrical resistivity of a rock containing a mixture of aqueous solution and either air or water vapor in the pores is not theoretically predictable. It strongly depends on the specific arrangement in the pores of the (conductive) fluid and (resistive) vapor or air and this arrangement can only be conjectured. Since such a situation might occur for dilatant rock at the LVT, experiments were carried out on actual rock samples, to get around this difficulty. Although, in the real situation water vapor may be involved, we studied mixtures involving air as the principal gaseous component. Presumably, since both are many orders of magnitude more resistive than the remaining aqueous solutions, behavior would be quite similar.

We previously observed resistivity changes in (fluid) saturated rocks during dilatancy [13]. Here, we wished to explore the effect of partial saturation. This was achieved either by starting with a rock sample which was partly saturated or by not allowing fluid to enter the sample and to restore saturation during dilatancy.

We studied the effects in Westerly granite, San Marcos Gabbro, and Maryland diabase at the same effective confining pressures as our previous work, so that we could directly compare results. The experimental arrrangement may be found in [13]. In Fig. 3 we compare two specimens of granite, stressed at the same effective confining

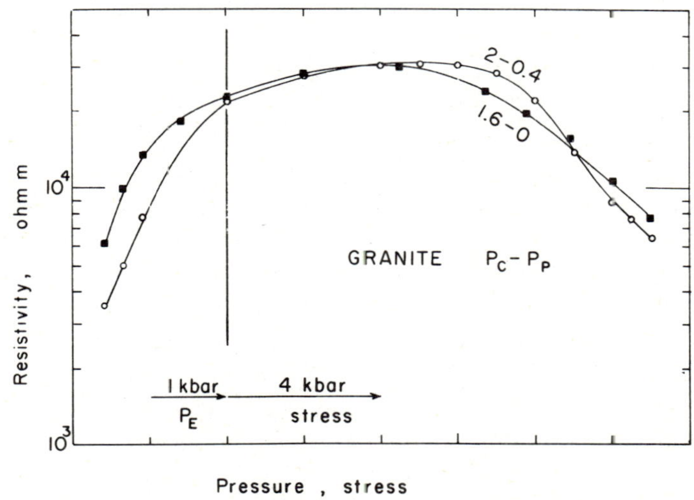

Figure 3
Resistivity of Westerly granite at constant effective confining pressure of 1.6 kbar. Open circles are from [13]. First number on curves in total pressure, P_C, second number pore pressure, P_P, in kbar

pressure of 1.6 kb. For the zero pore pressure sample shown by squares, above 5 or 6 kb stress, as the rock dilates, some of the new pore space is unsaturated; yet the resistivities are only slightly different from those of the saturated case shown by the circles. An even more extreme situation is shown in Fig. 4 for two samples of Westerly granite, one of which was fully saturated initially (squares) and the other (circles) initially at 56% saturation. The shapes of the two curves during hydrostatic compression, P_c, on the left side of Fig. 4 are markedly different, in agreement with our previous observations. Under stress, however, both curves show a very similar decrease in resistivity.

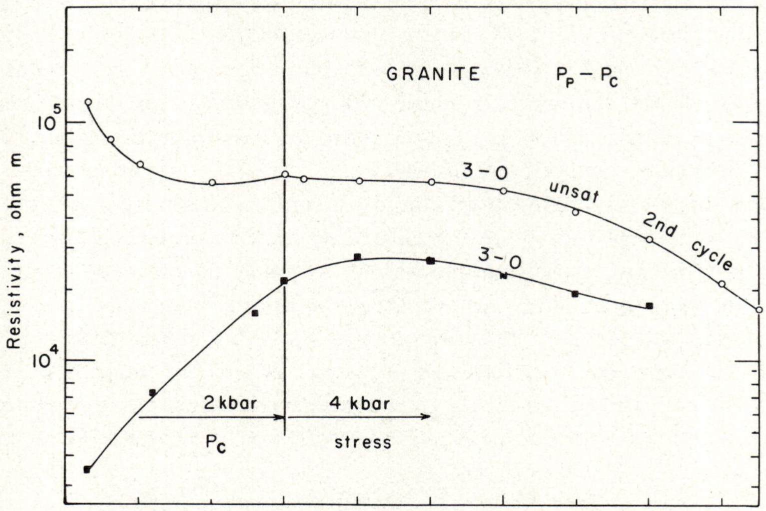

Figure 4
Same as Fig. 3. The lower curve was saturated before this first cycle of stress; the upper curve has 56% saturation and was in a second cycle

It is also important to note that for the partially saturated sample this was the second cycle. Work elsewhere [14, 15] suggests that dilatant cracking is largely accomplished during the first cycle with largely recoverable changes during later cycles. Additional results are shown in Fig. 5 for two other rocks, including a diabase which was also recycled.

Our results above apparently show that even relatively large amounts of vapor in the pore space do not have much effect. Conduction occurs nearly as effectively as it does when all pore space contains liquid. This is apparently in strong contrast to the elastic case studied by ANDERSON and WHITCOMB [1]; here, resistivity will not change appreciably as liquid is converted to vapor at the LVT.

Some explanation of this rather paradoxical result is needed. If conduction in rocks occurs principally through pore solution, how can conduction increase (and resistivity therefore decrease, as shown in Figs. 3, 4 and 5) if dry or partially dry cracks are opened during dilatancy? Recent microscopic studies of the structure of the cavities in rocks [16] may provide partial answers.

Grain boundaries are often the sites of interconnected rows of blunt, high aspect ratio cavities of irregular form and varying width. Dilatant cracks, in contrast, are straight, very narrow, and emanate from or connect the grain boundary features. Relatively small amounts of fluid are probably needed to fill or coat the surfaces of the dilatant cracks; this is probably availably from the grain boundary cavities which may serve as fluid reservoirs. MADDEN [17] has found that surface conduction is unusually important for the dilatant porosity, in accord with this model.

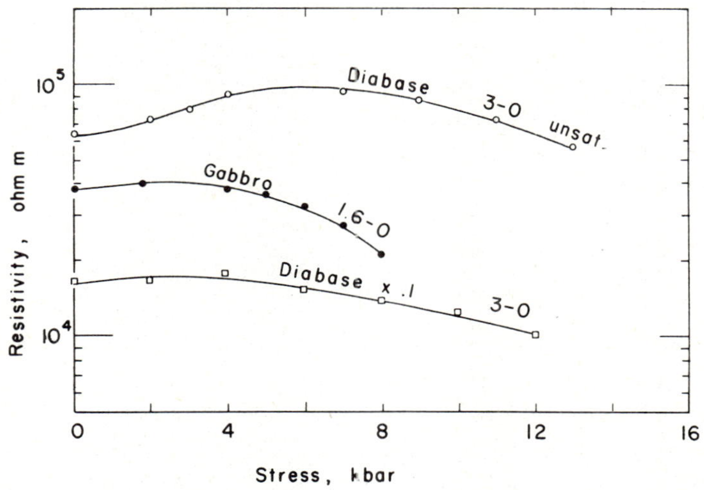

Figure 5

Same as Fig. 3. This contrasts two rocks. Maryland diabase and San Marcos gabbro, at different effective pressures. The lower two were saturated, on the first cycle; the upper one was 50% saturated on the second cycle. Bottom curve was shifted downward for clarity

In conclusion, we have observed the changes in resistivity which accompany the formation or re-opening of dilatant cracks for several cases in which insufficient fluid was present to fully saturate the dilatant cracks. Resistivity always decreases with stress by nearly the same amount as for full saturation. In only one or two situations has the reverse effect been observed [18, 19], namely resistivity increase with stress. These situations involve either unconfined compression tests or certain very high pressure deformation between solid anvils. It is still possible that for other ranges of pressure or temperature or for certain types of rock, such as the marble investigated earlier [13], or for very low initial saturation, resistivity may increase with stress. These possibilities should not be ruled out in the interpretation of resistivity anomalies preceding earthquakes.

4. Constraints on the dilatancy-diffusion model

HANKS [2] suggested constraints on the dilatancy–diffusion model imposed by energetic consideration and by observed uplift during the assumed period of dilatancy prior to an earthquake. Electrical resistivity changes prior to earthquakes of the sort noted by BARSUKOV [4] and MAZZELLA and MORRISON [5] impose additional, independent constraints.

Since condition in rocks at modest temperatures is almost entirely through pore fluids, changes in observed resistivity must correlate closely with changes in either pore volume V_P or pore fluid resistivity ρ_F. Above, we show that for typical ground water compositions, ρ_F is relatively insensitive to pressure and even to the presence of an appreciable fraction of air or water vapor. Accordingly, change in resistivity must correlate primarily with change in pore volume ΔV_P, which in this case would be the dilatant porosity. This correlation has been studied experimentally [13] for conduction parallel to the maximum compression. In terms of rock resistivity, ρ, in this direction, empirically,

$$\rho/\rho_F = (V_P/V_T)^{-\mathbf{m}} \tag{1}$$

where V_T is total volume, V_P/V_T is porosity, and $\mathbf{m}$ is a constant close to 1. ρ_F is the resistivity of pore fluids. For $\mathbf{m} = 1$, fractional change in porosity, $\Delta V_P/V_P$, equals minus the fractional change in resistivity.

HANKS [2] suggested that during the assumed drop in pore pressure to the LVT, $\Delta V_P/V_P$ might range from 50 to less than 1%. Synchronous changes in resistivity yield information which can narrow this range. The changes at Garm were 10 to 15% [5], and about 20% prior to an earthquake in central California [5]. In these cases, the increase in porosity, $\Delta V_P/V_P$, due to dilatancy was, based on (1), from 10 to 20%. This is close to the 15% suggested by NUR *et al.* [20] based on the analysis of velocity changes in regions such as Garm, but considerably less than the 50 to 100% of ANDERSON and WHITCOMB [1].

Actually, close agreement among these numbers would be fortuitous. Both dilatant strain and resistivity are anisotropic characteristics; their dependence can hardly be defined by a single scalar law such as (1). Also, both dilatant strain and resistivity must vary considerably in the vicinity of a locked fault, and a surface observation such as [4] presumably reflects some average over the dilatant volume. How average dilatancy and average resistivity are related is, of course, not known; again, the relation might not be the simple law (1). Thus, our estimate of $\Delta V_P/V_P$ must have an uncertainty of a factor of 5 or 10.

An independent estimate of dilatant porosity is given from the relation involving dilatant strain, $\Delta V_P/V_T$, and porosity, V_P/V_T, of rocks near the fault. Thus,

$$\Delta V_P/V_T = (V_P/V_T)(\Delta V_P/V_P) \tag{2}$$

Based on observed uplift, HANKS [2] estimated the dilatant strain associated with the San Fernando earthquake to be $\leqslant 2 \times 10^{-5}$. NUR [3] found the value 10^{-4} for the Matsushiro earthquake swarm. In laboratory experiments HADLEY [21] determined the amount of dilatant strain prior to slip on faults; this is a lot less than the dilatant strain at fracture, and is probably the appropriate quantity to use for earthquakes associated with well-established active faults. Hadley found values similar to those given above, namely, $\leqslant 10^{-4}$. Let us assume for the present a value of $\Delta V_P/V_T$ of about 10^{-4}.

Porosity is difficult to estimate for crustal rocks, but some recent observations may provide rough bounds. Porosity may be conveniently divided into pore, crack and fracture porosity [6]; the range of measured values is shown in Fig. 6. The crack and pore porosities of about 30 crystalline rocks are shown from various laboratory measurements [16, 22–24], the shaded portion shows the quartz-free rocks with the left-hand limit shown being the present resolution in porosity measurement. Fracture porosities are based on Snow's borehole study of rocks at 35 damsites [25]. The fracture porosity is due to joints, faults, cleavage or bedding plane openings; it varies with depth, as shown in Fig. 6, due to closure of these crack-like features under lithostatic pressure.

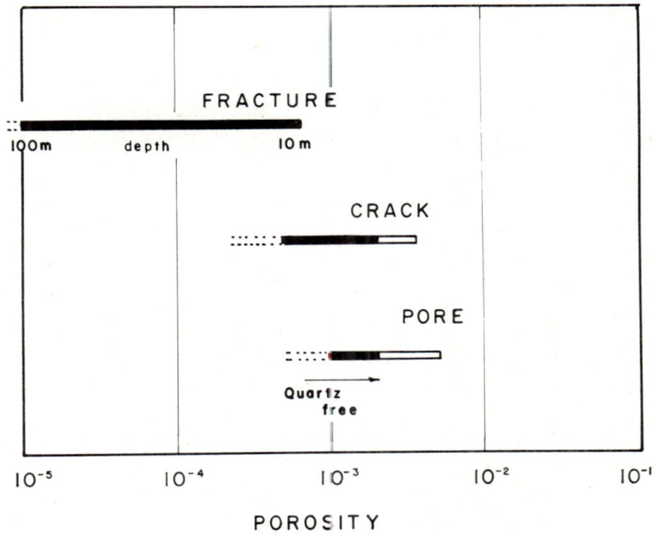

Figure 6
Magnitude of three types of porosity, from data in [16], [22–25]. See explanation in text

What is the porosity at depth in the crust? Based on [25] fracture porosity is relatively small and tends to become even smaller at depth. Crack porosity, too, disappears at high effective pressure, although probably at depths of kilometers rather than tenths of kilometers, as is the case with fracture porosity. Pore porosity, in contrast, may persist to depth. Most pores seem, from recent studies with the electron microscope, to have formed early in the history of at least some crystalline rocks [16]. The shape of these cavities often suggests formation at high temperature, in the presence of a high pressure gas phase [26]. For the present purposes, this suggests that pore porosity will have the same values deep in the crust as it does in samples in the laboratory shown in Fig. 6.

Accordingly, total porosity at depth can seldom fall much below the pore porosity, which for many rocks attains about 10^{-3}.

If we now calculate dilatant porosity according to (2) from $\Delta V_P/V_T$ of 10^{-4} and V_P/V_T of 10^{-3}, the result, namely 10^{-1}, is seen to be quite consistent with the value obtained through the resistivity and (1). Again, this second estimate of $\Delta V_P/V_P$ has a fairly large uncertainty; V_V/V_T may be as low as 10^{-4} and $\Delta V_P/V_T$ as low as 10^{-5}. Obviously both quantities must vary, with the geologic environment and perhaps with depth as well.

An interesting consequence of a $\Delta V_P/V_P$, of about 0.1 is, as Hanks points out, the low initial pore pressure required. Above about 250 the pore pressure drop to give an 0.1 cm^3/gm increase in $\Delta V_P/V_T$ is so small (Fig. 1) that pore pressure would have to be initially sub-hydrostatic. Hopefully this feature can be directly tested once deep holes become available in seismic areas.

Acknowledgements

This research was supported by the US Geological Survey under contract No. 14-08-0001-13229. Discussion of the problem with A. Nur, K. Hadley, T. Hanks, and T. Madden was particularly helpful.

References

[1] D. L. Anderson and J. H. Whitcomb, *The dilatancy–diffusion model of earthquake prediction*, in Proc. of Conf. on Tectonic Problems of the San Andreas Fault System, edited by R. L. Kovach and A. Nur (Stanford Univ. Press, California 1973), p. 417.

[2] T. C. Hanks, *Constraints on the dilatancy-diffusion model of the earthquake mechanism*, J. Geophys. Res. *79* (1974), 3023.

[3] A. Nur, *Matsushiro, Japan, earthquake swarm: Confirmation of the dilatancy-fluid diffusion model*, Geology *2* (1974), 217.

[4] O. M. Barsukov, *Relationship between the electrical resistivity of rocks and tectonic processes*, Izv. Earth Physics, No. 1 (1970), 84.

[5] A. Mazzella and H. F. Morrison, *Electrical resistivity variations associated with earthquakes on the San Andreas fault*, Science *185* (1974), 855.

[6] W. F. Brace, *Resistivity of saturated crustal rocks to 40 km based on laboratory measurements*, Geophysical Monograph Series Vol. 14 (Amer. Geophys. Un., Washington 1971), p. 243.

[7] T. Mitchell and W. F. Brace, *Electrical resistivity of partially saturated rocks* (Abstract), Trans. Amer. Geophys. Un. *54* (1973), p. 1209.

[8] A. S. Quist, *The ionization constant of water to 800° and 4000 bars*, J. Phys. Chem. *74* (1970), 3396.

[9] G. C. Kennedy and W. T. Holser, *Pressure–volume–temperature and phase relations of water and carbon dioxide*, *Handbook of Physical Constants*, Geol. Soc. Amer. Mem. *97* (1966), 371.

[10] A. S. QUIST and W. L. MARSHALL, *The electrical conductances of some alkali metal halides in aqueous solutions from 0 to 800° and at pressures to 4000 bars*, J. Phys. Chem. *73* (1969), 978.

[11] A. S. QUIST, W. L. MARSHALL, E. U. FRANCK and W. VON OSTEN, *A reference solution for electrical conductance measurements to 800° and 12 000 bars. Aqueous 0.01 demal potassium chloride*, J. Phys. Chem. *74* (1970), 2241.

[12] J. K. FOGO, S. W. BENSON and C. S. COPELAND, *The electrical conductivity of supercritical solutions of sodium chloride and water*, J. Chem. Phys. *22* (1954), 212.

[13] W. F. BRACE and A. S. ORANGE, *Electrical resistivity changes in saturated rocks during fracture and frictional sliding*, J. Geophys. Res. *73* (1968), 1433.

[14] M. ZOBACK and J. D. BYERLEE, *The effect of cyclic differential stress on dilatancy in Westerly granite under uniaxial and triaxial conditions*, J Geophys. Res. *80* (1975), 1526–1530.

[15] C. H. SCHOLZ and R. KRANZ, *Notes on dilatancy recovery*, J. Geophys. Res. *79* (1974), 2132.

[16] E. S. SPRUNT and W. F. BRACE, *Direct observation of microcavities in crystalline rocks*, Int. J. Rock Mech. Min. Sci. & Geomech. Abst. *11* (1974), 139.

[17] T. R. MADDEN, Personal communication (June, 1974).

[18] E. I. PARKHOMENKO, *Electrical Properties of Rocks* (Plenum Press, New York 1967).

[19] J. E. RUSSELL and E. R. HOSKINS, *Correlation of electrical resistivity of dry rock with cumulative damage*, Proc., 11th Sympos. on Rock Mechanics, Berkeley, June 1969 (SME & AIMMPE, New York 1970), p. 213.

[20] A. NUR, P. M. BELL and P. TALWANI, *Fluid flow and faulting, 1: A detailed study of the dilatancy mechanism and premonitory velocity changes*, in Proc. of Conf. on Tectonic Problems of the San Andreas Fault System, edited by R. L. KOVACH and A. NUR (Stanford University Press, California 1973), p. 391.

[21] K. HADLEY, *Laboratory investigation of dilatancy and motion on fault surfaces at low confining pressures*, in Proc. of Conf. on Tectonic Problems of the San Andreas Fault System, edited by R. L. KOVACH and A. NUR (Stanford Univ. Press, California 1973), p. 427.

[22] W. F. BRACE, A. S. ORANGE and T. R. MADDEN, *The effect of pressure on the electrical resistivity of water-saturated crystalline rocks*, J. Geophys. Res. *70* (1965), 5669.

[23] W. F. BRACE and A. S. ORANGE, *Further studies of the effect of pressure on electrical resistivity of rocks*, J. Geophys. Res. *73* (1968), 5407.

[24] A. NUR and G. SIMMONS, *The origin of small cracks in igneous rocks*, Int. J. Rock Mech. Min. Sci. *7* (1970), 307.

[25] D. T. SNOW, *Rock fracture spacings, openings and porosities*, J. Soil Mech. Found. ASCE *94* (1968), 73.

[26] C. MONTGOMERY and W. F. BRACE, *Micropores in plagioclase*, Contr. Min. Pet. (1974), in press.

(Received 28th October 1974)

Pageoph,
Vol. 113 (1975) Birkhäuser Verlag, Basel

Precursory and Coseismic Resistivity Changes

By Y. YAMAZAKI[1])

Summary – More than 20 coseismic resistivity steps were observed by a resistivity variometer at a station about 60 km south of Tokyo during a six-year period of observation. The logarithmic magnitude of steps normalized to an epicentral distance (Δ) of 100 km is linearly correlated to earthquake magnitude (M). The larger the earthquake magnitude is, the longer is the time required for completing a resistivity step.

Earthquakes, that produced a step, as plotted on an $M - \Delta$ graph are clearly separated by a straight line

$$M = -12.5 + 2.5 \log_{10} \Delta$$

from those that did not produce a step, Δ being measured in units of km. The straight line is approximately equivalent to a resolution of strain steps of the order of 10^{-9}. As the smallest rate of resistivity change that can be observed by the variometer is 10^{-5} or so, it is apparent that mechanical strain is amplified by a factor of 10^4 or thereabouts when it is measured in terms of resistivity change.

Some of the steps are preceded by a premonitory effect. When long-period changes, mostly due to tidal loadings, are eliminated by a numerical filtration technique, precursors are brought out very clearly in a number of cases. The precursor time ranges from 1 to 7 hours and appears to be independent of magnitude.

1. Introduction

A resistivity variometer (YAMAZAKI [1–5]; RIKITAKE and YAMAZAKI [6–7]) has been at work at a station about 60 km south of Tokyo, Japan, since 1968. Twenty-three coseismic resistivity steps have been recorded during the six-year period of observation. The results added recently since the last report on coseismic steps (YAMAZAKI [5]) will be briefly described in Section 2. As the number of observed steps increases it has also become possible to examine the lowest limit of magnitude of earthquakes which give rise to a step in relation to epicentral distances.

It now seems to be certain that a precursory change in resistivity occurs in some cases. Although it is not as yet quite clear which kind of earthquake is associated with a precursor, a number of examples of precursory change will be presented in Section 3.

2. Coseismic step

Many of the coseismic resistivity steps which have been observed by the present resistivity variometer have been presented by YAMAZAKI [5], so that only one recent

[1]) Earthquake Research Institute, University of Tokyo, Tokyo, Japan.

example of such steps is shown in Fig. 1. The step was associated with an earthquake of magnitude (M) 6.9 that occurred at the tip of Izu Peninsula, about 100 km southwest of the variometer station.

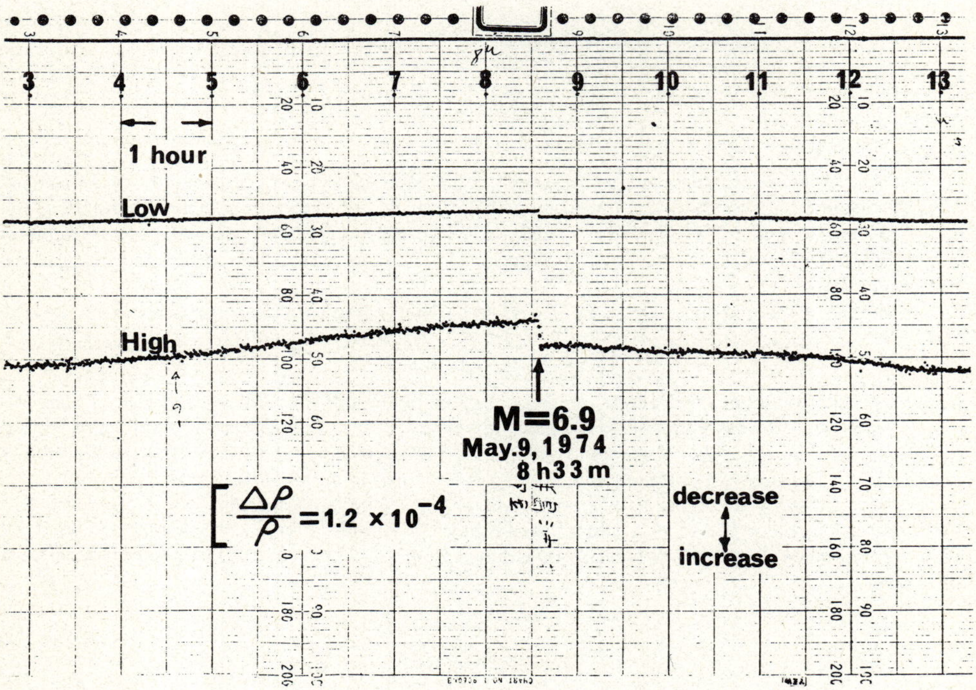

Figure 1
Resistivity changes associated with an earthquake of $M = 6.9$, that occurred at an epicentral distance of 100 km southwest of the observation point on May 9, 1974, as recorded by the high- and low-sensitivity channels. The sensitivity of the resistivity variometer for the high channel is indicated by the $\Delta\rho/\rho$ value

The step is not quite instantaneous as usual. As changes in resistivity are recorded by a multi-channel recorder having a plotting interval of 30 sec, it is seen from the figure that the step was completed within 90 sec.

2.1. Step–magnitude relationship

A hypothetical resistivity step $(\Delta\rho/\rho)_{100\,\mathrm{km}}$ that should have been observed at an epicentral distance (Δ) of 100 km is estimated for each step observed on the assumption that the step decreases proportionally to $\Delta^{-3/2}$ as has been done in the previous papers. $\Delta\rho/\rho$ is then converted to mechanical strain by means of the relationship between the rate and the resistivity change obtained by YAMAZAKI [5].

Figure 2 shows the $(\text{step})_{100\,\text{km}}$ or $(\Delta\rho/\rho)_{100\,\text{km}}$ vs. M relationship thus obtained for the 20 steps. A linear relation between logarithmic amplitude of step and earthquake magnitude as proposed in the previous papers seems to be confirmed, even though no adjustment for source mechanism has been made.

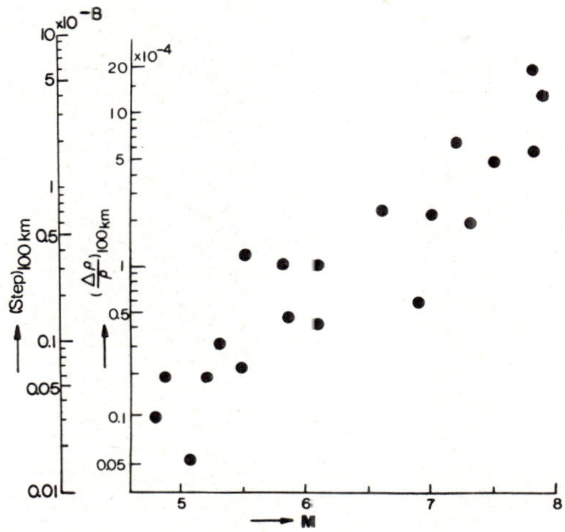

Figure 2
$(\Delta\rho/\rho)_{100\,\text{km}}$ or $(\text{step})_{100\,\text{km}}$ vs. M relationship

2.2. Rise-time

We define a rise-time τ by the time required for completing a resistivity step. $\text{Log}_{10}\tau$ is plotted against M in Fig. 3. The previous speculation (YAMAZAKI [5]) that the larger the energy released by an earthquake, the longer becomes the time needed for completing the strain step seems to hold.

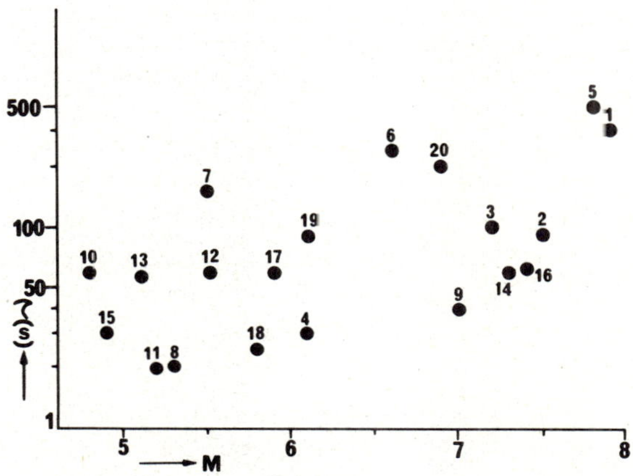

Figure 3
Rise-time (τ) vs. M relationship. Numerals indicate the sequence in which the earthquake occurred

2.3. *Distribution of earthquakes for which a resistivity step is observed*

As the data has increased, it now becomes possible to see the distribution of earthquakes which accompany a resistivity step. Figure 4 is an epicenters map of earthquakes of $M \geqq 5$ in and around Japan during the period from May, 1968, to March, 1974.

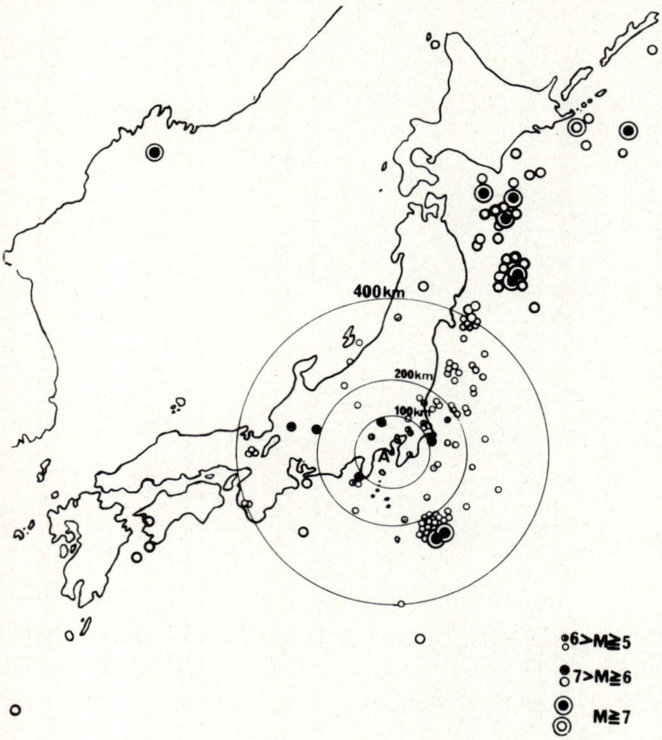

Figure 4
Epicenters of earthquakes of $M \geqq 6$ in and around Japan and of $6 > M \geqq 5$ in a circular area having a radius of 400 km centered at the observation point indicated by *A* during a period from May, 1968, to March, 1974. Solid and half-solid circles indicate earthquakes that accompanied a resistivity step

Within a circular area having a radius of 400 km centered at the observation point indicated by *A* earthquakes with $M \geqq 5$ are shown. Solid and half-solid circles indicate that these earthquakes accompanied a resistivity step at *A*. In addition to these earthquakes, two earthquakes of $M = 4.8$ and 4.9 in the Tokyo area gave rise to a step. It can be said that all earthquakes of $M \geqq 5$ within an epicentral distance of 100 km from *A* accompanied a resistivity step.

It appears to the author that earthquakes in southwestern Japan or those related to the Philippine Sea plate do not give rise to a resistivity change at *A* although such a conclusion would be premature because the number of conspicuous earthquakes in the area is not large.

Figure 5 is a plot of earthquake magnitude against logarithms of epicentral distance. Earthquakes accompanied by a resistivity step are indicated by solid circles. It is interesting to note that almost all earthquakes accompanied by a resistivity step are

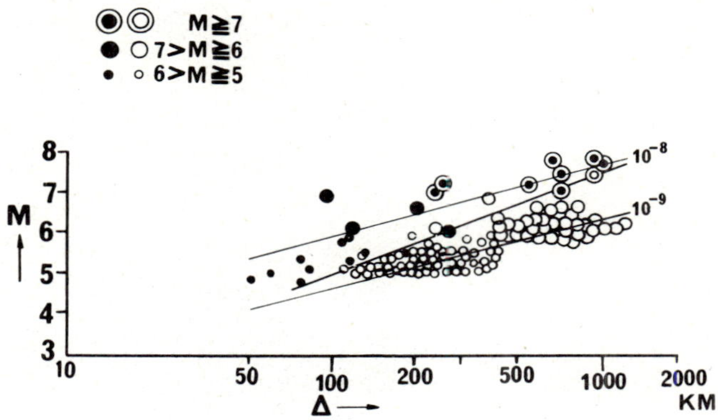

Figure 5

Earthquake magnitude (M) as plotted against epicentral distance (Δ). Solid circles indicate earthquakes that accompanied a resistivity step. The thick line shows the lowest limit of magnitude of earthquakes that give rise to a resistivity step. Thin lines indicate the Wideman-Major relationships between M and Δ for earthquakes which result in a mechanical step of the order of 10^{-9} and 10^{-8}, respectively

plotted in the upper half of the $M - \Delta$ plane bounded by a thick straight line as shown in the figure. When Δ is measured in units of km, the line is expressed as

$$M = -12.5 + 2.5 \log_{10} \Delta \tag{1}$$

According to WIDEMAN and MAJOR [8], empirical relationships between M and Δ which indicate that the magnitude of the smallest earthquake, from which a strain step of the order of 10^{-8} and 10^{-9} may be expected, are given by the thin straight lines which are also drawn in Fig. 5. It is thus seen that the resistivity variometer can detect a strain change of the order of 10^{-9} or thereabout.

The smallest rate of change in resistivity that can possibly be detected is of the order of 10^{-5} as can be seen in Fig. 1. On the other hand it was estimated by YAMAZAKI [5], who relied on a calculation of earth strain on the basis of the dislocation theory, that small earth strains of the order of 10^{-8}–10^{-9} can be picked up by the resistivity variometer with a magnification factor of 10^4 or so although nothing certain about the magnifying mechanism is known. Taking the magnification factor into account, the smallest strain that can be detected must be of the order of 10^{-9}. It may therefore be said that the relationship as shown in Fig. 5 is in good harmony with the fact that mechanical strain is amplified by a factor of 10^4 or so when it is measured in terms of resistivity change.

3. Precursory change

If one looks at Fig. 1 carefully, it seems likely that a change in resistivity started a few hours prior to the step. As has been mentioned in YAMAZAKI [5], there are a number of cases for which some sort of premonitory effects were observed.

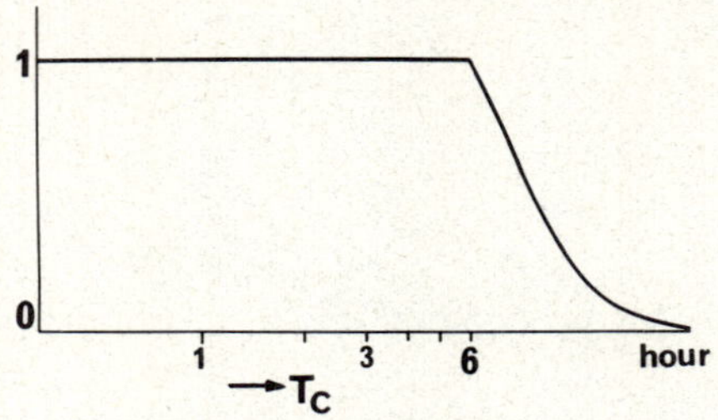

Figure 6
The characteristics of numerical high-pass filter

In order to bring out the precursory change more clearly, the long period changes mostly due to tidal loading were eliminated by applying a numerical filter to the records. For example, the high-sensitivity trace in Fig. 1 is read off with an interval of 12 minutes over a period of four days centered by the time of the earthquake. A high-pass filtration procedure (MASON [9]) having a cut-off period (T_c) of six hours is then applied to the readings. The characteristics of the numerical filtration is shown in Fig. 6.

Figure 7 shows the original readings (A) of the resistivity–change trace that was shown in Fig. 1 along with the altered record (B). The figure clearly demonstrates that

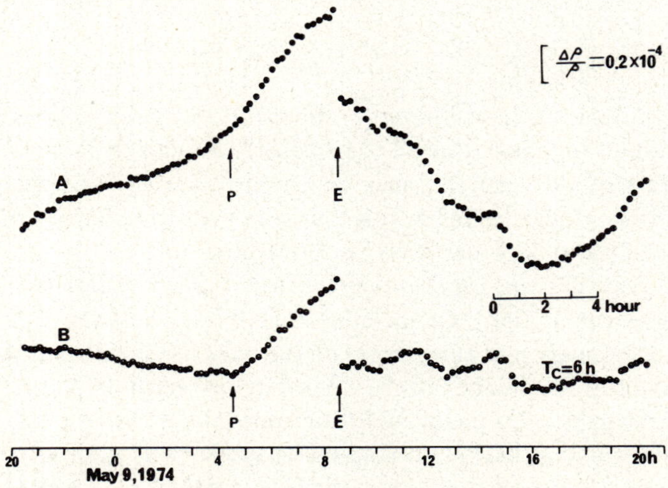

Figure 7
Original record of resistivity changes (A) and the filtered ones (B) for an earthquake of $M = 6.9$ at $\Delta = 100$ km on May 9, 1974. The precursor time is defined by the time-interval from P to E. The resistivity decreases upwards in the present and following figures

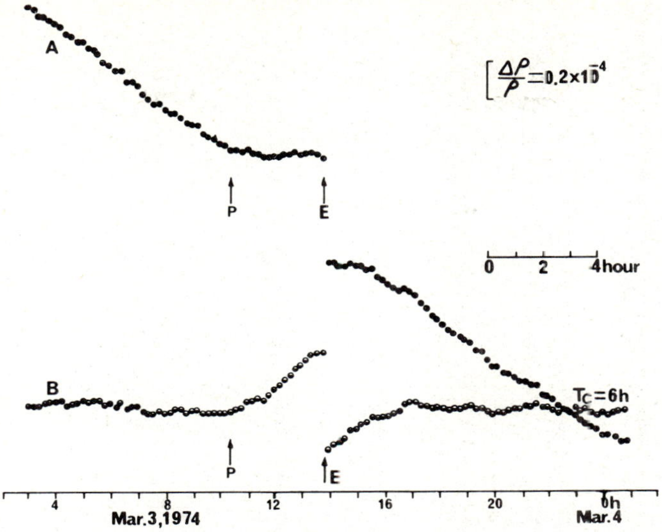

Figure 8
Precursory and coseismic changes for an earthquake of $M = 6.1$ at $\Delta = 119$ km on March 3, 1974

a precursory change starts at about the time denoted by P. The step at the time of the shock is marked by E.

A few more examples of precursory change in resistivity are also illustrated for earthquakes of $M = 6.1$ at $\Delta = 119$ km on March 3, 1974, and $M = 4.9$ at $\Delta = 50$ km on May 27, 1973, respectively, in Figs. 8 and 9. It seems highly likely that these earthquakes were preceded by a resistivity decrease although the records sometimes suffer from noise.

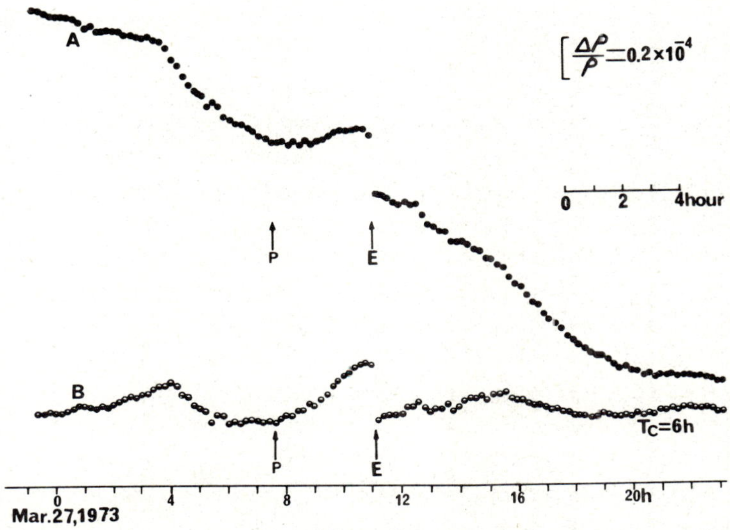

Figure 9
Precursory and coseismic changes for an earthquake of $M = 4.9$ at $\Delta = 50$ km on March 27, 1973

The nature of precursors as revealed by the resistivity variometer is not clear. Figure 10 shows the relationship between presursor time T_P and earthquake magnitude. Although there may be a slight tendency for the larger earthquakes to have larger precursor times, it may as well be said that T_P is independent of M as far as the data accumulated are concerned.

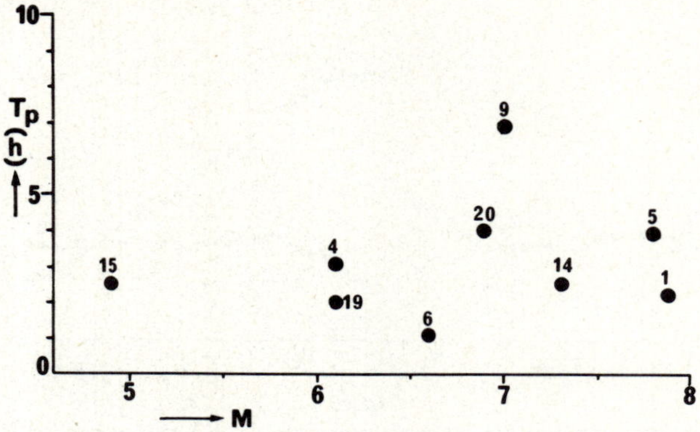

Figure 10
Precursor times (T_p) in hours as plotted against earthquake magnitudes (M). Numerals are numbers of earthquakes analyzed

4. Conclusion

The nature of coseismic steps of resistivity as observed by a resistivity variometer becomes fairly clear. The logarithmic magnitude of the steps normalized to a constant epicentral distance, is linearly correlated to earthquake magnitude. It seems that the larger the earthquake magnitude is, the longer is the time required for completing a step.

Resistivity steps are observed for earthquakes having a magnitude larger than a certain lowest limit, which is expressed by a straight line on a magnitude (M) – epicentral distance (Δ) plot. The limit seems to approximately coincide with a line which indicates that a mechanical strain step of the order of 10^{-9} is to be observed on $M - \Delta$ plane.

Some of the resistivity steps are preceded by a precursory change. The precursor time ranges from 1 to 7 hours and appears to be independent of magnitude. The precursor times observed in terms of resistivity change do not fit the precursor time – magnitude relation obtained by SCHOLZ *et al.* [10], WHITCOMB *et al.* [11] and TSOBOKAWA [12]. It seems likely that the present precursors have nothing to do with the models suggested by the above authors. They might be associated with preliminary failure immediately before the main rupture of the earth's crust. The precursors are brought out very clearly after eliminating long period changes due to tidal loading by applying a numerical high-pass filter to the original records. Resistivity precursors thus observed may have an important bearing on short-range earthquake prediction.

Acknowledgements

The writer thanks Professor T. RIKITAKE for his advice in the course of this study. He also thanks Dr. T. WATANABE and Dr. Y. HONKURA for their help in designing a numerical filter used in this paper. The writer wishes to express thanks to Mr. T. TAKAHASHI at the Aburatsubo Crustal Movement Observatory, Earthquake Research Institute. The present work could not have been accomplished without his collaboration in the actual observation.

REFERENCES

[1] Y. YAMAZAKI, *Electrical conductivity of strained rocks (the 1st paper), Laboratory experiments on sedimentary rocks*, Bull. Earthquake Res. Inst. *44* (1965), 783–802.

[2] Y. YAMAZAKI, *Electrical conductivity of strained rocks (the 2nd paper), Further experiments on sedimentary rocks*, Bull. Earthquake Res. Inst. *44* (1966), 1553–1570.

[3] Y. YAMAZAKI, *Electrical conductivity of strained rocks (the 3rd paper), A resistivity variometer*, Bull. Earthquake Res. Inst. *45* (1967), 849–850.

[4] Y. YAMAZAKI, *Electrical conductivity of strained rocks (the 4th paper), Improvement of the resistivity variometer*, Bull. Earthquake Res. Inst. *46* (1968), 957–964.

[5] Y. YAMAZAKI, *Coseismic resistivity steps*, Tectonophysics *22* (1974), 159–171.

[6] T. RIKITAKE and Y. YAMAZAKI, *Electrical conductivity of strained rocks (the 5th paper), Residual strains associated with large earthquakes as observed by a resistivity variometer*, Bull. Earthquake Res. Inst. *47* (1969), 99–105.

[7] T. RIKITAKE and Y. YAMAZAKI, *Strain step as observed by a resistivity variometer*, Tectonophysics *9* (1969), 197–203.

[8] C. J. WIDEMAN and M. W. MAJOR, *Strain steps associated with earthquakes*, Bull. Seism. Soc. Amer. *57* (1967), 1429–1444.

[9] R. G. MASON, *Spatial dependence of time variations of the geomagnetic field in the range 24 hr to 3 min on Christmas Island*, Geophys. Dept. Imp., Coll. Sci. Technol., London, Publ. *63–3* (1963), 1–20.

[10] C. H. SCHOLZ, L. R. SYKES and Y. P. AGGARWAL, *Earthquake prediction: a physical basis*, Science *181* (1973), 803–810.

[11] J. H. WHITCOMB, J. D. GARMANY and D. L. ANDERSON, *Earthquake prediction: variation of seismic velocities before the San Fernando earthquake*, Science *181* (1973), 632–635.

[12] I. TSUBOKAWA, *On relation between duration of precursory geophysical phenomena and duration of crustal movement before earthquake*, J. Geod. Soc. Japan *19* (1973), 116–119 (in Japanese).

(Received 2nd October 1974)

Pageoph,
Vol. 113 (1975) Birkhäuser Verlag, Basel

Application of Electric Method to the Tentative Short-Term Forecast of Kamchatka Earthquakes

By G. A. Sobolev[1])

Summary – Tentative forecasts of strong earthquakes on Kamchatka have been carried out continuously since 1972 with scientific purpose. Three methods are being used: electric, ratio V_p/V_s and seismostatistical. The main scientific result of the two-year study is the statement that it is possible to predict earthquakes of magnitude ≥5.5 some days in advance; eight earthquakes had been predicted from a total number of twelve.

In the course of investigations for earthquake prediction on Kamchatka, a number of short-term precursors were found that appeared 5–20 days before the earthquakes (Sobolev *et al.*, 1972; Sobolev, 1973; Fedotov *et al.*, 1972). These precursors became the basis of the program of tentative short-term forecasts, created in December 1971 by S. A. Fedotov and the author, of Kamchatka earthquakes of magnitude ≥5.5. Three principal methods are being used: electric, V_p/V_s ratio, and seismostatistical. Tentative forecasts have been carried out continuously since 1 January 1972.

Recordings of the electrotelluric field in search of precursors of strong earthquakes began in 1966 along the eastern coast of Kamchatka. The process of preparation of a large earthquake manifests itself by cracking, accompanied by a change in deformation velocities and stress in certain parts of the seismoactive area. It would be natural to expect that, in this case, in the preparation zone, changes would appear in the electric state of the ground. Observations on the mechanics of failure in heterogeneous medium allow us to suggest that, during the process of preparation of the earthquake, precursors of different duration can appear. The summarized data of precursors (Miachkin and Zubkov, 1973) shows the presence of a wide range of precursor times, from several minutes to several years, for different geophysical fields. We are inclined to believe that there can exist an almost uninterrupted frequency spectrum of precursors that reflects the fracturing of different-size blocks of rock. In the present work, we report anomalous changes in electric state of the order of several days. The absence of precursors of shorter duration can be accounted for because the frequency of observations is one hour. Long-term precursors are not seen because the period of research is not sufficient to remove the effects of seasonal variations and other long-period noise.

Observations of electric precursors are conducted by a network of stations spaced at distances of 100–200 km. As transducers, lead electrodes and electrodes with special

[1]) Institute of Earth Physics, Moscow, USSR.

low self-polarization film fixed at a depth of 2 m are used. At all the stations along the eastern coast of Kamchatka, potential differences are recorded on main and doubling lines oriented in north–south and east–west directions. The distance between the electrodes is 200 m. At some stations additional lines of different length are employed with which it is possible to exclude the influence of contact phenomena on the electrodes

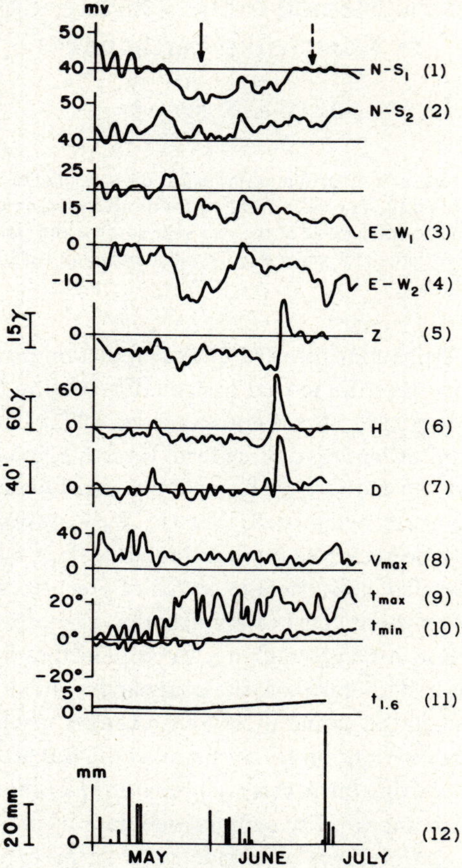

Figure 1
Example of electric state of ground change before earthquake with magnitude 6, at station 'Cape Schipunsky'. 1–4 are daily mean values of the difference of voltage between electrodes along main and doubling lines; 5–7 are for daily mean values of geomagnetic field; 8 is for wind velocity; 9-10 are for soil temperature at the surface; 11 is for rock temperature at a depth of 1.6 m; 12 is for precipitation

and the disturbances due to induced currents. This is achieved by means of cross-correlation of the data of many measuring lines of several stations and comparison with geomagnetic variations. Such analysis allow us to distinguish unusual changes of the electrotelluric field, presumably connected with earthquakes.

An example of anomalous change of the electric state of ground before an earthquake is shown in Fig. 1. On 27 May 1972 a deep earthquake occurred with magnitude 6, its source situated under Kamchatka at the epicentral distance of 250 km from the station 'Cape Schipunsky'. Approximately one week before the earthquake, synchronous changes of electrotelluric field level, including all components, were observed (graphs 1–4, Fig. 1). Meteorological factors in this region and the geomagnetic field did not change noticeably before the earthquake.

If we assume the values of the electrotelluric field on 20 May to be normal, then, at the moment of the earthquake, the changes of the field on both east–west lines were 10–13 mV and on north–south lines 8–10 mV. Hourly records during that period, given in Fig. 2, show the identity of azimuth change of the field along the lines.

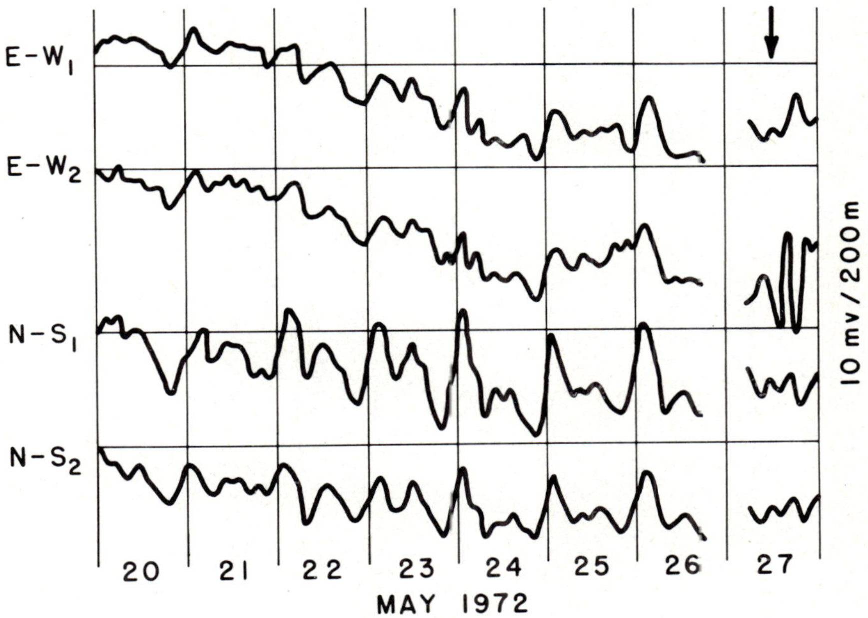

Figure 2
Hourly records of mean and doubling lines showing the change of electric state of ground before earthquake with magnitude 6

In the majority of cases, the changes of electrotelluric field before an earthquake do not exceed the background noise. This makes it impossible to distinguish precursors through the recording of one measuring line. As a result, standardization of methods is needed to conduct the joint processing of several synchronous records.

The recognition of anomalous changes of the electrotelluric field was made by computer analysis (Fig. 3). The algorithm employed for this purpose decreases the high-frequency noise and creates a 'zero' line of seasonal variations for each measuring line. The deflection of real changes of the field from the 'zero' line are then calculated. To increase the signal-noise ratio, the data are averaged on all the measuring lines for each station. To increase stability of this analysis, we used multiple averaging on variable time windows. Maximums of the outlet function correspond to an alarm period. Figure 3 shows the unprocessed records of four components for one station situated in the central part of the Kamchatka seismoactive area, and the results of analysis of these records by means of two different programs. The arrows show the moments of all earthquakes with magnitude ⩾5.5 that occurred closer than 250 km from the station. The numbers near the arrows correspond to the numbers of the earthquakes in Fig. 4. One can see that most earthquakes were preceded by an anomalous increase of the processed

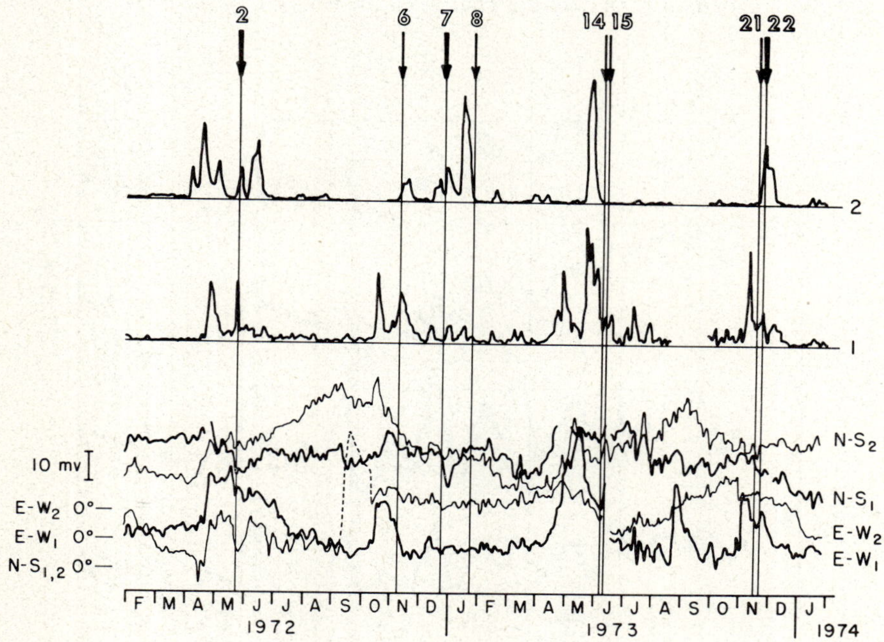

Figure 3
Detection of the Kamchatka earthquakes with magnitude ⩾5.5 by the computer programs exemplified by the electric station 'Cape Schipunsky'. The numbers near the arrows serve to identify the corresponding earthquakes in Fig. 4

signal. The programs are not good enough to exclude all seasonal variations, as those for example which appeared in May 1972 and 1973. But the main maxima before the earthquakes are not connected with such variations.

The tentative short-term forecast of the time and place of Kamchatka earthquakes was based on data from electrotelluric stations, and changes of the ratio of *p*- and *s*-wave velocity calculated for each seismological station according to the method developed by L. B. Slavina.

Individual values of the parameter $x = V_p/V_s$ for each station as a function of time were estimated using the relation $x = (Ts - p)/(Tp - o) + 1$, where $Ts - p$ is for the difference in the time of arrival of *p*- and *s*-waves to the station; $Tp - o$ is for the travel

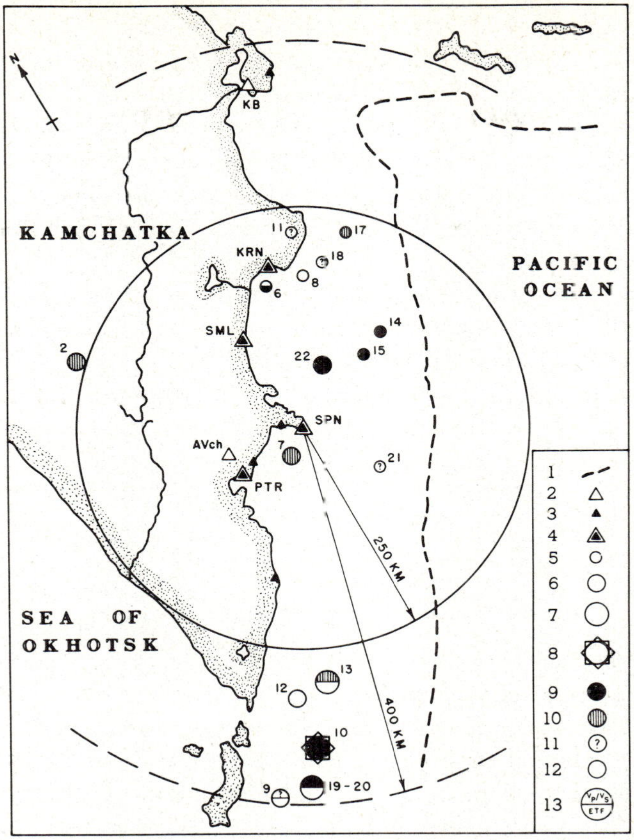

Figure 4

Map of epicenters of the Kamchatka earthquakes for which the computer forecast was made for the station 'Cape Schipunsky'. 1 is the axis of deep-sea trench; 2 is the seismic station; 3 is the electric station; 4 is the joined seismic and electric station. Magnitude of earthquakes: 5 is approximately 5.5; 6 is approximately 6; 7 is greater than 6; 8 is greater than 7; 9 is the epicenters of earthquakes before which considerable anomalies were observed; 10 those with less-distinct anomalies; 11 those with questionable anomalies; 12 those with no anomalies. The presence of anomalies for the electric and V_p/V_s method is depicted by symbols, according to the scheme in the right lower corner

time of p-waves from the source to the station. The moment of earthquake occurrence was determined with the help of data from all Kamchatka seismological stations.

The analysis of time dependence of parameter x was made by computer (SOBOLEV and SLAVINA, 1974). The algorithm includes the calculation of differences between the current value of x and the average value $\Delta x = x_i - \bar{x}$, the introduction of weight coefficients p_i to reduce the effect of accidental errors, and separate summing of positive and negative Δx along variable time intervals. At the outlet we have the function

$$|\bar{x}|^2 = \sum_{i=1}^{n} (\overline{\Delta x}_{\text{pos}} \cdot p_i) \cdot \sum_{i=1}^{n} (\overline{\Delta x}_{\text{neg}} \cdot p_i).$$

The maximum of the outlet function correspond to the alarm period as in the case of electric forecast. In Fig. 4, the map of computer forecasts of Kamchatka earthquakes, according to data from only one station (Cape Shipunsky), is shown. The solid and lined circles denote the earthquakes before which both electrotelluric (ETF) and V_p/V_s outlet functions exceeded the alarm levels. One can see that most earthquakes with precursors occurred within 250 km of the recording stations. Some of the earthquakes in the south, for example number 10, were predicted on the basis of the southern stations records.

It is possible to estimate the success of the electric method for the tentative short-term forecast. 235 separate forecasts were made and recorded in informal proceedings of the Institute from 1 January 1972 to 20 April 1974. Thus the forecasts were made for 3 to 4 days in advance, twice a week. The data of five electrotelluric stations were used.

Table 1

Method	N	N_p	$L(\%)$	I
ETF	12	8	29	2.30
V_p/V_s	14	10	56	1.28
ETF + V_p/V_s	12	7	22	2.64

The numerical results are summarized in Table 1, where the data from the V_p/V_s method are also given for comparison. The forecasts were made in Petropavlovsk-Kamchatskij by A. A. Chromov for the ETF method, and by L. B. Slavina for the V_p/V_s method. In addition, these two methods were combined for short-term forecasts. The conditional alarm forecasts were given when alarm periods of both methods coincided. In Table 1 N is the number of earthquakes with magnitude $\geqslant$5.5 which occurred during the period for which forecasts had been made by means of a given method; N_p is the number of predicted earthquakes; L is the ratio of the sum of products of alarm time and the square of alarm region to the product of total time of forecast and total square of region; I is the efficiency of the forecast, i.e. the ratio of the density of earthquakes during alarms to the average one. The main result of the two-year study is the statement that it is possible to predict earthquakes with magnitude $\geqslant$5.5 some days in advance with a probability two times higher than chance guessing. It is very interesting that I tends to increase as a result of a combination of the two methods. This increase is accounted for by reduction of time of false alarms.

From these results it would seem that the attempt to conduct a continuous forecast program on Kamchatka is, at this time, of scientific interest only. The efficiency of forecast is too low for practical purposes.

References

Fedotov, S. A. *et al.* (1972), *Progress of earthquake prediction in Kamchatka*, Tectonophysics *14* (3/4).

Miachkin, V. J., and Zubkov, S. J. (1973), *Svodnij graphik predvestnikov zemletriasenij*, Izvestia AN SSSR, No. 6.

SOBOLEV, G. A. (1973), *Perspectivy operativnogo prognoza zemletriacenij po electrotelluricheskim nabludeniam*, Sbornick, predvestniky zemletriacenij, VJNJTJ.

SOBOLEV, G. A., MOROSOV, V. N., AND MIGUNOV, N. J. (1972), *Electrotelluricheskoe pole i silnoe zemletriasenie na Kamchatka*, Izvestia AN SSSR, Physica Zemly, No. 2.

SOBOLEV, G. A., and SLAVINA, L. B. (1974), *Bistrie izmenenia electricheskich i seismicheskich svoistv sredy v seismoactivnom raione*, DAN SSSR *215*, No. 5.

(Received 22nd November 1974)

Pageoph,
Vol. 113 (1975) Birkhäuser Verlag, Basel

The Nature of Surface Tilt Along 85 km of the San Andreas Fault – Preliminary Results from a 14-Instrument Array

By Carl E. Mortensen and Malcolm J. S. Johnston[1])

Summary – The continuous monitoring of surface deformation near active faults is clearly necessary for an understanding of elastic strain accumulation and elastic and anelastic strain release associated with earthquakes. Fourteen 2-component tiltmeters have been installed in shallow boreholes along 85 km of the currently most active section of the San Andreas fault in the western United States. These instruments operate at a sensitivity of 10^{-8} radians. Five of these tiltmeters, extending along one 35 km section of the fault, have been in operation since June 1973. The results indicate that regional tectonic tilting has occurred before more than ten individual earthquakes or groups of earthquakes with epicenters within ten earthquake source dimensions of one or more instruments. This tilting has a time scale of up to a month depending on earthquake magnitude. The amplitude of these tilts exceeds by almost an order of magnitude that expected from a dislocation model of the source using seismically determined parameters. No indication of rapid or accelerated tilt just prior to these earthquakes has been seen.

1. Introduction

The accepted explanation of the earthquake mechanism, proposed by Reid [15], is the rapid release of slowly accumulated elastic strain energy by fault slip or breakage; however, the details of stress accumulation in the earth's crust are not well understood. This simple explanation, however, is unlikely to be useful in the problem of earthquake prediction where visco-elastic behavior could dominate in the period immediately before an earthquake. In particular, this includes the expected and observed variability of material properties and behavior, both in time (Dieterich [3], Griggs *et al.* [4] Orowan [12], Scholz [19]) and space. The time scale of seismic stress change will probably also be nonlinear (Stacey [20]).

Precise determination of surface displacement and displacement gradients near active faults may contribute to understanding the earthquake mechanism. Unfortunately, observations of this type are difficult to make and are easily contaminated by instrumental, meteorological, and site effects. Progress in the last ten years has been slow. To obtain useful data, it is still necessary to test the instrument thoroughly under simulated site conditions before installation, to obtain simultaneous meteorologic records, and to place strict checks on the data produced to ensure authenticity, e.g., simultaneous recording on at least two independent instruments.

[1]) U.S. Geological Survey, Menlo Park, California 94025.

By far the easiest method for monitoring surface deformation is to record the associated surface tilt. The best evidence of tilts accompanying earthquakes have come from several observers in Japan (HAGAWARI and RIKITAKE [5], SASSA and NISHIMURA [17]), though some observations of possible anomalous tilting preceding earthquakes have been made in the USSR (OSTROVSKII [13]), Italy (CALOI [2]), and the USA (Wood and Allen [24]). There does not appear from the reported data to be any systematic tilt behavior associated with earthquakes.

To examine the relationship of crustal tilts and earthquakes, the installation of a network of 2-component shallow borehole tiltmeters was initiated in 1973. This network extends along an 85 km section of the San Andreas fault from Mt. Madonna (23°01′N, 121°42′W) to Dry Lake (36°29′N, 121°05′W). This is currently the most seismically active section of the fault. Fourteen instruments have now been installed and five have been in operation since June 1973. The purpose of this paper is to present the records from several of these instruments and discuss some possible relation to earthquake tectonics.

2. *The instruments and their installation*

Although observations of crustal tilting data back to the early 1900's (OMORI [11]), the basic problems plaguing this kind of measurement are not easy to minimize. The most important of these are:

1) meteorological loading effects,
2) thermal and mechanical instabilities both in the instrument and the site,
3) the heterogeneous nature of crustal rock and surface strain fields.

Deep mine installations may reduce some surface noise problems but introduce others such as tunnel-shape effects, as discussed by KING [8], complex fracturing due to mining, and cost. Few deep mines exist in seismically active areas. If reliable data on tilt before earthquakes are to be obtained, it appears necessary to successfully operate near-surface installations at as many closely spaced sites as possible.

The instrument used in this array is the model 541B manufactured by the Autonetics Division of Rockwell International. It consists of a bubble in a conducting electrolyte contained under a concave quartz lens. The position of the bubble is sensed by two A.C. bridge circuits which have a common electrode at the bottom of the fluid cavity and orthogonal pairs of sensing electrodes deposited on the lens surface (KOHLENBERGER *et al.* [10]).

Prior to installation the calibration, linearity, thermal, and mechanical stability were measured by JOHNSTON [6]. The most important of these parameters is the mechanical stability, which in the worst case was 2×10^{-7} radians in a two-month period. This measurement was made by suspending the instrument, with viscous damping, on a long, thin stainless steel wire. The output depends then only on the gravitational constant and the mechanical and electronic stability. With the viscous damping removed, this was also a convenient way to measure impulse response.

Sensitivity to temperature was high ($5 \times 10^{-8}/°C$ near the null balance position). The range of temperature at 3 m depth is not expected or observed to exceed a few degrees and operation is kept near the null balance position. Sensitivity to thermal gradients was also high; however, in the borehole installations used, thermal gradients have been minimized ($<0.001°C/cm$).

The installation consists of a pit 1 m in diameter and 2 m deep, lined with a fiberglass culvert surrounded by a layer of tar and gravel to prevent surface runoff from seeping down the outside and underneath the culvert. Extending down about 2 m from the bottom of the pit is a 15 cm diameter borehole lined with aged steel tubing, closed at the bottom. The sensor is wellpacked with fine grain silica sand near the bottom of the borehole. The closely packed sand appears to have better stability than the surrounding fault zone materials. We are currently testing a more rigid mount using an expansive grout to cement the sensor in the borehole. The tiltmeter electronics are located in the bottom of the pit, which is filled with marble-sized styrofoam pellets. Battery power and recording facilities are remotely located in a second pit. This method of installation is similar to that described by ALLEN *et al.* [1].

As a check on the whole system, two installations, 20 m apart, were completed at one site. For various reasons, the site selected for this test has proven to be a bad choice. However, the two instruments have tracked one another since installation within 0.8 microradians prior to the rainy season. As an additional check on the system, surveyed level lines have been established across or near the instrument sites and will be resurveyed periodically. Preliminary results of the first resurvey at the Bear Valley site show close agreement in direction and magnitude with recorded instrumental tilt (KINOSHITA [9]).

Site selection was found to be a crucial factor in the shallow borehole installation. Figure 1 shows the sites selected and the tiltmeters that have been installed. The instruments are located between 1 and 4 km from the fault with a spacing of about 6 km on alternate sides along the fault. The Nutting and Libby instruments, the two instruments at the Sage Range (North and South), and an instrument near Tres Pinos which has given records of varying quality due to installation and recording problems, were installed in June 1973. The Aromas and San Juan Bautista sites have been operational since December 1973, and the Stone Canyon and Bear Valley instruments were installed in January 1974. Other instruments have been installed more recently.

Sites selected on the basis of radial symmetry, homogeneity of material, and remoteness from topographical changes show no apparent diurnal temperature effects or ground loading effects due to rainfull at a sensitivity of 10^{-7} radians. It was not always possible to satisfy these criteria. The primary tilt effects observed at installations near steep topography were thermoelastic in origin. Meteorological effects also were more obvious at these sites.

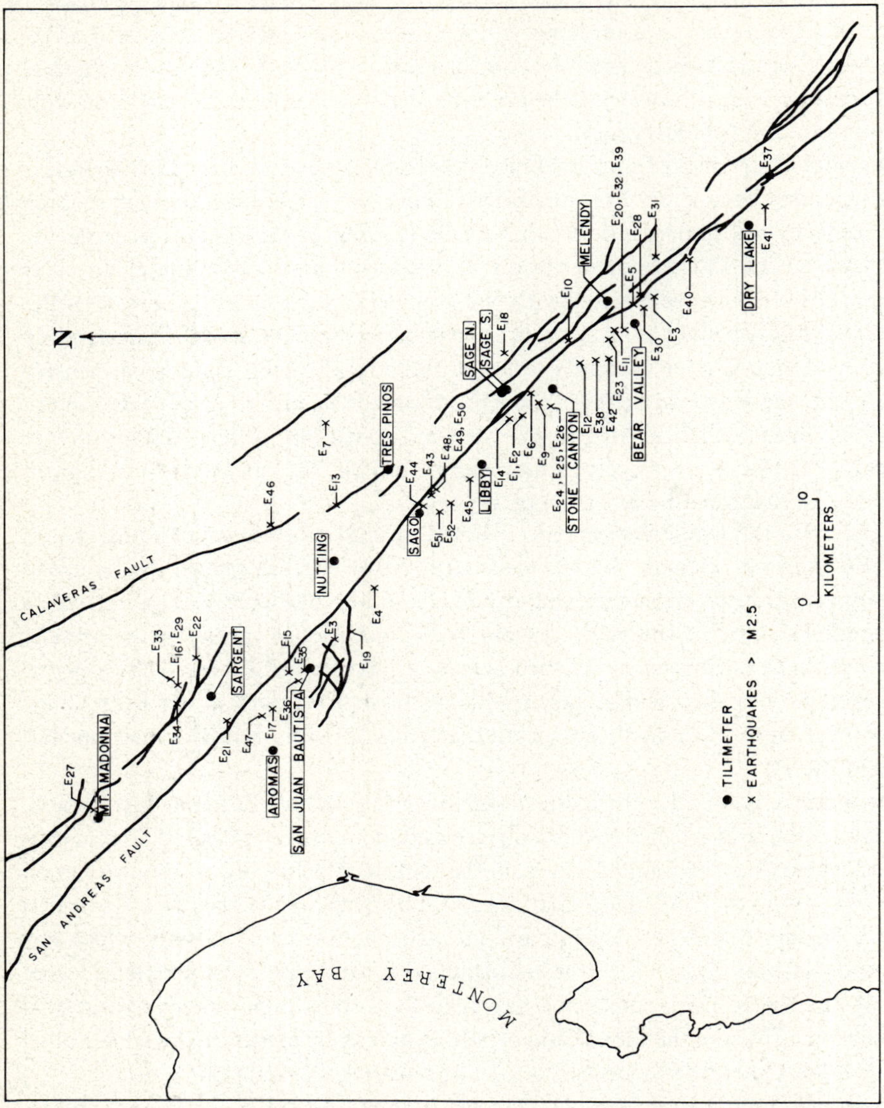

Figure 1
Tiltmeter sites (circles) along an 85 km section of the San Andreas fault from Dry Lake (36°29′N, 121°05′W) to Mt. Madonna (37°01′N, 121°42′W), and earthquakes (×'s) of $M > 2.5$, listed in Table 1, which have occurred near the instrument sites

3. The earthquake source region

Various models of the earthquake source region imply particular, though sometimes highly complex, spatial distributions of surface tilting prior to and following earthquakes. The possibility of large-scale dilatant behavior in the crust associated with earthquakes, which has been suggested by SCHOLZ *et al.* [19] to explain most precursive phenomena, implies, in its simplest form, symmetric uplift. The detection of this uplift should be possible with a sensitive tiltmeter array. In contrast, a slip model of the earthquake source region implies an asymmetric tilt field. In particular, this is the case for strike-slip and thrust faulting. Using a simple dislocation model in an elastic homogeneous half-space to represent fault slip, Press [14] calculated the tilt field at distances large compared to the earthquake source dimension. More recently, ROSENMAN and SINGH [16] have calculated the surface tilt fields for several elastic and viscoelastic models. These fields are highly dependent on viscosity and other model parameters. In terms of such slip models, tiltmeters operating at tidal sensitivity should resolve tilts as distant as 8.5 source dimensions from the epicentral region or about 30 km for a magnitude 4.0 earthquake. Source dimension (L) as a function of magnitude (M_L) for San Andreas earthquakes has been plotted by THATCHER and HANKS [22] and appears to best fit a relation (WYSS and BRUNE [25]) which can be approximated by

$$M_L = 1.9 \log L + 2.9. \tag{1}$$

With the instrument separation selected, therefore, some indication of regional deformation around the fault is expected with the tiltmeter array. Far-field and perhaps also near-field tilt effects from earthquakes with magnitudes greater than 4 should be observed on at least one instrument.

A crude but convenient way of ranking the effects of various earthquakes was to consider a weighting function, W, dependent on source dimension (L), depth (D), and distance (Δ):

$$W = K\frac{LD}{\Delta} \tag{2}$$

where K is a constant. This function probably will overestimate the effects of distant earthquakes due to the assumption of linear fall-off with distance. For central California earthquakes the observed range of depths is limited to between 2 and 10 km.

After carefully examining the tilt records at the times of small to moderate earthquakes in central California, very few cases of anomalous tilt behavior were observed for earthquakes that occurred at distances greater than 10 L. Preliminary data analysis has been concentrated, therefore, on earthquakes with less than a 10L distance to an instrument.

4. Summary of tilt observations

The general character of the raw data tends to show a long-term systematic tilt which can be as much as 2×10^{-6} radians per month, depending on the site, on which

Table 1

Occurrence times, locations, and magnitudes (>2.5) of the local earthquakes E_1 to E_{52} that have occurred since June 1973 in central California near the tiltmeter installations

No.	Date	GMT	Lat. N.	Long. W	Depth	Mag.
E_1	July 5	1147	36 39.2	121 18.2	4.8	2.7
E_2	July 5	1702	36 39.2	121 18.1	5.5	3.0
E_3	July 9	2100	36 48.7	121 32.9	6.2	3.3
E_4	August 7	0417	36 46.8	121 29.3	5.0	3.0
E_5	September 17	0531	36 33.6	121 11.4	4.8	2.8
E_6	October 4	0537	36 38.8	121 17.6	6.7	2.9
E_7	October 6	0918	36 49.0	121 18.4	8.3	2.7
E_8	October 12	1919	36 32.5	121 10.5	7.6	2.8
E_9	October 14	1657	36 38.6	121 17.8	5.2	2.5
E_{10}	October 27	2008	36 36.9	121 13.6	2.43	2.9
E_{11}	November 15	0948	36 34.7	121 12.9	6.2	2.8
E_{12}	November 24	0108	36 36.6	121 15.0	5.8	2.8
E_{13}	December 13	0217	36 48.6	121 24.0	7.7	2.7
E_{14}	December 14	1204	36 39.9	121 18.5	9.0	2.6
E_{15}	December 14	2356	36 50.6	121 35.3	6.0	2.9
E_{16}	January 10	1122	36 57.1	121 35.8	7.7	4.3
E_{17}	January 23	1556	36 52.2	121 37.3	6.0	3.0
E_{18}	January 27	1922	36 35.2	121 14.0	6.8	2.7
E_{19}	February 1	0327	36 47.6	121 32.6	8.5	3.6
E_{20}	February 7	1035	36 34.2	121 12.8	5.0	3.2
E_{21}	February 8	0004	36 54.5	121 37.7	1.5	2.7
E_{22}	February 8	0215	36 56.2	121 33.3	4.8	2.7
E_{23}	February 20	1055	36 35.3	121 13.7	12.8	2.8
E_{24}	March 8	1855	36 38.3	121 17.6	4.0	2.8
E_{25}	March 8	1856	36 38.6	121 17.5	4.1	2.9
E_{26}	March 8	1910	36 38.3	121 17.2	4.2	3.1
E_{27}	March 16	1624	37 1.1	121 43.6	9.7	3.0
E_{28}	March 17	0029	36 33.5	121 10.6	3.5	2.5
E_{29}	March 31	2306	36 57.2	121 35.8	5.1	3.4
E_{30}	April 7	1047	36 33.1	121 11.4	3.7	3.0
E_{31}	April 7	2207	36 32.6	121 8.1	9.1	2.8
E_{32}	April 9	1637	36 34.5	121 12.9	5.0	2.8
E_{33}	April 10	1713	36 56.9	121 35.7	4.7	2.5
E_{34}	April 17	1930	36 57.1	121 36.0	5.0	3.3
E_{35}	April 22	0821	36 50.5	121 34.4	5.0	2.7
E_{36}	April 22	0824	36 50.5	121 34.9	5.8	3.0
E_{37}	April 25	0440	36 27.0	121 3.3	8.9	3.0
E_{38}	April 25	1905	36 35.5	121 14.6	9.6	2.5
E_{39}	April 30	0545	36 34.4	121 12.8	5.3	2.8
E_{40}	May 4	0347	36 30.9	121 8.0	11.0	3.1
E_{41}	May 5	2013	36 27.2	121 5.1	3.3	2.6
E_{42}	June 10	0003	36 35.0	121 14.5	8.1	2.8
E_{43}	June 12	1850	36 44.0	121 23.1	6.8	2.9
E_{44}	June 12	1921	36 44.1	121 23.6	7.3	3.3
E_{45}	June 12	1922	36 42.0	121 22.5	7.1	3.3
E_{46}	June 14	0249	36 52.2	121 25.0	5.7	3.1
E_{47}	June 15	1410	36 52.7	121 37.7	6.2	2.6
E_{48}	June 15	1749	36 43.9	121 23.7	6.9	3.3

Table 1—*continued*

No.	Date	GMT	Lat. N.	Long. W	Depth	Mag.
E_{49}	June 15	1816	36 43.9	121 24.0	6.4	2.5
E_{50}	June 15	1818	36 43.6	121 23.9	7.4	2.7
E_{51}	June 15	1900	36 43.3	121 24.1	5.5	2.8
E_{52}	June 15	1929	36 43.5	121 23.9	6.6	2.9

are superimposed shorter-term perturbations with time spans of weeks to a month. Some very short-term tilt changes (minutes to hours) have also been observed.

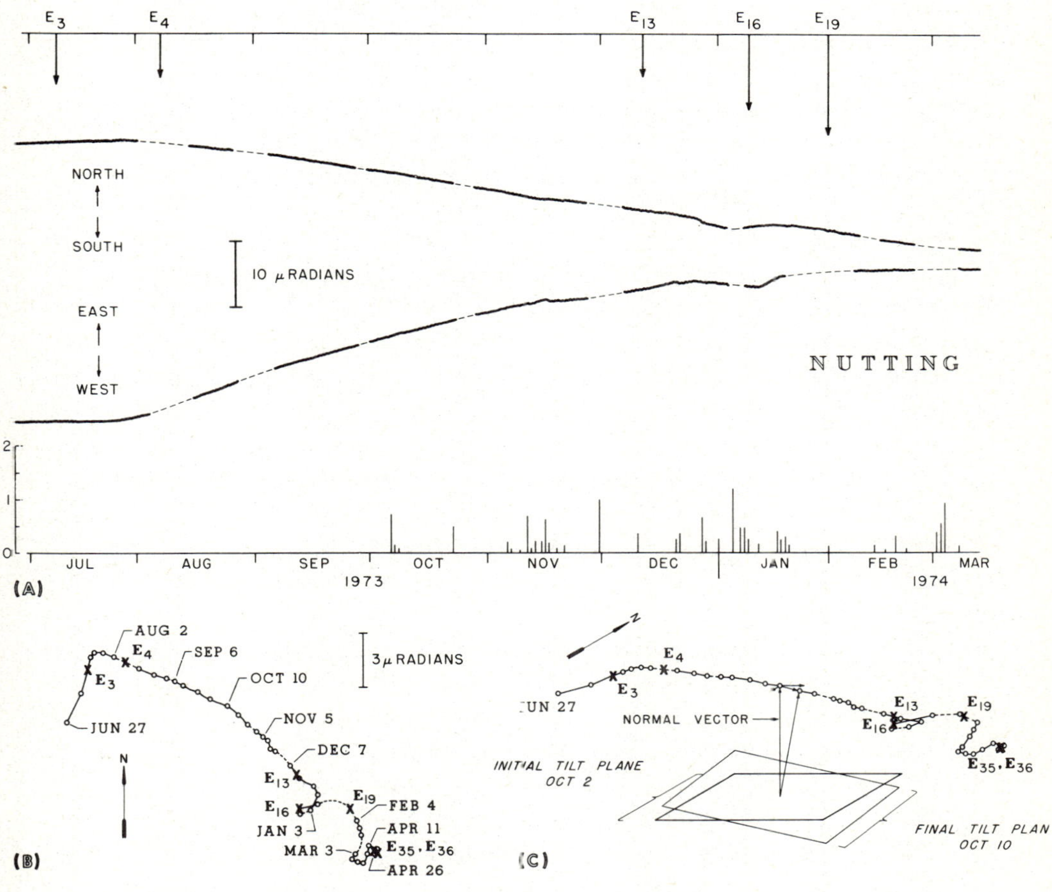

Figure 2

a) Time compressed raw record, with rezeros removed, from the Nutting Ranch site. The occurrence times of earthquakes are noted by arrows having length proportional to W. b) Cumulative plot of weekly average tilt-change vectors for the Nutting site. c) The tilt-change vector related to the tilt-plane

Many earthquakes having magnitudes M_L in the range $2.5 < M_L < 4.3$ have occurred in the area of installation since the array has been in operation. The intermediate-term tilt perturbations of up to a month appear to be associated with these earthquakes. In this paper the emphasis will be on these effects, which are best illustrated by filtering the data to remove frequencies of greater than one cycle/day. This averaging removes the diurnal and semi-diurnal solid-earth tidal perturbations and, in some cases, a diurnal thermoelastic effect. It is important to point out that no indication has yet been found of any accelerating tilt immediately before (hours to seconds) any of the 52 earthquakes that have occurred within 25 km of an instrument since the array has been operating.

A typical example of the raw data from the Nutting site is illustrated in Fig. 2a. Here the north–south and east–west components are shown with electrical offsets removed. Tilt vectors generated from these data were monitored with time at each instrument. The change in tilt behavior from week to week can be displayed by plotting a weekly average tilt-change vector, as shown in Fig. 2b, where each vector represents the amount and direction of tilting during a given week. The physical meaning of such a plot is illustrated in Fig. 2c. The amplitude and azimuth of running weekly average tilt-change vectors are plotted in Fig. 3 for the six instruments that have been longest in operation.

The earthquakes that have occurred within a 10-source dimension distance from an instrument are listed in Table 1 and are plotted, together with the site locations, in Fig. 1. Their occurrence is noted in Fig. 3 by arrows. The length of the arrows representing these earthquakes is proportional to the weighting function W.

Temperature and rainfall records have also been included in Fig. 3. It was not expected that temperature changes could explain the observations. However, since the instruments do have a thermal response, the possibility of some thermal contamination is real. At the installation depth, the diurnal temperature range is less than 0.1°C and the annual range is less than a few degrees. No obvious temperature effect has been found.

Careful examination shows some correlation between tilt and rainfall, particularly at sites lacking symmetric topography. However, some of the dramatic changes in the tilt vector appear to be independent of rainfall and in each case there are some periods of rainfall not accompanied by significant changes in tilt. There does appear to be some relation between earthquakes and surface tilt. Curiously, it is the direction of the tilt-change vector that appears most unambiguously related to subsequent earthquake occurrence. This can be emphasized, as in Fig. 4, by differentiating the azimuth plot. Weighted earthquakes and their occurrence times, together with rainfall and temperature records, are also included. This plot is useful for identification of general anomalous periods and coherency between independent instruments, as between Nutting and Libby during January 1974. The peaks have, of course, little physical significance.

The amplitude information included in Fig. 3 is complicated. Details of the tilt amplitude behavior and the relation this has to the earthquake source mechanism are currently being investigated.

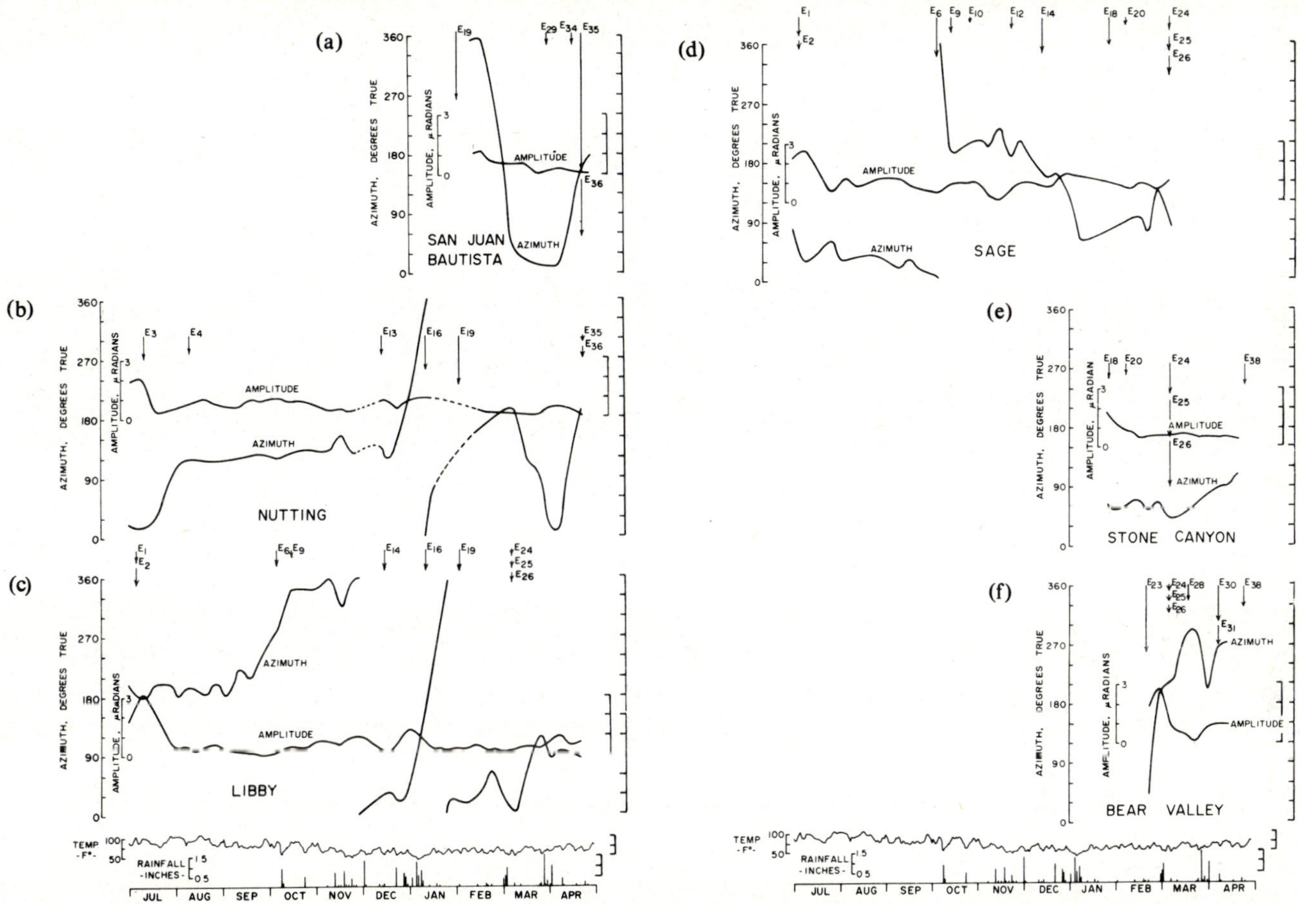

Figure 3a–f
Azimuth and amplitude of the weekly average tilt-change vectors for the six tiltmeter sites longest in operation. Temperature and rainfall are included along the abscissa

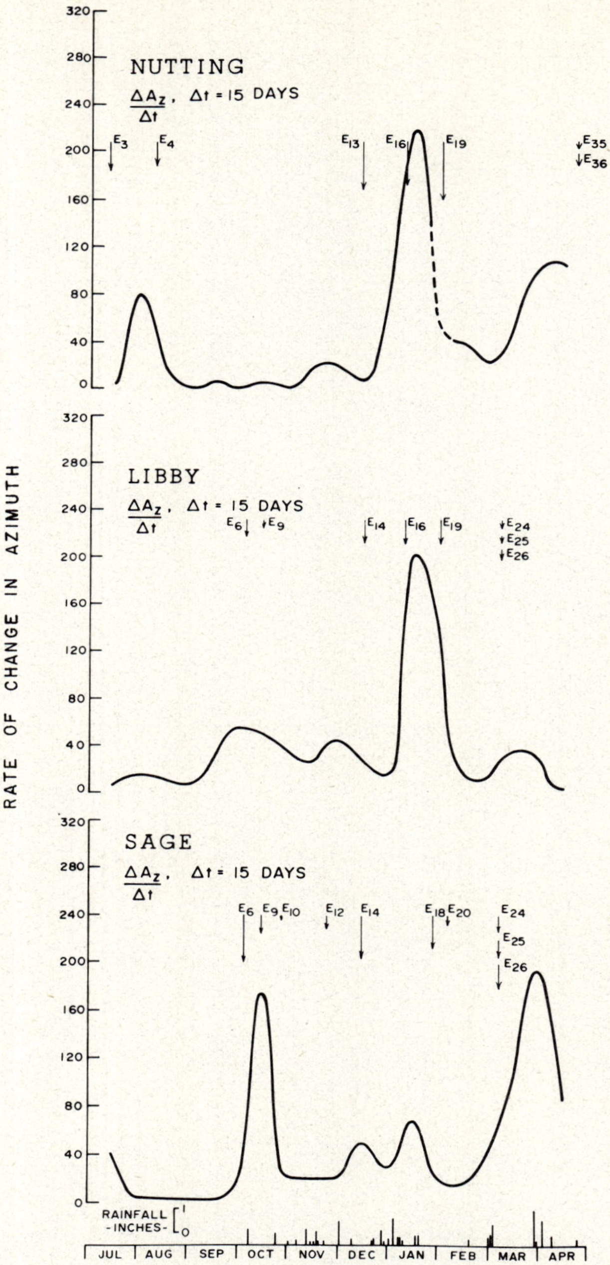

Figure 4
Differentiated average tilt-change vector azimuth records for the Nutting, Libby, and Sage sites

In Fig. 3 a significant slow rotation of the resultant weekly tilt-change vector (azimuth) at the Nutting site can be seen, beginning on about 17 December. On 10 January, a magnitude 4.3 earthquake (E_{16}) occurred 17 km, or roughly four source dimensions, north-west of the Nutting Ranch site, at a depth of 7.7 km, some 24 days after the onset of anomalous tilt behavior. This behavior is also apparent in Fig. 2. On 1 February, earthquake E_{19} ($M_L = 3.6$) occurred 7.4 km west of the Nutting site, or roughly 3.7 source dimensions distant. These earthquakes appear also to have affected the Libby Ranch site, at distances of nearly 8 and 10 source dimensions respectively. Other incidences of this type of precursory behavior can be seen in Fig. 3, but they are not as large as the above events and precursor times are correspondingly shorter. The range of magnitudes of the earthquakes that have occurred so far is too small to test fully the idea of WHITCOMB *et al.* [23] that the relation between magnitude (M) and presursor time (t) takes the form

$$\log t = 0.8M - 1.9 \qquad (3)$$

although the data do not violate the equation.

An important aspect of the data is the long term or secular tilting which takes place at each site and which can amount to several microradians per month, depending on the site and, in some cases, time. This secular trend has been consistent at a site such as Nutting, while at other locations, such as the Sage site, the trend has changed direction markedly. In general, the secular trend increases with seismic activity in the surrounding region.

Preliminary results indicate that if the secular trends are removed from the data, then the resultant tilt vectors prior to and following earthquakes do not systematically point away from the earthquake as would be expected for the simplest dilatancy model. Rather, the observed vector amplitude and direction are dependent on where the earthquake occurs with respect to the instrument. This position dependence would be expected for slip models of the earthquake source. However, the amplitudes are larger by almost an order of magnitude than those expected from a dislocation model of the earthquake source using seismically determined source parameters.

Because of its amplitude it is unlikely that the secular trend is of instrumental origin. Since all installations are identical, and since all sites are picked for apparent stability, it is unlikely also that site effects are solely responsible. It is possible that there is some contribution from a seasonal effect. However, at the Nutting site in particular, the trend has been the same through both summer (no rain) and winter (20 inches of rain). The most probable cause of the secular trend is a long-term tectonic effect on the fault. Several new installations set along two lines perpendicular to the fault should indicate whether this is the case.

For sites where seismicity is high it is difficult to isolate the effects of individual earthquakes. This is particularly evident for the Bear Valley record in Fig. 3 where seven earthquakes, ranging from $M_L = 2.5$ to $M_L = 3.1$, occur in only two months, with epicenters quite near the instrument. New instrument sites have been located further west of the fault at Bear Valley in order to find the variation of the tilt fields with increasing distance from the fault.

5. Conclusions

A tight array of tiltmeters has been used to minitor active crustal deformation. The results indicate that regional tilting does occur before earthquakes near these instruments and that a real possibility now exists for answering many questions concerning fault tectonics and the earthquake source mechanism. The prerequisites for successful operation appear, in order of importance, to be

a) site selection,
b) mechanically and thermally stable instruments capable of at least 10^{-8} radians sensitivity,
c) stable installation,
d) a sufficiently tight array in order that some coherency is obtained between instruments,
e) installation at a distance from the fault such that minor localized tectonic behavior does not confuse the observations.

Preliminary results appear to indicate that broad scale symmetric uplift associated with earthquakes does not occur on the San Andreas fault (JOHNSTON and STUART [7], STUART and JOHNSTON [21]). Rapid non-linear tilt just prior to earthquakes, indicating the onset of catastrophic failure, has not been observed. This may be due to attenuation with distance of non-linear strain effects in the immediate source region. Tilt offsets ($>10^{-8}$ radians) associated with earthquakes occur very rarely.

Acknowledgements

REX ALLEN made many useful technical suggestions and contributions.

REFERENCES

[1] R. V. ALLEN, M. D. WOOD and C. E. MORTENSEN, *Some instruments and techniques for measurements of tidal tilt*, Phil. Trans. Roy Soc. Lond. A *274* (1973), 219–222.

[2] P. CALOI, *About some phenomena preceding and following the seismic movements in the zone characterized by high seismicity*, International Series of Monographs on Earth Sciences – Contributions in Geophysics *I* (1958), 44–56.

[3] J. H. DIETERICH, *Time-dependent friction as a possible mechanism for aftershocks*, J. Geophys. Res. *77* (1972), 3771–3781.

[4] D. T. GRIGGS and D. W. BAKER, *The origin of deep-focus earthquakes*, in H. MARK and S. FERNBECH, *Properties of Matter Under Unusual Conditions* (Wiley-Interscience, New York, 1968), p. 23.

[5] T. HAGAWARI and T. RIKITAKE, *Japanese program on earthquake prediction*, Science *157* (1967), 761–768.

[6] M. J. S. JOHNSTON, in preparation (1974).

[7] M. J. S. JOHNSTON and W. D. STUART, *The form of pre- and post-earthquake crustal deformation for moderate earthquake on the San Andreas fault* (abstract), EOS, *56* (1974), 1196.
[8] G. C. P. KING, *The siting of strainmeters for teleseismic and tidal studies*, Roy. Soc. of New Zealand Bull. *9* (1971), 239–247.
[9] W. T. KINOSHITA, personal communication (1974).
[10] C. W. KOHLENBERGER, G. L. COOPER and W. T. SCHMARS, *Dynamic properties of a new biaxial tiltmeter* (abstract), Earthq. Notes *44* (1973), 11.
[11] F. OMORI, *Tilting of the ground during a storm*, Bull. Imp. Earthq. Invest. Com. *1* (1907), 167–171.
[12] E. OROWAN, *Mechanism of seismic faulting*, in D. GRIGGS and J. HANDIN, *Rock deformation*, Geol. Soc. Am. Memoir *79* (1960), 323.
[13] A. Y. OSTROVSKII, *On variations in tilts of the earth's surface before strong near earthquakes*, Akad. Nauk SSSR Inst. Fiziki Zemli (1970), 58–62.
[14] F. PRESS, *Displacements, strains, and tilts at teleseismic distances*, J. Geophys. Res. *70* (1965), 2395–2412.
[15] H. F. REID, *The elastic rebound theory of earthquakes*, Bull. Dept. Geology, Univ. California *6* (1911), 413.
[16] M. ROSENMAN and S. J. SINGH, *Quasi-static strains and tilts due to faulting in a viscoelastic half-space*, Bull. Seism. Soc. Am. *63* (1973), 1737–1742.
[17] K. SASSA and E. NISHIMURA, *On phenomena forerunning earthquakes*, Bull. of Disaster Prevention Res. Instit., Kyoto Univ. *13* (1956), 1.
[18] C. H. SCHOLZ, *Crustal movement in tectonic areas*, Tectonophysics *14* (3/4) (1972), 201.
[19] C. H. SCHOLZ, L. R. SYKES and Y. P. AGGARWAL, *Earthquake prediction: a physical basis*, Science *181* (1973), 803–810.
[20] F. D. STACEY, *Physics of the Earth* (John Wiley and Sons, Inc. 1969), p. 217.
[21] W. D. STUART and M. J. S. JOHNSTON, *Tectonic implications of anomalous tilt before Central California earthquakes* (abstract) EOS, *56* (1974), 1196.
[22] W. THATCHER and T. C. HANKS, *Source parameters of Southern California earthquakes*, J. Geophys. Res. *78* (1973), 8547–8576.
[23] J. H. WHITCOMB, J. D. GARMANY and D. L. ANDERSON, *Earthquake prediction: variation of seismic velocities before the San Fernando earthquake*, Science *180* (1973), 632–635.
[24] M. D. WOOD and R. V. ALLEN, *Anomalous microtilt preceding a local earthquake*, Bull. Seism. Soc. Am. *61* (1971), 1801–1809.
[25] M. WYSS and J. N. BRUNE, *Seismic moment, stress, and source dimension for earthquakes in the California-Nevada region*, J. Geophys. Res. *73* (1968), 4681–4694.

(Received 2nd December 1974)

Pageoph,
Vol. 113 (1975) Birkhäuser Verlag, Basel

Possible Premonitory Strain Changes Associated with an Earthquake Swarm Near Mina, Nevada

By KEITH F. PRIESTLEY[1])

Summary – Rapid strain changes at a station near Mina, western Nevada, occurred during a time of increased local seismic activity. The recordings indicate a period of crustal dilation followed by a contraction and a swarm of earthquakes in May, 1972, and the process may have been repeated during the next month. This pattern is consistent with precursory changes that have been proposed in connection with pre-earthquake dilatancy.

1. *Introduction*

A three-component quartz-tube strainmeter installation was constructed in a mine tunnel in the Garfield Hills near Mina during the fall and winter of 1970. This station operated from January to June, 1971, when it became inoperative due to instrument problems. It was repaired in February, 1972, and has operated almost continuously to the time of this writing. The site, instrumentation and data have been discussed in some detail by PRIESTLEY (1974a, b); station data are listed in Table 1.

Table 1
Characteristics of the strainmeters at Mina, Nevada

Latitude	Longitude	Elevation	Strainmeter Length	Orientation†
38.44°N	118.15°W	1670 m	24 m	23.0°
			24 m	74.0°
			12 m	137.0°

† Orientation is in degrees east of north.

A comparison of long-term strain observations for stations at Mina and Round Mountain, Nevada, is given by PRIESTLEY (1974b). The strain pattern at Round Mountain for 1970–1973 was in general agreement with focal mechanisms for nearby earthquakes found by SMITH *et al.* (1971), and is consistent with strain accumulation in that area. At Mina, the strain pattern is much more variable, and is interpreted as alternating strain buildup and release by inelastic processes (microearthquakes and creep). Noting

[1]) Seismological Laboratory, University of Nevada, Reno, Nevada 89507. Present address: Dept. of Geodesy and Geophysics, Univ. of Cambridge, Madingley Rise, Madingley Road, Cambridge, England.

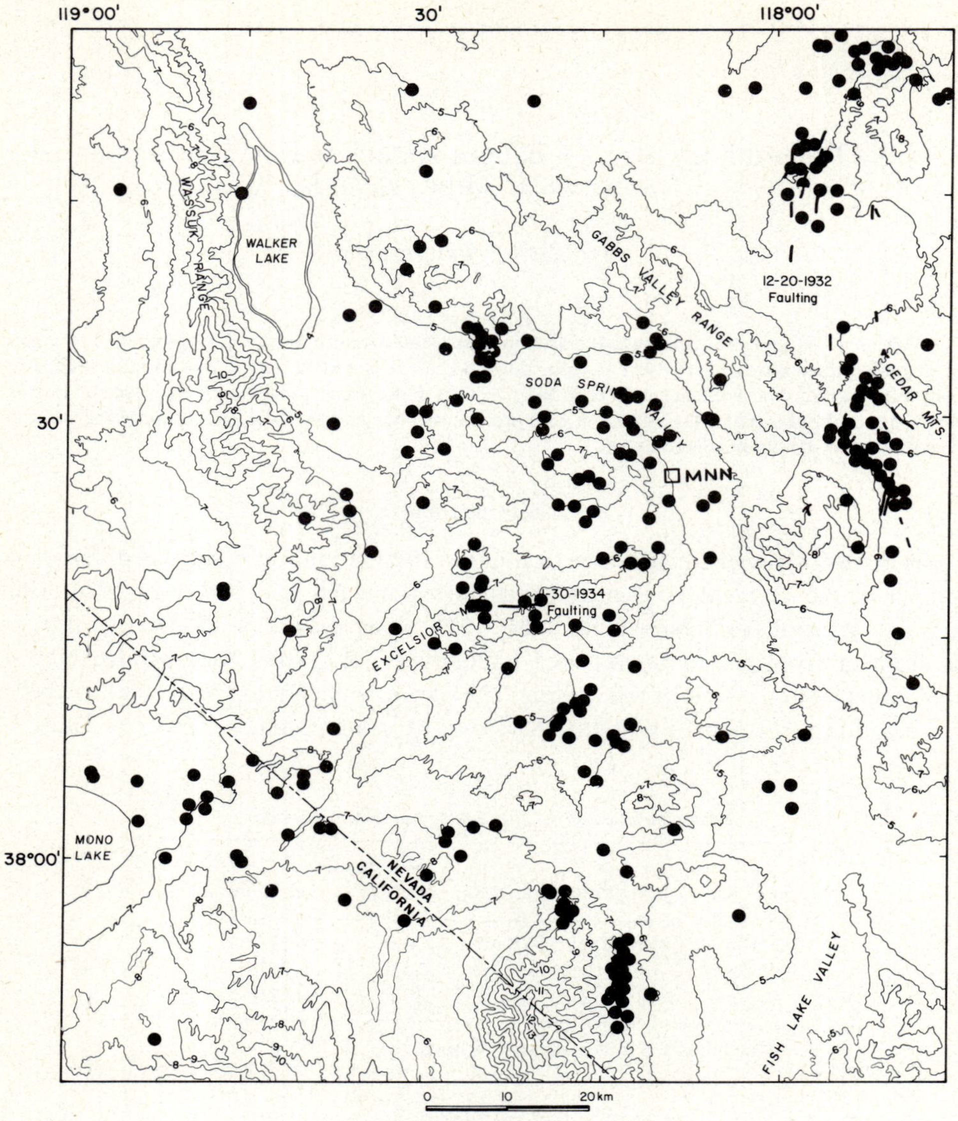

Figure 1
Map of epicenters in the Excelsior Mountains area, western Nevada, for 1970–1972. MNN: Mina strain station

that Round Mountain is in an aseismic part of the Nevada region, while Mina is in a very active area (Fig. 1), Priestley compared the secular strain rates for the two stations with SCHOLZ' (1970) four stages of rock deformation. He suggested that the higher strain rates and low seismic activity at Round Mountain might correlate with Scholz' stage II, which is characterized by linear elastic strain; and that the variable strain rates and high level of seismicity at Mina might indicate a later stage (III or IV) of deformation. RYALL and PRIESTLEY (1975) compared the character of strain observations at the Mina station with characteristics of earthquakes and faulting in the area, and suggested that a relatively low level of tectonic stress there appeared to be generating strain release by a continuing series of moderate earthquakes and fault creep.

2. *Analysis and results*

During the period of observation two episodes of rapid strain change occurred at the Mina station (Fig. 2). The first of these was during May and June, 1972, and the

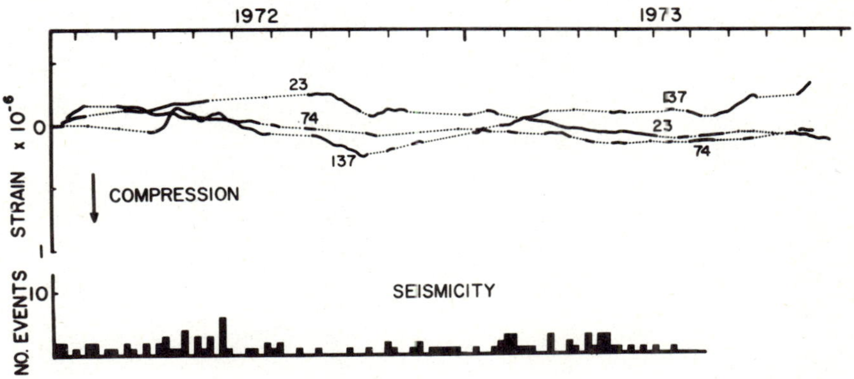

Figure 2
Secular strain observed at Mina from February, 1972, to October, 1973. Lower part of the figure shows number of earthquakes located using Nevada network data, within 50 km of the Mina station. Dotted lines: periods when secular strain was obtained by bi-weekly system calibration rather than continuous recording

second in September and October of the same year. The first strain change may have been related to an earthquake swarm, and will be discussed below. The second occurred when the level of seismic activity was low and at a time when the weather began to turn cool. However, this strain change appears not to have been related to air temperature changes in the mine tunnel, for two reasons: First no changes in air temperature were recorded on a thermograph capable of detecting changes of $\pm 0.02°C$. Second, if the air temperature decreased by an amount too small to measure, this decrease would have had the same effect on all the instruments, leading to an apparent ground extension on all components. This was not the case: both the 137° and the 23° components showed compression during this period, while the 74° component showed no change.

Similarly, this rapid change appears not to have been the result of seasonal temperature changes on the mountain block in which the instruments are located. BENIOFF (1959) showed that diurnal heating of the mountain block containing the Dalton Dam strainmeter installation had significant effects on the observed strain. Although the expansion or contraction of the surface layer due to diurnal or seasonal temperature changes is a small effect, propagating to a depth of only a few inches or a foot, in areas of extreme topography such effects can lead to strain changes at considerable depth. The effect on a buried strainmeter will be greatest if the instrument is oriented in the direction of the largest topographic gradient. For the Mina station, the instrument that should be most affected by surface temperature changes is the 74° component, but this instrument showed the least strain change during the period in question. Further, if the strain changes during this period were due to seasonal cooling of the surface layer, we should expect to see the reverse pattern during the following spring and summer, and this reversal did not occur. We therefore conclude that the rapid strain change was tectonic strain in the earth.

As mentioned above, the strain change during May and June, 1972, occurred during a time of increased seismic activity. The strain variation and rate of earthquake occurrence for this period are shown in Fig. 3. During the early part of May, each of

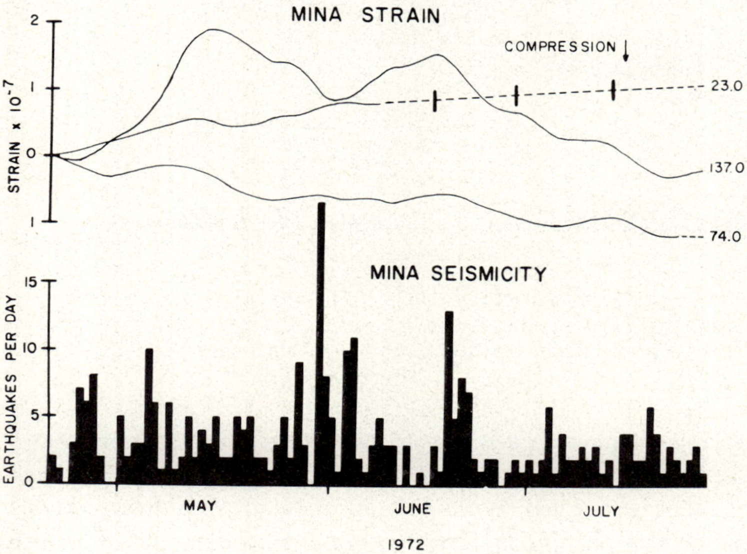

Figure 3
Secular strain at Mina from May to July, 1972, together with rate of microearthquake activity recorded at the Mina seismographic station

the strain components showed an extensional strain, followed by a period of compressional strain. This took place principally on the NW–SE component, but also to a lesser degree on the other two components. In other words, the crustal sample under observation initially underwent a dilatation, followed by a contraction. During the period of expansion, 2×10^{-7} extensional strain accumulated in the NW–SE direction.

The bargraph in Fig. 3 shows the rate of occurrence of local earthquakes recorded at the Mina station. Most of these events were too small to give accurate locations using data of the Nevada network; however, many of them were identical in character to several earthquakes of magnitude 1–3 which were located in the Garfield Hills, 12–15 km north-west of the strainmeter site.

The bargraph shows an initial decrease in seismic activity during the period when the strain recordings indicate dilatation was taking place. The earthquake rate then increases rapidly and a large number of events occurs at the time when the maximum contraction occurs.

This may have been followed by a similar occurrence during the first part of June. The rate of seismic activity followed the same pattern as in May, with an initial decrease followed by a swarm of earthquakes. Although one component (23°) did not record during this period, the behavior of the other two is essentially identical to the strain change that took place in May, except that the June change was followed by a contraction that continued until October (Fig. 2).

There have been no records kept of the precipitation at the Mina strainmeter installation. The annual precipitation in this area is 3.45 inches with monthly averages varying from 0.15 to 0.48 inches. The strain variations described above are unlikely due to precipitation because of the depth of burial of the instrument, and because of the size and duration of the strain variation.

3. Discussion

Rapid strain changes at the Mina station in May and June, 1972, occurred during a time of increased seismic activity. The Mina recordings indicate a period of crustal dilatation followed by a contraction and a swarm of earthquakes in May, and this process may have been repeated in June. During the interval of strain observations shown in Fig. 2, these were the only such occurrences. Scholz *et al.* (1973) have proposed this type of deformation as one of the precursory effects that would be associated with pre-earthquake dilatancy. According to them, the amount of dilatancy required to produce the 10–20% V_P/V_S anomalies observed at Garm in the USSR (Nersesov *et al.*, 1971), the New York Adirondacks (Aggarwal *et al.*, 1973), and the San Fernando region of California (Whitcomb *et al.*, 1973) would lead to crustal strains of the order of 10^{-6} to 10^{-5}. They also compiled data on the duration of precursor phenomena as a function of the magnitude of the resulting earthquake. The magnitude of the strain change at Mina was 2×10^{-7}. This is less than the strains proposed by Scholz *et al.*, but could be smaller due to the distance of the station, about 15 km, from the source area. The duration of the strain change, 20–25 days, is longer than would be expected for an earthquake of magnitude 3.0, which was the largest event in the May–June sequence. However, this was a swarm, with several events of comparable magnitude, and if we consider these events together the time duration of the strain change is brought into better agreement with the findings of Scholz *et al.* Thus, while the rapid strain changes in May and June might have been unrelated to the earthquake swarm, the observations are at least consistent with precursory changes proposed in connection with the dilatancy model. A more definitive conclusion will have to be deferred until this relationship, between strain changes and earthquake activity, is either confirmed or rejected by additional observations.

Acknowledgements

This research was partly supported by the Advanced Research Projects Agency and was monitored by the Air Force Office of Scientific Research under grant AFOSR-71-2041 and contract F44620-72-C-0069. It was also partly supported by the National Science Foundation under grant GA-38732.

References

AGGARWAL, Y. P., SYKES, L. R., ARMBRUSTER, J., and SBAR, M. L. (1973), *Premonitory changes in seismic velocities and prediction of earthquakes*, Nature *241*, 101–104.

BENIOFF, H. (1959), *Fused-quartz extensometer for secular, tidal, and seismic strains*, Bull. Geol. Soc. Am. *70*, 1019–1032.

NERSESOV, I. L., SEMENOV, A. N., and SIMBIREVA, I. G. (1971), *Prostransvenno-vre mennoe raspredelenie otnoshenii vremen probega poperechnykh i prodol'nykh voln v garmskom raione* (*Space-time distribution of the traveltime ratio of transverse and longitudinal waves in the Garm region*), Experimental'naya Seismologiya, Akad. Nauk SSSR, 334–345.

PRIESTLEY, K. F. (1974a), *Earth Strain Observations in the Western Great Basin*, Univ. Nevada Ph.D. Dissertation, 185 pp.

PRIESTLEY, K. F. (1974*b*), *Crustal strain measurements in Nevada*, Bull. Seis. Soc. Am. *64*, 1319–1328.

RYALL, A., and PRIESTLEY, K. F. (1975), *Seismicity, secular strain and earthquake potential in the Excelsior Mountains area, Nevada*, Bull. Geol. Soc. Am., in press.

SCHOLZ, C. H. (1970), *The role of microfracturing in rock deformation, Proc. Second Congress of the Int. Soc. for Rock Mechanics*, 323–327.

SCHOLZ, C. H., SYKES, L. R., and AGGARWAL, Y. P. (1973), *Earthquake prediction: a physical basis*, Science *181*, 803–810.

SMITH, B. E., HAMILTON, R. M., and JACKSON, W. H. (1971). *Seismicity of the Central Nevada Test Area, 26 September 1969–1 October 1970*, U.S. Geol. Survey report.

WHITCOMB, J. H., GARMANY, J. D., and ANDERSON, D. L. (1973), *Earthquake prediction: variation of seismic velocities before the San Fernando earthquake*, Science *180*, 632–635.

(Received 4th December 1974)

Pageoph,
Vol. 113 (1975) Birkhäuser Verlag, Basel

Crustal Structure and Temporal Velocity Change in Southern California[1])

By HIROO KANAMORI and DAVID HADLEY

Summary – About twenty blasts are used to determine crustal structure and to monitor temporal seismic velocity changes in southern California. The shot time is determined up to 10 msec by using a disposable pick-up placed directly on the explosive. About 17 permanent stations and 20 temporary stations are used for the recordings. With a fast paper speed (typically 1 cm/sec) and the WWVB radio signals superposed on the seismic trace, absolute timing accuracy of up to 10 msec is achieved. A representative structure thus determined consists of a 4 km thick 5.5 km/sec layer underlain successively by 23.4 km thick 6.3 km/sec layer, 5.0 km thick 6.8 km/sec layer and 7.8 km/sec half space. The details of the lower crust are somewhat uncertain. This structure can explain the travel time data, corrected for the station and source elevations and for the station delays, to ±0.15 sec. Small but systematic temporal velocity changes up to 3% have been found for some of the profiles. If the effect of the migration of the shot point is small enough, these changes are larger than experimental errors and represent real temporal change in the material property between the shot point and the stations.

1. *Introduction*

Besides its obvious importance in studying regional geology and tectonics, detailed knowledge of crustal structure is becoming increasingly important in relation to earthquake location, mechanism and prediction. Pioneer works on this subject in southern California include GUTENBERG (1944, 1951, 1952, 1955), RICHTER (1950), SHOR (1955), PRESS (1956, 1960), and ROLLER and HEALY (1963).

The structure determined by Press (1960) has long been used in the earthquake location program by the Seismological Laboratory, California Institute of Technology (HILEMAN *et al.*, 1973) (Press's model has an upper layer 26.2 km thick with *P* velocity of 6.11 km/sec; a second layer 25.6 km thick with *P* velocity of 7.66 km/sec and an underlying half-space with *P* velocity of 8.11 km/sec). The recent increase in the number of seismic stations as well as the improved timing accuracy warrant a more detailed study of the crust.

In addition to this static aspect of the problem, temporal changes in seismic wave velocity are now becoming an increasingly important subject of study in relation to earthquake prediction. Temporal variations in local seismic wave velocities precursory to earthquakes have been reported for many events, and have been explained in terms of rock dilatancy and subsequent fluid diffusion. Since the hypotheses of temporal fluctuations and dilatancy–diffusion are still a matter of controversy, it is very important

[1]) Contribution No. 2530, Division of Geological Sciences, California Institute of Technology, Pasadena, California.

to determine with a reliable method if seismic wave velocities do indeed change in such a tectonically active area as southern California. For this purpose, man-made explosions have several advantages over natural events as the former provide accurate location and origin times, impulsive onsets and, in many cases, repeatability of the measurements. These advantages remove practically all of the uncertainties associated with results obtained from natural events. The present paper makes extensive use of large explosions located at various mines and quarries within the Caltech seismic network in southern California. Some of these blasts have already been used by ALLEN and HELMBERGER (1973) and KANAMORI and CHUNG (1974) in attempts to detect temporal velocity changes.

2. Data

Figure 1 shows the locations of the explosions used in the present study. The coordinates and nature of the blasts are listed in Table 1. In most cases, the shot was timed with a disposable pick-up placed directly on the explosive. The signal was recorded on a Kinemetrics PS-1 portable seismograph unit together with the WWVB radio signal. This method permits a determination of the shot time to within ±10 msec. The shot point location was determined to 1/100 minute from US Geological Survey

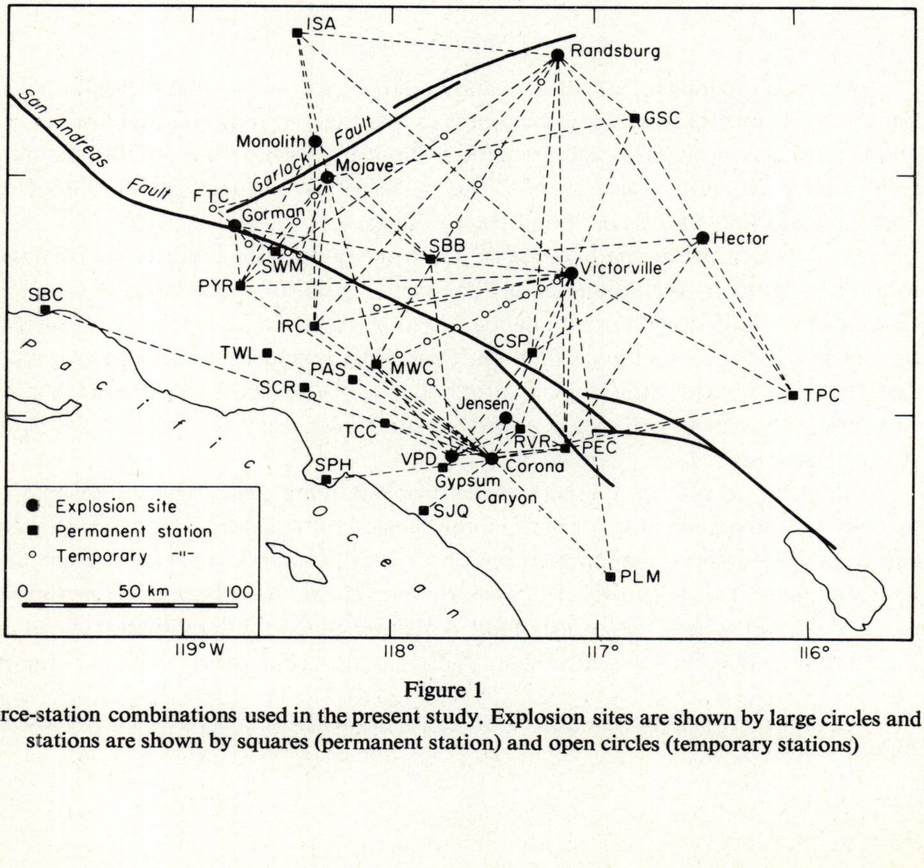

Figure 1
Source-station combinations used in the present study. Explosion sites are shown by large circles and stations are shown by squares (permanent station) and open circles (temporary stations)

Table 1
Explosion data

Location	Approximate Coordinates	Elevation (m)	Note
Victorville	34.63°N, 117.11°W	1280	Cement Quarry
Corona	33.84°N, 117.51°W	300	Dacite Porphyry
Jensen	34.03°N, 117.43°W	360	Cement Quarry
Gypsum Canyon	33.86°N, 117.71°W	150	Sand and Gravel
Randsburg	38.52°N, 117.17°W	750	Navy Testing
Gorman	34.83°N, 118.76°W	1290	Cement Quarry
Mojave	35.03°N, 118.32°W	1190	Cement Quarry
Monolith	35.15°N, 118.39°W	1350	Cement Quarry
Hector	34.75°N, 116.42°W	600	Bentonite Quarry

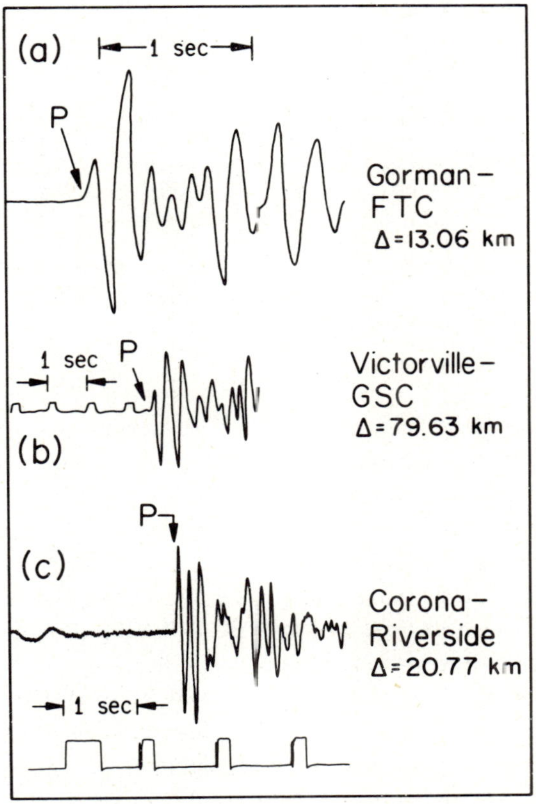

Figure 2
Examples of the records. (A) Kinemetrics (PS-1) portable unit (paper speed: 1 cm/sec) used at temporary stations. (B) Telemetered record on Develocorder from permanent stations (film speed: 1 cm/sec on a viewer). (C) Strip chart recording at temporary stations (paper speed: 2 cm/sec)

Table 2a
Victorville blast

Station	Δ (km)	Azimuth (deg)	$T-\Delta/6$ (sec)	Corrected $T-\Delta/6$ (sec)	Grade
Navajo	8.24	237.8	0.21	0.21	C
UICT	16.18	241.1	0.28	0.28	B
Trash	24.39	245.1	0.52	0.39†)	B
Baldy Mesa	34.82	242.6	0.68	0.47†)	C
EXP-O	44.55	244.8	0.74	0.29	A
Table Mt.	59.08	242.4	0.60	0.24	A
Crystal Lake	74.63	243.1	0.37	0.06	B
Charlton Flats	89.55	245.0	0.36	0.07	C
MWC	98.03	242.9	0.26	−0.05	C
Mt. Emma	90.62	258.8	0.34	0.08	B
CSP	43.12	211.6	0.46	0.20	A
SBB	65.64	275.9	0.54	0.33	A
RVR	74.64	199.0	0.44	0.28	B
GSC	79.63	20.5	0.62	0.39	B
PEC	81.97	183.2	0.54	0.35	B
VPD	108.43	213.7	0.33	0.18	C
TPC	113.92	120.5	0.66	0.46	C
IRC	121.29	257.7	0.19	0.00	C
PLM	143.44	170.7	0.25	−0.05	C
PYR	149.71	267.8	−0.04	−0.30	C
ISA	167.58	312.5	−0.49	−0.71	C

For station coordinates, see Hileman *et al.* (1973) for permanent stations (three-letter abbreviation) and Table 2j for temporary stations.
† Corrected for local structure.

quandrangles. These events were recorded by permanent seismic stations belonging to the southern California seismic network and temporary stations consisting of the Kinemetrics portable units and Caltech seismographic trailers (Fig. 1). Except on a few occasions, the signals recorded by the permanent stations were telemetered through telephone lines to Pasadena and recorded on 16 mm films (Develocorder) with the WWVB radio signal superimposed on the seismic trace. The portable Kinemetrics units are run with 1 cm/sec paper speed with the WWVB radio signal superimposed on the trace. In some cases a strip-chart recorder was used with 1 to 2 cm/sec paper speed. The trailer recordings are also timed with either WWVB or WWV radio signals. Figure 2 shows typical examples of the seismograms. With these recording systems, a timing accuracy of up to 10 msec is achieved when the signal onset is clear.

Figure 1 also shows the source-station combinations used in the present study and several profiles which were established specifically for the determination of crustal structure.

The travel time data, summarized in Table 2, are assigned grades A (± 10 msec) B (± 30 msec), and C (± 100 msec).

Table 2b
Corona blast

Station	Δ (km)	Azimuth (deg)	$T-\Delta/6$ (sec)	Corrected $T-\Delta/6$ (sec)	Grade
RVR	20.77	35.8	0.23	0.23	A
VPD	23.81	263.0	0.72	0.72	B
PEC	32.48	80.0	0.32	0.32	B
SJQ	39.85	232.0	0.77	0.77	B
Dalton Lake	45.85	322.3	0.47	0.39	A
CSP	52.53	15.2	0.40	0.24	B
MWC	66.27	309.9	0.44	0.23	B
PAS	70.25	299.2	0.25	0.19	B
SPH	77.42	262.4	0.37	0.30	B
PLM	80.71	132.0	0.38	0.19	B
UCLA-1	86.67	290.2	0.16	0.09	—
SCR	92.37	288.8	0.11	0.05	C
IRC	102.46	306.7	0.02	−0.07	B
TWL	111.53	296.1	0.41	0.35	C
TPC	137.89	77.4	−0.08	−0.17	B
PYR	139.48	305.6	0.25	0.10	C
GSC	174.3	21.5	−0.52	−0.70	B
IKP	185.7	135.0	−1.55	−1.72	C
SBC	214.15	288.7	−1.69	−1.74	C
ISA	218.74	336.3	−2.36	−2.52	C

Table 2c
Jensen blast

Station	Δ (km)	Azimuth (deg)	$T-\Delta/6$ (sec)	Corrected $T-\Delta/6$ (ec)	Grade
RVR	6.45	124.3	0.14	0.14	B
PEC	29.23	120.6	0.33	0.33	B
CSP	30.99	13.0	0.54	0.54	B
VPD	38.40	232.5	0.74	0.74	B

Table 2d
Gypsum Canyon

Station	Δ (km)	Azimuth (deg)	$T-\Delta/6$ (sec)	Corrected $T-\Delta/6$ (sec)	Grade
VPD	7.52	223.3	0.83	0.83	A
SJQ	30.01	205.5	1.18	1.18	B
TCC	31.84	297.1	1.82	1.82	B
RVR	33.71	65.1	0.57	0.57	C
PEC	50.55	86.4	0.60	0.52	A
MWC	51.42	320.8	0.91	0.71	C
CSP	57.94	33.7	0.82	0.67	B
SBB	92.07	353.3	0.74	0.64	B

Table 2e
Randsburg blast

Station	Δ (km)	Azimuth (deg)	$T-\Delta/6$ (sec)	Corrected $T-\Delta/6$ (sec)	Grade
Navy Test Range—1	13.90	237.8	0.47	0.47	B
Galileo Hill	63.29	237.6	0.58	0.40	B
Kramer Hill	71.11	213.1	0.22	0.06	A
Jackrabbit Hill	92.46	212.8	0.30	0.14	A
SBB	110.14	213.1	0.10	−0.06	B
Lovejoy Buttes	121.01	211.3	−0.04	−0.21	A
GSC	40.95	126.5	0.58	0.48	A
CLC	50.84	310.2	0.23	0.07	C
ISA	119.47	276.9	0.32	0.15	B
CSP	136.82	187.3	0.24	−0.03	B
SWM	156.67	235.8	0.97	0.77	C
MWC	165.49	209.8	0.22	−0.04	C
PEC	180.81	179.8	−0.64	−0.85	C

Table 2f
Mojave blast

Station	Δ (km)	Azimuth (deg)	$T-\Delta/6$ (sec)	Corrected $T-\Delta/6$ (sec)	Grade
Willow Springs	11.71	211.0	0.73	0.73	A
Rosamond	26.93	214.2	1.13	1.13	B
Lake Hughes	41.95	194.1	0.85	0.63	B
SWM	43.16	214.0	0.90	0.66	A
FTC	55.47	250.8	0.79	0.57	B
SBB	59.78	130.4	0.61	0.41	A
PYR	65.24	216.6	0.90	0.66	A
ISA	68.54	348.1	0.51	0.31	C
IRC	72.37	185.8	0.77	0.59	B
MWC	93.62	165.1	0.66	0.36	B
GSC	141.09	77.6	0.32	0.10	C

Table 2g
Gorman blast

Station	Δ (km)	Azimuth (deg)	$T-\Delta/6$ (sec)	Corrected $T-\Delta/6$ (sec)	Grade
FTC	12.88	288.9	0.78	0.78	A
West Liebre Mt.	13.21	153.2	0.82	0.82	B
SWM	20.89	128.6	0.88	0.88	A
Pine Canyon	27.81	123.1	0.82	0.82	A
PYR	29.76	176.8	1.01	1.01	A
Lake Hughes	35.20	120.5	0.92	0.92	A

Table 2h
Monolith blast

Station	Δ (km)	Azimuth (deg)	$T-\Delta/6$ (sec)	Corrected $T-\Delta/6$ (sec)	Grade
ISA	55.76	351.7	0.64	0.41	B
PYR	71.81	206.9	0.68	0.41	B
SBB	72.36	134.4	0.54	0.32	C
IRC	83.88	180.7	0.62	0.42	C

Table 2i
Hector blast

Station	Δ (km)	Azimuth (deg)	$T-\Delta/6$ (sec)	Corrected $T-\Delta/6$ (sec)	Grade
GSC	70.71	330.7	1.16	1.00	B
TPC	79.12	154.0	1.01	0.88	B
CSP	98.98	240.1	0.77	0.58	B
PEC	116.53	215.8	0.72	0.60	B
SBB	128.35	267.5	0.60	0.46	B

Table 2j
Station coordinates

Station	Latitude	Longitude	Height
Navajo	34°35.41′N	117°11.25′W	910
UICT	34°33.54′N	117°15.95′W	900
Trash	34°32.22′N	117°21.15′W	870
Baldy Mesa	34°29.11′N	117°26.88′W	1012
EXP-O	34°27.47′N	117°33.00′W	1130
Table Mt.	34°22.91′N	117°40.84′W	2280
Crystal Lake	34°19.39′N	117°50.07′W	1736
Charlton Flats	34°17.09′N	117°59.53′W	1556
Mt. Emma	34°28.07′N	118°04.74′W	1305
SBB	34°41.31′N	117°49.43′W	829
Dalton Lake	34°10.20′N	117°48.56′W	523
UCLA—1	34°06.52′N	118°23.27′W	347
Navy Test Range—1	35°27.29′N	117°17.79′W	867
Galileo Hill	35°12.87′N	117°45.23′W	945
Kramer Hill	34°59.03′N	117°35.52′W	774
Jackrabbit Hill	34°49.21′N	117°42.89′W	830
Lovejoy Buttes	34°35.27′N	117°51.14′W	879
Willow Springs	34°57.03′N	118°23.04′W	1180
Rosamond	34°50.40′N	118°29.00′W	872
Lake Hughes	34°40.43′N	118°25.76′W	1000
Lake Hughes—2	34°40.59′N	118°26.07′W	1024
West Liebre Mt.	34°43.77′N	118°41.70′W	1627
Pine Canyon	34°41.93′N	118°30.34′W	1146

3. *Upper crust*

All of the travel time data obtained for the source-station combinations shown in Fig. 1 are plotted, regardless of the azimuth, on the reduced travel time curve given by Fig. 3. Since the altitude of the shot points and the stations varies from 350 to 2280 m,

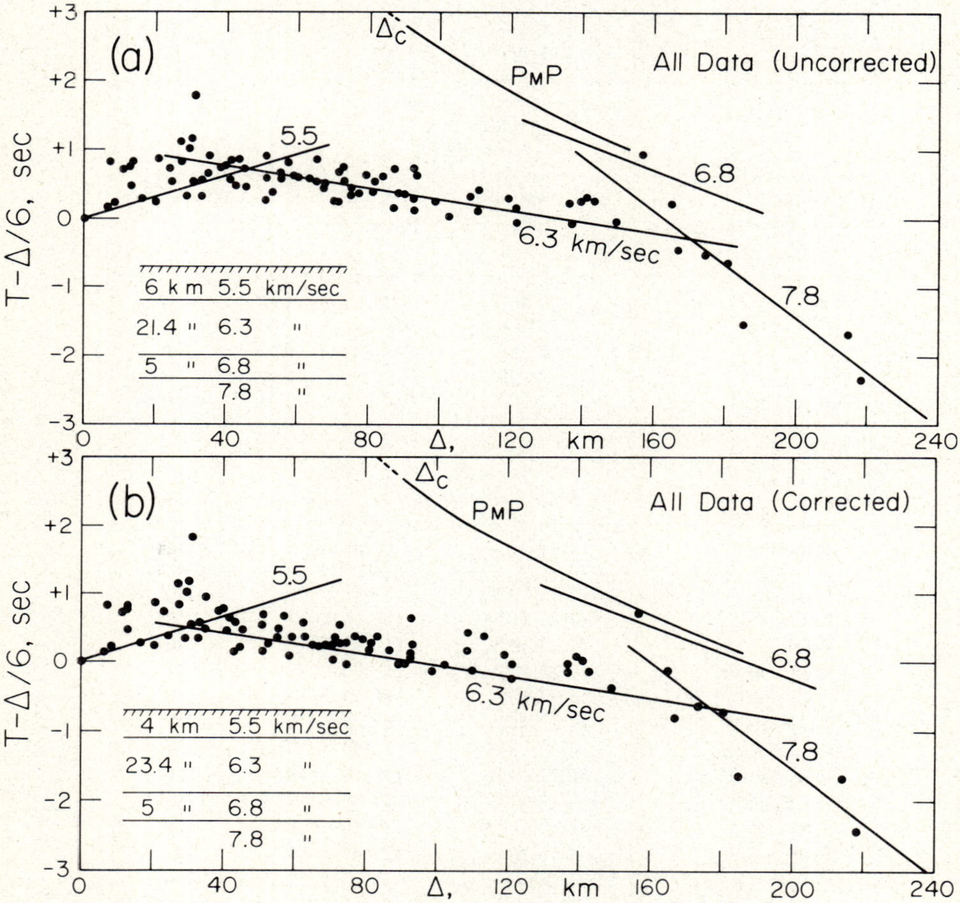

Figure 3

(a) Reduced travel times for all data except those for the Hector blast. Travel time curves are for the structure shown in the inset. P_MP branch shows the Moho reflection. (b) Reduced travel times corrected for the elevations at both source and stations. The structure is shown in the inset

and also the local geologic structure differs beneath individual shot points and stations, these data are not homogeneous. Nevertheless, in view of the fairly random coverage of the paths as shown in Fig. 1, this figure depicts the overall variation of the short-distance ($\Delta < 220$ km) travel times in southern California. The scatter is surprisingly small in view of the vast variability of geology in this area. With a few exceptions, the scatter is within ±0.3 sec of the average. The straight-line segments shown in Fig. 3 correspond to a structure determined for a more restricted profile which will be discussed in the following.

Victorville blast

The Victorville blasts have been used to establish a fairly complete profile extending from the quarry to the station MWC (Fig. 1). Figure 4a shows the data uncorrected for station elevation and local geology. The data along the profile (closed circles) can be fit very well by two segments having apparent velocities of 5.32 and 6.35 km/sec, the crossover being at 37.5 km. This indicates a structure as shown in the inset. The off-profile stations (open circle) are within ±0.3 sec of the above travel time curves indicating a lateral uniformity of the structure.

Next we will consider corrections for elevation and local structure at both the source and the station. These corrections are very difficult to make precisely. Utilizing the structure given in the inset of Fig. 4a, the station and the source height corrections can

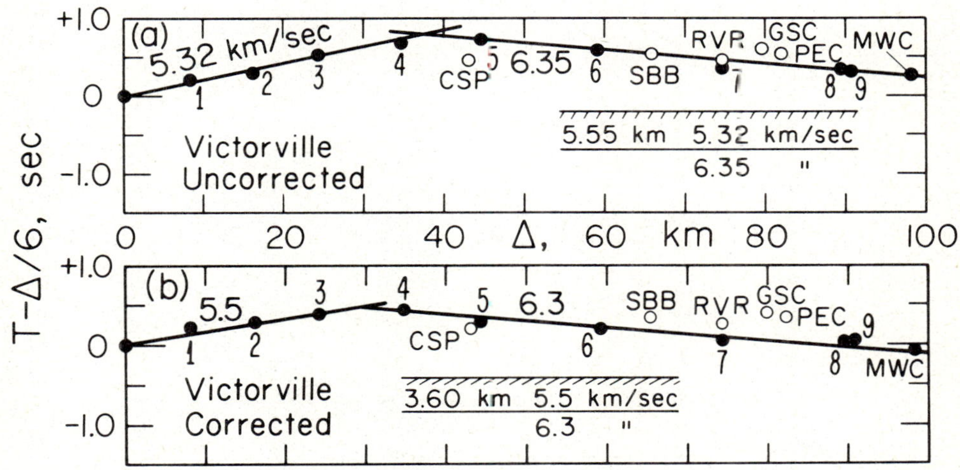

Figure 4

Uncorrected (a) and corrected (b) travel times for the Victorville blast. The corresponding structures are given in the inset. Numbered stations are; 1: Navajo; 2: UICT; 3: Trash; 4: Baldy Mesa; 5: EXP-O; 6: Table Mt.; 7: Crystal Lake; 8: Charlton Flats; 9: Mt. Emma

be given by $\Delta t = -h \cos i_0/v_0$ where h is the height, v_0 is the velocity at the surface, here 5.32 km/sec, and i_0 is the incidence angle at the station or the take-off angle at the source. For P_g, $i_0 = 90°$, for P^*, $i_0 = 57°$, and for P_n, $i_0 = 42°$.

The effect of the local geology cannot be estimated unless the structure is known. The correction is probably negligible for stations on bed rocks, but it can become considerably large for stations on alluvium. For estimating this correction we have

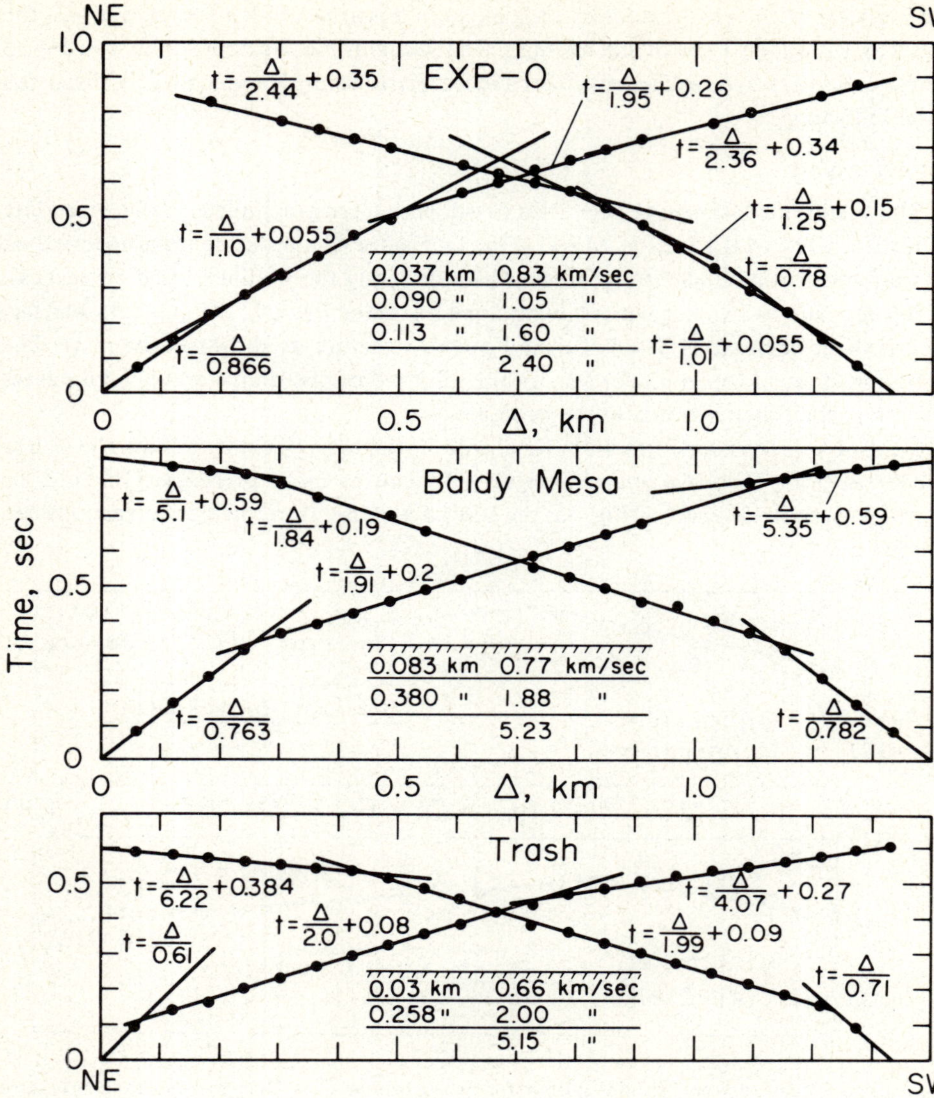

Figure 5

Short-range refraction measurements at temporary stations EXP-O (a), Baldy Mesa (b) and Trash (c). Structures (averaged) are given in the inset

made short range (up to 1.4 km) refraction measurements at stations 3 (Trash), 4 (Baldy Mesa) and 5 (EXP-O) shown in Fig. 4. These profiles were recorded with a 24-channel refraction system. Small charges were used as seismic sources. The profiles are shown in Figs. 5a, b, and c. Since these structures are used only for correction purposes, the dip of the layers is not considered and the insets represent the average structure. The corrections are then computed as the difference of the travel times for a vertical path between this structure and a homogeneous 5.32 km/sec layer. The largest height correction amounts to 0.23 sec (Table Mountain, 2280 m) and the largest station correction amounts to 0.21 sec (EXP-O). Although these corrections are significant, they are still small enough so that the errors arising from the crude correction method are probably insignificant. Figure 4b shows the corrected travel times. These data can be fit very well by two straight segments having apparent velocities of 5.5 km/sec, and 6.3 km/sec; the cross-over is at 28.0 km. The corresponding structure is given in the inset. In this structure, any feature above the sea level has been stripped off. Considering the source-station combinations shown in Fig. 1, this structure can be considered to represent the structure north of the San Andreas fault in this area. We will use these travel time curves as master curves for interpreting the data obtained from other explosions.

Corona, Jensen and Gypsum Canyon blast

For these blasts (Fig. 6a), no attempt has been made to align stations to make up a profile. However, it is remarkable that the combined travel times for the Corona and Jensen blasts are very consistent with those determined by the Victorville blast. If the 6.3 km/sec branch is shifted upward by 0.1 sec, the agreement would be even better. This indicates that the 5.5 km/sec layer is 0.5 km thicker in this region than the Victorville profile. It should also be noted that the arrivals at stations VPD, SJQ, SPH and TWL for the Corona blast and at VPD for the Jensen blast are delayed by 0.3 sec to 0.7 sec, the average being 0.5 sec. All of these stations are located on the alluvium of the Los Angeles basin. Although the details of the local structure beneath these stations are unknown, a delay of 0.5 sec seems reasonable for the alluvium. If we use this value for the time delay at these stations, the data for the Corona and the Jensen blasts becomes very consistent with the structure obtained for the Victorville profile.

For the Gypsum Canyon blast the data, though scattered considerably, show systematically late arrivals. However, if we assume a delay of 0.5 sec for the Los Angeles basin alluvium, these delays can readily be explained. Since in this case, the shot point itself is on the alluvium, allowance should be made for an additional delay of 0.5 sec. Thus if we subtract 1.0 sec from the times at stations on the Los Angeles basin alluvium (VPD, TCC, SJQ) and 0.5 sec from other stations (RVR, MWC, PEC, CSP, SBB), the general trend for the Gypsum Canyon blast becomes consistent with that for the Corona and the Jensen blasts. These source-station combinations represent primarily the structure south of the San Andreas fault. We therefore conclude that, except for the delay of about 0.5 sec beneath the stations on the Los Angeles basin alluvium, the crustal structure to the south of the San Andreas fault is nearly the same, velocity-wise, as that to the north of it.

Randsburg blast

Figure 6b shows the travel time data for the Randsburg blast. The height corrections

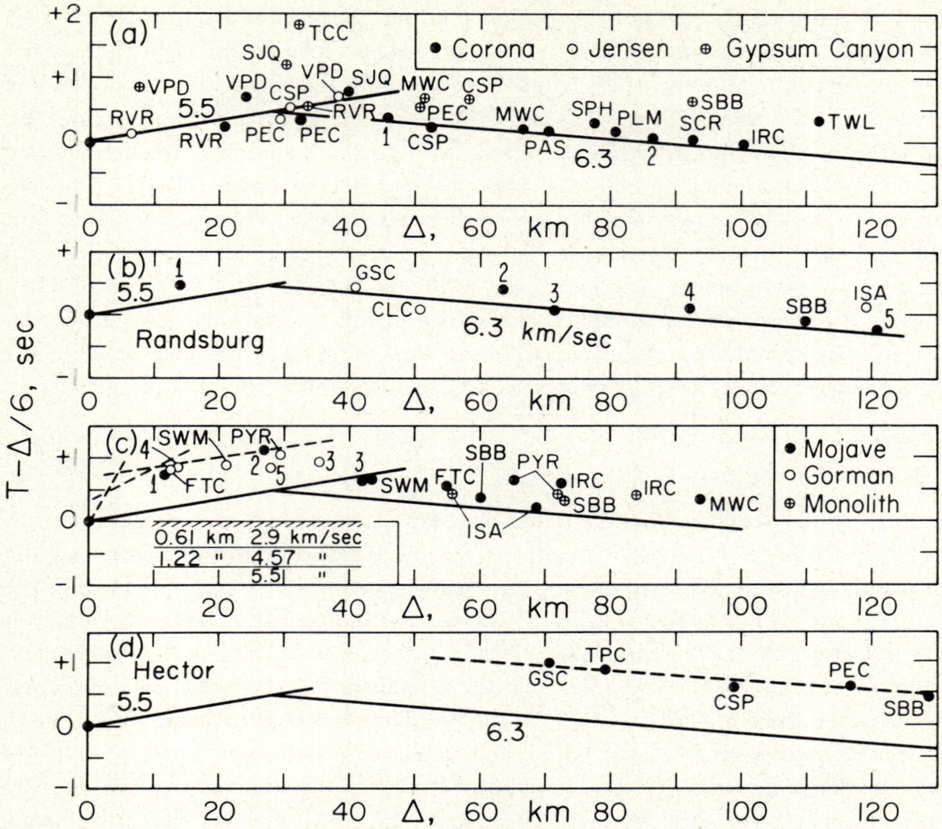

Figure 6

Travel times for various blasts. Height corrections are applied for both source and stations. Solid travel time curves are for the structure fitted to the corrected Victorville data. (a) Corona, Jensen and Gypsum Canyon blast. Numbered stations are; 1: Dalton Lake; 2: UCLA-1. (b) Randsburg blast. Numbered stations are; 1: Navy Test Range—1; 2: Galileo Hill; 3: Kramer Hill; 4: Jackrabbit Hill; 5: Lovejoy Buttes. (c) Mojave, Gorman and Monolith blasts. Dotted travel time curves correspond to the structure shown in the inset. Numbered stations are; 1: Willow Springs; 2: Rosamond; 3: Lake Hughes; 4: West Liebre Mt.; 5: Pine Canyon. (d) Hector blast. The dotted line shows the 6.3 km/sec line displaced upward by 0.9 sec from the Victorville curve

are made, but no corrections are made for the local structure. The closed circles represent the stations along the profile to Mount Wilson, and the open circles, off-profile stations. There is no obvious difference between these two groups of stations, and they are, as a whole, consistent with the Victorville travel time curves. If the whole curve is shifted upward by 0.1 sec, the agreement would become even better. It is possible that this 0.1 sec delay is due to the local structure beneath the source. However, owing to the lack of short-distance data, it is not resolvable whether the delay is due to the source or to a thickening of the top, presumably 5.5 km/sec, layer.

Mojave, Monolith and Gorman blast

Figure 6c shows the composite travel-time data for the Mojave, Monolith and Gorman blasts together with the travel-time curves obtained for the Victorville profile. The observed arrivals at $\Delta < 40$ km are delayed by nearly 0.5 sec. Most of these data are obtained from the Gorman blast for which the paths are essentially within the fault zone of either the Garlock or the San Andreas faults. Refraction measurements made in 1965 in this area (DON L. ANDERSON, personal communication, 1973) suggest low-velocity, presumably highly shattered, surface layers to a depth of 600 m as shown in the inset of Fig. 6c. The dotted lines in Fig. 6c are the travel time curves corresponding to this structure; the delay can be accounted for by this low-velocity surface layer.

The arrival times at $\Delta > 40$ km are delayed by 0.25 sec. Both Mojave and Monolith are cement quarries. Usually the *in situ* velocity in limestones is relatively high, about 4 km/sec, so that a large source delay is unlikely. It seems that the above delay can be explained most reasonably by a thickening of about 1.5 km of the 5.5 km/sec layer.

Hector

Figure 6d shows the travel time data for the Hector blast. Unlike other blasts, all the arrivals are consistently late by 0.9 sec. Although the lack of short-distance data does not permit the determination of the structure, the above delay seems to be mostly due to the source. Hector mine is quarrying bentonite, a very soft clay mineral (ultrasonic velocity ~1.7 km/sec), and an excessively large delay is quite possible. In view of this possible large source delay, the data obtained by this blast may not be adequate for crustal studies unless the structure beneath the quarry can be determined. In the interpretation that follows, these data will not be used directly. However, the consistent 6.3 km/sec branch strengthens the evidence for the existence of a layer with this velocity.

In summary, if we remove all the structures above the sea level, and are to find a single upper crustal structure that represents the whole area, we obtain a model consisting of a 4 km thick 5.5 km/sec layer underlain by a 6.3 km/sec layer. There is an indication that the 5.5 km/sec layer is about 1 to 1.5 km thinner along the Victorville to Mount Wilson path than elsewhere. The well-defined 6.3 km/sec branch up to 150 km suggests a substantial thickness of about 10 km for this layer. This model will be used in the following.

4. *Lower crust*

Since most of the blasts are not large enough to be recorded at distances beyond 150 km, the data for the lower crust are incomplete. As shown in Figure 3, there are arrivals at $\Delta > 160$ km which indicate the P_n branch, but the slope cannot be well defined. It is also not clear whether there is another branch intermediate between the 6.3 km/sec and the P_n branch. However, there are several constraints which are useful for determining the lower crustal structure. These constraints come from the following

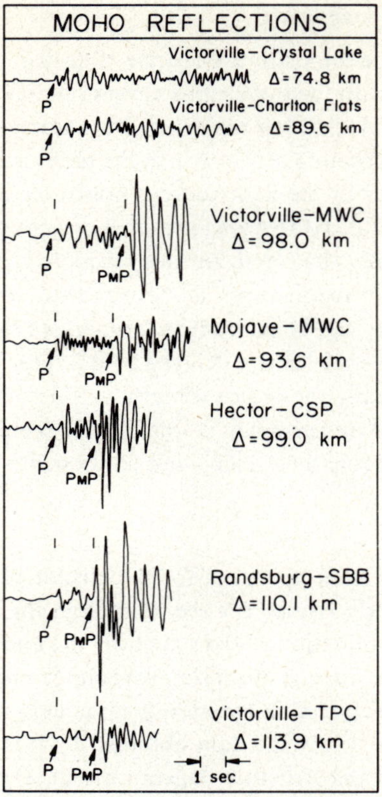

Figure 7

Examples of the Moho reflection (P_MP) near the critical distance. Note the absence of the P_MP phase at Crystal Lake ($\Delta = 74.8$ km) and Charlton Flats ($\Delta = 89.6$ km)

observations: 1) The round-trip travel time of the clear near-vertical Moho reflections, P_MP, observed at Corona is 10.5 sec (Shor, 1955; see also Richter, 1958). 2) For the Victorville blast, a very large second arrival was observed at Mount Wilson ($\Delta = 98$ km), but it was not observed at Crystal Lake ($\Delta = 75$ km) nor Charlton Flats ($\Delta = 90$ km) (see Fig. 7); we consider this strong phase as the Moho reflection near the critical distance Δ_c, and set Δ_c at 90 to 98 km. 3) This strong phase is also observed for other blasts at stations from $\Delta = 90$ to 110 km. $P_MP - P$ times are 1.5 sec to 3 sec depending on the distance (see Fig. 7). 4) The P_n velocity is determined as 7.66 km/sec by Press (1960), while 8.0 to 8.2 km/sec has been suggested by Gutenberg (1955). We examined a Nevada event to determine the P_n velocity over a distance range 200 to 400 km beneath the area considered. The origin time and the location of this event are known sufficiently well for this purpose. Figure 8 shows the reduced travel time curve. The station corrections suggested in the earlier section are applied. The P_n velocity of about 7.8 km/sec seems appropriate. This value which is intermediate between Press (1960) and Gutenberg (1955) will be used in the following.

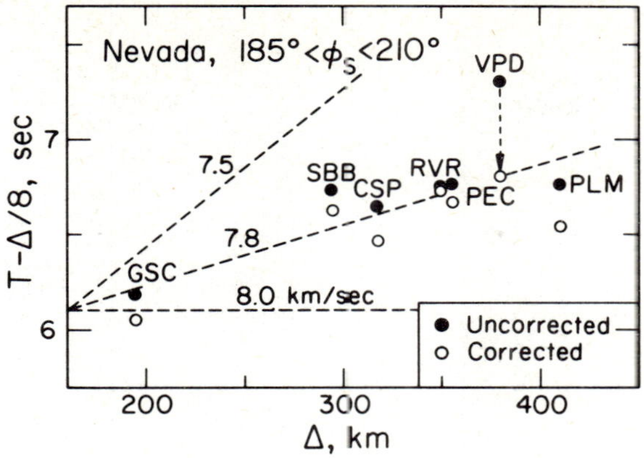

Figure 8
Travel times for a Nevada event. The stations within the azimuth range from 185° to 210° are used

First, we assume that the 6.3 km/sec layer extends to the Moho. Then to satisfy 1), the thickness of the 6.3 km/sec layer should be approximately 28 km, the height correction having been considered. However, simple calculation shows that this structure brings Δ_c at 85 km which is slightly too small. One possible way of increasing Δ_c is to insert an intermediate layer at the bottom of the crust. For example, replacing the bottom 4.6 km of the 6.3 km/sec layer by a 5 km thick 6.8 km/sec layer brings Δ_c up to 90 km and still satisfies 1).

This model gives $P_M P - P$ times of 2.7 sec and 2.1 sec, at $\Delta = 90$ km and 110 km respectively, and therefore satisfies 2) and 3). Of course, this is not the only possibility. However, since the structure down to 15 km is fairly well constrained by the data for $\Delta < 150$ km, the range of the possible models is not large. The least certain is the velocity of the intermediate layer. The travel time curves for this structure are given in Fig. 3b which shows the overall consistency of this model with the data. The phase refracted in the 6.8 km/sec layer never appears as the first arrival for shallow focus events.

The structure obtained here differs from that of PRESS (1960). In particular, the 4 km thick upper layer (5.5 km/sec) found in the present study does not exist in PRESS (1960). Since Press's emphasis was on the regional structure rather than the local structure, he had only two stations at $\Delta \leqslant 50$ km, which are too few to determine this layer. Press's model includes a 8.11 km/sec layer at about 50 km depth. None of our profiles is long enough to detect this layer.

5. Temporal change

For Mojave, Gorman, Victorville and Gypsum Canyon blasts, measurements were made repeatedly for the purpose of monitoring possible temporal changes in seismic velocity. For this purpose, only the readings with grade *A* or *B* were used. Table 3 and Figs. 9a, b, c and d show the results. As the shot point migrates as much as 800 m within each quarry, the travel time varies slightly. Therefore instead of the travel

Table 3a
Gorman blast

Source		Stations											
		FTC			SWM			Lake Hughes			PYR		
Date	Coordinates 34°N, 118°W	Δ (km)	T (sec)	$\bar{v}$ (km/sec)	Δ	T	$\bar{v}$	Δ	T	$\bar{v}$	Δ	T	$\bar{v}$
June 27, 1973	49.62′, 45.72′	13.06	3.01 (A)	4.34	20.44	4.35 (A)	4.70	—	—	—	28.83	5.72 (A)	5.04
Sept. 10, 1973	49.52′, 45.80′	13.03	2.96 (A)	4.40	20.43	4.38 (A)	4.66	34.89	6.84 (A)	5.10	28.62	5.75 (A)	4.98
Nov. 14, 1973	50.15′, 45.60′	12.88	2.93 (A)	4.40	20.89	4.36 (A)	4.79	35.20	6.79 (A)	5.18	29.76	5.97 (B)	4.99
Feb. 25, 1974	50.25′, 45.11′	13.54	3.01 (A)	4.50	—	—	—	—	—	—	29.91	5.93 (B)	5.04
Apr. 29, 1974	50.10′, 45.57′	12.96	2.91 (A)	4.45	20.80	4.45 (A)	4.67	34.74†)	6.68 (B)	5.20	29.67	5.86 (A)	5.06

†) Lake Hughes—2.

Table 3b
Mojave blast

Source		Stations														
		FTC			PYR			SWM			Lake Hughes			SBB		
Date	Coordinates 35°N, 118°W	Δ (km)	T (sec)	$\bar{v}$ (km/sec)	Δ	T	$\bar{v}$	Δ	T	$\bar{v}$	Δ	T	$\bar{v}$	Δ	T	$\bar{v}$
July 13, 1973	1.80′, 19.19′	—	—	—	64.20	11.59 (A)	5.54	—	—	—	—	—	—	59.10	10.46 (B)	5.65
Aug. 8, 1973†)	2.34′, 19.22′	55.48	9.71 (A)	5.71	—	—	—	42.89	8.05 (A)	5.33	—	—	—	59.80	10.50 (A)	5.70
Sept. 6, 1973†)	2.22′, 19.08′	55.60	9.81 (A)	5.67	64.89	11.59 (A)	5.60	42.79	7.98 (A)	5.36	41.52	7.71 (A)	5.39	59.47	10.37 (A)	5.74
Nov. 14, 1973	2.46′, 19.08′	55.75	9.89 (B)	5.64	65.24	11.77 (A)	5.54	43.16	8.09 (A)	5.34	41.95	7.84 (B)	5.35	59.76	10.57 (B)	5.65
Jan. 23, 1974	2.34′, 19.22′	—	—	—	64.94	11.80 (B)	5.50	42.85	8.04 (B)	5.33	41.68	7.85 (B)	5.31	59.78	10.57 (A)	5.66
Mar. 14, 1974	2.36′, 19.20′	—	—	—	64.99	11.75 (A)	5.53	—	—	—	41.73†)	7.85 (A)	5.32	59.78	10.62 (A)	5.63
May 14, 1974	2.41′, 19.07′	—	—	—	—	—	—	—	—	—	41.87‡)	7.88 (A)	5.31	59.69	10.60 (A)	5.63

†) Shot time is uncertain to 0.1 sec.
‡) Lake Hughes—2.

Table 3c
Victorville blast

Source		Stations														
		CSP			PEC			Table Mt.			SBB			Mt. Emma		
Date	Coordinates 34°N, 117°W	Δ (km)	T (sec)	$\bar{v}$ (km/sec)	Δ	T	$\bar{v}$	Δ	T	$\bar{v}$	Δ	T	$\bar{v}$	Δ	T	$\bar{v}$
Oct. 11, 1973	37.78′, 06.68′	43.12	7.65 (B)	5.64	81.97	14.20 (B)	5.77	59.08	10.45 (A)	5.65	65.64	11.48 (B)	5.72	90.62	15.44 (B)	5.87
Jan. 29, 1974	37.78′, 06.66′	43.13	7.68 (A)	5.62	81.98	14.26 (B)	5.75	59.10	10.48 (A)	5.64	65.67	11.51 (A)	5.71	—	—	—
June 26, 1974	37.87′, 06.71′	43.24	7.65 (A)	5.65	82.14	14.20 (A)	5.78	59.11	10.49 (A)	5.63	65.58	11.52 (A)	5.69	90.61	15.50 (B)	5.85

Table 3d
Gypsum Canyon blast

Source		Stations												
		VPD			PEC			CSP			SBB			
Date	Coordinates 33°N, 117°W	Δ (km)	T (sec)	$\bar{v}$ (km/sec)	Δ	T	$\bar{v}$	Δ	T	$\bar{v}$	Δ	T	$\bar{v}$
Feb. 13, 1974	51.86′, 42.36′	7.52	2.08 (A)	3.62	50.55	9.02 (A)	5.60	57.94	10.48 (B)	5.53	92.07	16.08 (B)	5.73
July 10, 1974	51.83′, 42.30′	7.54	2.18 (B)	3.46	50.46	9.05 (B)	5.58	57.94	10.57 (B)	5.48	92.13	16.15 (B)	5.70

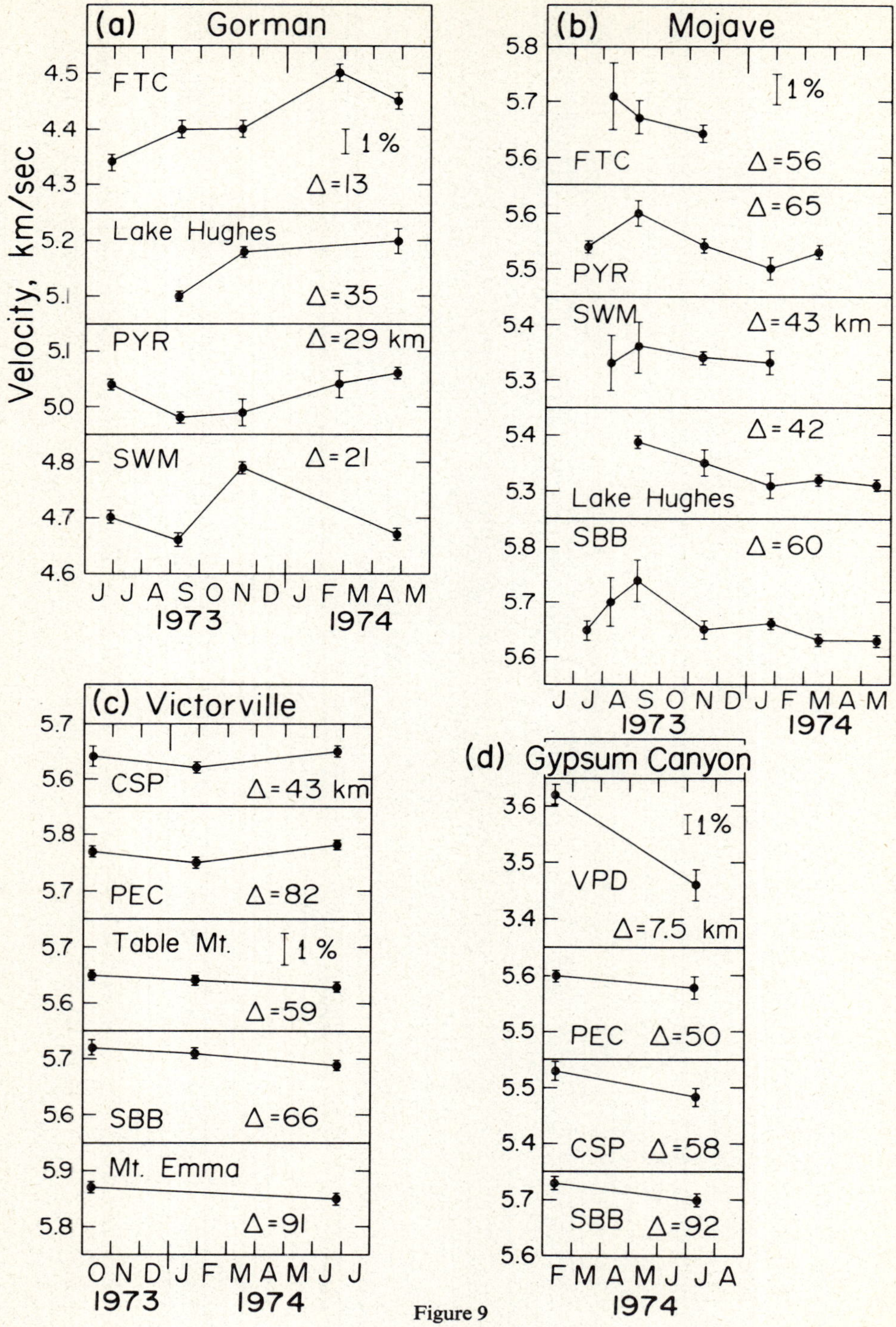
(a) Gorman
FTC
Δ=13
Lake Hughes
Δ=35
PYR
Δ=29 km
SWM
Δ=21
1%
Velocity, km/sec
J J A S O N D J F M A M
1973
1974
(b) Mojave
FTC
Δ=56
Δ=65
PYR
SWM
Δ=43 km
Δ=42
Lake Hughes
SBB
Δ=60
1%
J J A S O N D J F M A M
1973
1974
(c) Victorville
CSP
Δ=43 km
PEC
Δ=82
Table Mt.
Δ=59
SBB
Δ=66
Mt. Emma
Δ=91
1%
O N D J F M A M J J
1973
1974
(d) Gypsum Canyon
VPD
Δ=7.5 km
PEC Δ=50
CSP Δ=58
SBB Δ=92
1%
F M A M J J A
1974

Figure 9

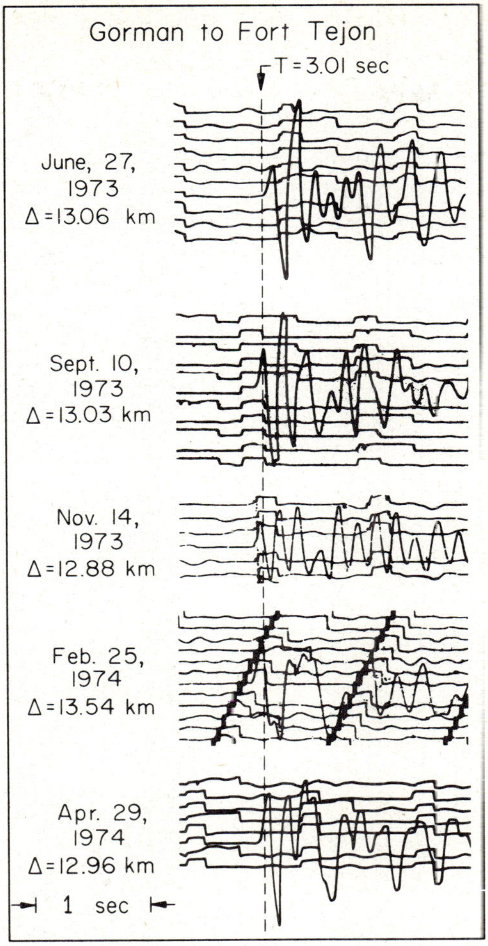

Figure 10
Seismograms of the Gorman blast recorded at FTC. The origin time of the ith event is shifted by $\Delta t_i = -(\Delta_i - \Delta_1)/v_i$ with respect to the 1st event, where Δ_i and v_i are epicentral distance and the average velocity of the ith event. If there is no velocity change, the onsets should be aligned

times, the average velocities, distance/travel time, are plotted. If the structure is homogeneous, the average velocity is a perfectly adequate parameter. However, when the structure is vertically and/or laterally heterogeneous some uncertainty is unavoidable. Although we think the error resulting from this effect is small, this is the only ambiguity in the present experiment.

Figure 9
Temporal variation of the average velocity. (a) Between Gorman and FTC, Lake Hughes, PYR and SWM. (b) Between Mojave and FTC, PYR, SWM, Lake Hughes and SBB. (c) Between Victorville and CSP, PEC, Table Mountain, SBB and Mount Emma. (d) Between Gypsum Canyon and VPD, PEC, CSP and SBB

Figure 9a shows the results obtained for the Gorman blasts. The error bars indicate the uncertainty corresponding to the grade assignment A or B. It is interesting to note that there seems to be a change exceeding the error limit; in particular, the paths from Gorman to Fort Tejon (FTC) and to Lake Hughes, which are both along the San Andreas fault zone, show a more or less steady increase. In order to show the quality of the data, the seismograms at FTC are shown in Fig. 10. The suggested temporal change appears evident, if the effect of the change in the distance is insignificant. If the structure in the quarry is relatively homogeneous, then the effect would be very small. However, if, for some reason, the velocity is locally very low, the effect may become significant. In order to pursue this problem, we made two short-range refraction profiles at the quarry site. Figure 11 shows the travel time curves and the structure. Except for a very thin soil layer, the *in situ* velocity is more or less what one might expect for competent limestones, and is about the same as the average velocity along the path.

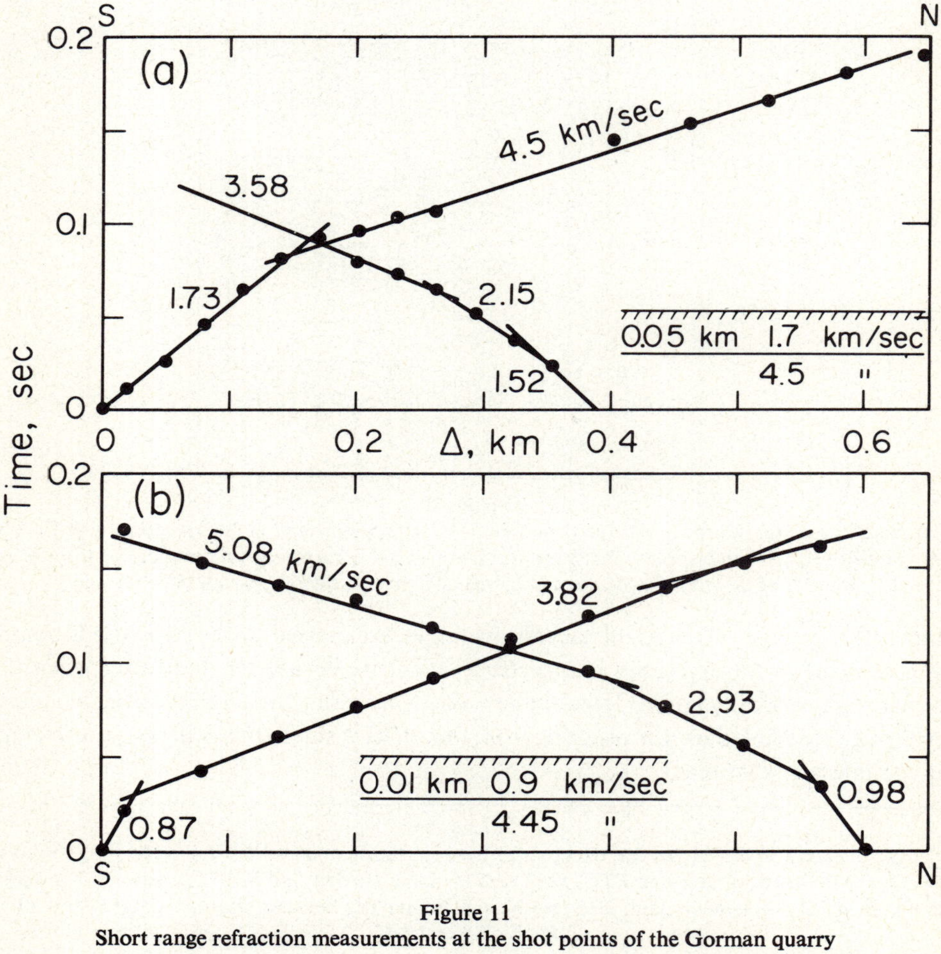

Figure 11
Short range refraction measurements at the shot points of the Gorman quarry

Although the profile coverage in the quarry is not complete, this result seems to reinforce the conclusion that the velocity did change. Figure 9b shows the results for the Mojave blasts. The change is not large, but a steady decrease is indicated since Septemper, 1973. Figure 9c shows the results for the Victorville blasts. The paths to the south (to CSP and PEC) show a shallow bay-shaped anomaly similar to each other, while the paths to the west (Table Mt., SBB) show more or less monotonic decrease.

Figure 9d shows the results for the Gypsum canyon blasts. Although only two measurements have been taken, a decrease larger than those observed for other profiles is found. This change may be related to the large velocity change near Riverside reported earlier (KANAMORI and CHUNG, 1974; WHITCOMB *et al.*, 1974).

As a whole, the change in the velocity seems systematic although the magnitude is much smaller than those reported elsewhere as precursory to earthquakes. During the period of these experiments, no earthquake larger than $M_L = 5$ has occurred in this area. In view of the complete lack of any previous data as to the background fluctuation of seismic velocity in a tectonically active area, we do not attempt, at present, to interpret these data in relation to impending earthquakes. It seems most important to establish a base-line by which we can judge what is and is not anomalous. Obviously, a change having a time constant larger than several years is most important in relation to major earthquakes. From this point of view, the present data are still very preliminary, but it is hoped that they serve as a nucleus for a longer and more definitive data set which can be used for actual prediction purposes.

6. Discussion and conclusions

The crustal structure obtained here (see the inset of Fig. 3b) is not unique, particularly the lower crust, but it fits the travel–time data within ±0.3 sec. If corrections are made for station and the source delays, the fit would be better than ±0.15 sec, and therefore it will serve as a useful model for the location of local earthquakes. It may be suggested that, if no corrections for station height and delay are included in the location program, use of the structure given in Fig. 3a is appropriate. If the station corrections for height and possibly for station delays are included in the location program the structure shown in Fig. 3b would be more appropriate. (0.5 sec may be tentatively used for the stations in the Los Angeles basin. This delay is consistent with that calculated for the velocity–depth functions obtained from sonic log and check shock surveys for oil explorations. The calculated delay ranges from 0.47 to 0.69 sec for several locations in the Los Angeles basin.)

The present data do not preclude a possibility of a velocity reversal in the lower crust. More detailed analyses of the amplitude and the wave form of the first as well as later arrivals are obviously necessary for the determination of the details of the lower crust.

It is rather surprising that the structure is very homogeneous in the entire area. PRESS (1960) also pointed out this fact. The velocity of the upper 3 to 5 km thick layer

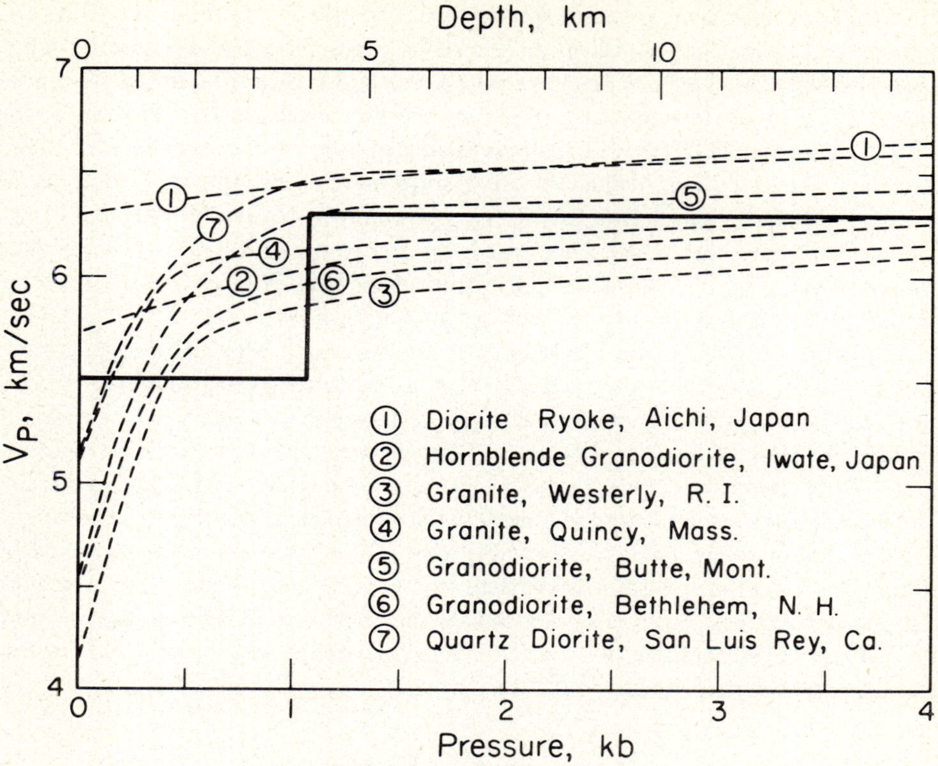

Figure 12

Comparison of crustal velocity with velocities of granites. Data for rocks are from BIRCH (1960) ((3), (4), (5), (6)), and KANAMORI and MIZUTANI (1965) ((1) and (2))

is fairly well defined at 5.5 km/sec (see Figs. 3b and 4b). It is instructive to compare the crustal velocities with laboratory velocity–pressure data for typical granites. Figure 12 shows such comparison. The crustal velocity to a depth of 4 km is significantly lower than that for granites. If we accept the prevalent notion that the upper crustal layer consists of granitic rocks, this discrepancy requires explanation. Most obvious is that the *in situ* cracks remain open because of either a high degree of fracturing or large deviatoric tectonic stress. The latter case is usually referred to as dilatancy, and is a much debated subject in relation to earthquake prediction. Whether the relatively low southern California crustal velocity is due to 'regional' dilatancy or not may be resolved by examining the *in situ* anisotropy. This is an important subject for the future.

Although the present results indicate a temporal velocity change, it is much smaller than those believed to take place precursory to earthquakes (10 to 20%). The observed change may be reflecting a gradual change in the tectonic stress in southern California, but it could be also due to some minor local effects such as the change in the ground water level, change in the water content in the rocks near the source or the station, etc. The present method has demonstrated that we now have the capability of detecting a change in the velocity as small as 1% over the most important part of southern California. If the dilatancy–diffusion model is correct and a precursory velocity change as large as 10 to 20% does take place over an area 100 km or larger in extent, it would hardly escape detection. If the present monitoring system is continued at least until the next major earthquake, we will have a very complete pre-seismic record which we can use as a guideline for future predictions.

Acknowledgements

We are very much indebted to RALPH GILMAN, FRANCIS LEHNER and his associates for the field operations. SHAWN BIEHLER and his colleagues at the University of California, Riverside, LEON TENG and his colleagues at the University of Southern California, and many gradute students at the Seismological Laboratory at Caltech helped us in the various aspects of the field program. DAVID JACKSON at the University of California at Los Angeles provided us with a reading at a UCLA station. HSI-PING LIU kindly measured the ultrasonic velocity of a bentonite sample from the Hector quarry.

Our thanks go to Minnesota Mining and Manufacturing Co., General Portland Inc., California Portland Cement Co., Southwestern Portland Cement Co., Owl Rock Products Co., L. S. Hawley Corp., Baroid Division, N. L. Industries, Inc., Riverside Division, American Cement Corp., and Dr. Pierre St. Amand at U.S. Naval Ordnance Test Station at China Lake for kind cooperation in recording these blasts.

This research was supported by Earthquake Research Affiliates, National Science Foundation Grant GA 29920, and US Geological Survey Contract 14-08-001-14563.

REFERENCES

ALLEN, C. R. and D. V. HELMBERGER, *Search for temporal changes in seismic velocities using large explosions in southern California*, in *Proceedings of Conference on Tectonic Problems of the San Andreas Fault System, June 20–23, 1973*, edited by R. B. KOVACH and A. NUR (Stanford, California 1973), pp. 436–445.

BIRCH, F., *The velocity of compressional waves in rocks to 10 kilobars, 1*, J. Geophys. Res. *65* (1960), 1083–1102.

GUTENBERG, B., *Travel times of principal P and S waves over small distances in southern California*, Bull. Seismol. Soc. Am. *34* (1944), 13–32.

GUTENBERG, B., *Revised travel times in southern California*, Bull. Seismol. Soc. Am. *62* (1951), 427–439.

GUTENBERG, B., *Waves from blasts recorded in southern California*, Trans. Am. Geophys. Union *33* (1952), 427–431.

GUTENBERG, B., *Wave velocities in the earth's crust*, Geol. Soc. Am. Special Paper 62 (1955), 19–34.

HILEMAN, J. A., C. R. ALLEN and J. M. NORDQUIST, *Seismicity of the southern California region, 1 January 1932 to 31 December 1972* (Seismological Laboratory, California Institute of Technology, Pasadena 1973), p. 83.

KANAMORI, H. and W. Y. CHUNG, *Temporal changes in P-wave velocity in southern California*, Tectonophysics, *23* (1974), 67–78.

KANAMORI, H. and H. MIZUTANI, *Ultrasonic measurement of elastic constants of rocks under high pressure*, Bull. Earthq. Res. Inst. Tokyo Univ. *43* (1965), 173–194.

PRESS, F., *Determination of crustal structure from phase velocity of Rayleigh waves, 1.*, Bull. Geol. Soc. Am. *67* (1956), 1647–1658.

PRESS, F., *Crustal structure in the California-Nevada region*, J. Geophys. Res. *65* (1960), 1039–1051.

RICHTER, C. F., *Velocities of P at short distances*, Bull. Seismol. Soc. Am. *40* (1950), 281–289.

RICHTER, C. F., *Elementary Seismology* (W. H. Freeman and Company, San Francisco 1958), pp. 44–45.

ROLLER, J. C. and J. H. HEALY, *Seismic-refraction measurements of crustal structure between Santa Monica Bay and Lake Mead*, J. Geophys. Res. *68* (1963), 5837–5849.

SHOR, G. G., *Deep reflections from southern California blasts*, Trans. Am. Geophys. Union *36* (1955), 133–138.

WHITCOMB, J. H., H. KANAMORI and D. HADLEY, *Earthquake prediction: Variation of seismic velocities in southern California, abstract*, EOS, *55* (1974), 355.

(Received 2nd December 1974)

Pageoph,
Vol. 113 (1975) Birkhäuser Verlag, Basel

Time Variations in Teleseismic Residuals Prior to the Magnitude 5.1 Bear Valley Earthquake of February 24, 1972

By C. H. CRAMER and R. L. KOVACH[1])

Summary – Teleseismic arrivals for large earthquakes occurring in the Circum-Pacific seismic source regions have been analysed for *P*-velocity variations prior to the February 24, 1972, Bear Valley earthquake at several stations in the vicinity of Bear Valley, California. The teleseismic arrivals have been analysed by the method of two-station residuals and corrected for observed azimuthal variations. The data covers the time period of July, 1971, through April, 1972, and suggests that during part of January, 1972, a *P*-velocity anomaly occurred beneath station BVL 2 km from the epicenter of the magnitude 5.1 Bear Valley event. A maximum vertical travel time delay of 0.15 seconds is observed. No other anomalous behavior associated with this event is suggested by the data for the other stations ranging from 7 to 19 km from the event's epicenter. These results support an anomalous zone of limited size with a maximum horizontal extent of less than 5 km perpendicular and about 10 km parallel to the San Andreas fault relative to the epicenter and confined within a 5 to 10 km portion of the uppermost crust. Also a *P*-velocity delay for waves travelling essentially along the intermediate stress axis would imply in this case that the *P*-velocity anomaly is caused by a bulk modulus mechanism such as that proposed by the dilatancy–fluid flow theory.

1. *Introduction*

Since the first papers on the premonitory changes in seismic velocities prior to thrust fault type earthquakes (SEMENOV [1], NERSESOV *et al.* [2], WHITCOMB *et al.* [3], AGGARWAL *et al.* [4], STEWART [5]), several studies of premonitory changes before strike-slip events have been undertaken. Local source region studies have been the dominant approach in the search for seismic wave velocity anomalies along strike-slip faults (BAKUN *et al.* [6], BROWN [7], MCEVILLY and JOHNSON [8,–9], and ROBINSON *et al.* [10]). Of these studies, Robinson *et al.*'s use of seismic waves originating outside the anomalous zone and passing through the zone on the way to the receiving station has been most successful in detecting precursory velocity anomalies before moderate sized earthquakes in central California (ROBINSON *et al.* [11]). Some studies have been made using teleseismic waves to search for *P*-velocity anomalies before large and some moderate sized events (WYSS and HOLCOMB [12], UTSU [13], CRAMER and KOVACH [14], and WYSS and JOHNSTON [15]).

This paper presents some results on the use of the two-station teleseismic residual technique to search for travel-time anomalies at seismograph stations of the National Center of Earthquake Research (NCER) network along the San Andreas fault in central California. In particular, teleseismic arrivals for large events occurring in the Circum-Pacific source regions have been analysed (prior to the February 24, 1972, magnitude 5.1

[1]) Geophysics Department, Stanford University, Stanford, California 94305.

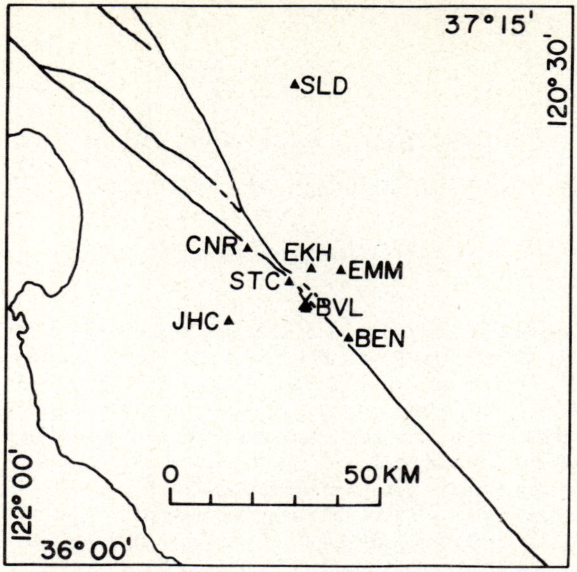

Figure 1
A map of the Bear Valley region showing the location of the Bear Valley event (×) and the seismograph stations used in this study (triangles). Monterey Bay is shown to the west and the juncture of the Calaveras Fault with the San Andreas Fault is also shown

Bear Valley earthquake) for *P*-velocity variations at several stations in the vicinity of Bear Valley, California (Fig. 1). (This is the same event for which ROBINSON *et al.* [10] examined local *P*-arrivals.) Besides differencing station residuals with a presumably stable station, a correction for observed azimuthal variations has been applied to the two-station residuals to remove the effects of upper mantle inhomogeneities from the data. This enables one to use a larger range of source azimuths and hence more data without increasing the observational scatter.

2. Method

Figure 1 shows the locations of the seismograph stations used in this study. Stations Johnson Canyon (JHC), Bear Valley (BVL), Benito (BEN), Elk Horn Ranch (EKH), Emmet (EMM), and Cienega Road (CNR) are part of the NCER telemetered network of seismograph stations. Stations Stone Canyon (STC) and San Luis Dam (SLD) are presently also part of the NCER network but prior to mid-1973 were part of a telemetered network maintained by the National Oceanic and Atmospheric Administration's Earthquake Mechanism Laboratory (EML) in San Francisco, California. The specific source areas for this study are seismic zones associated with the Aleutian, Alaskan, Kurile, Japanese, Philippine and New Guinea through Tonga–Kermadec island arc systems plus the seismic zones along the west coasts of South and Central America, including Mexico. These areas provide a relatively frequent source of

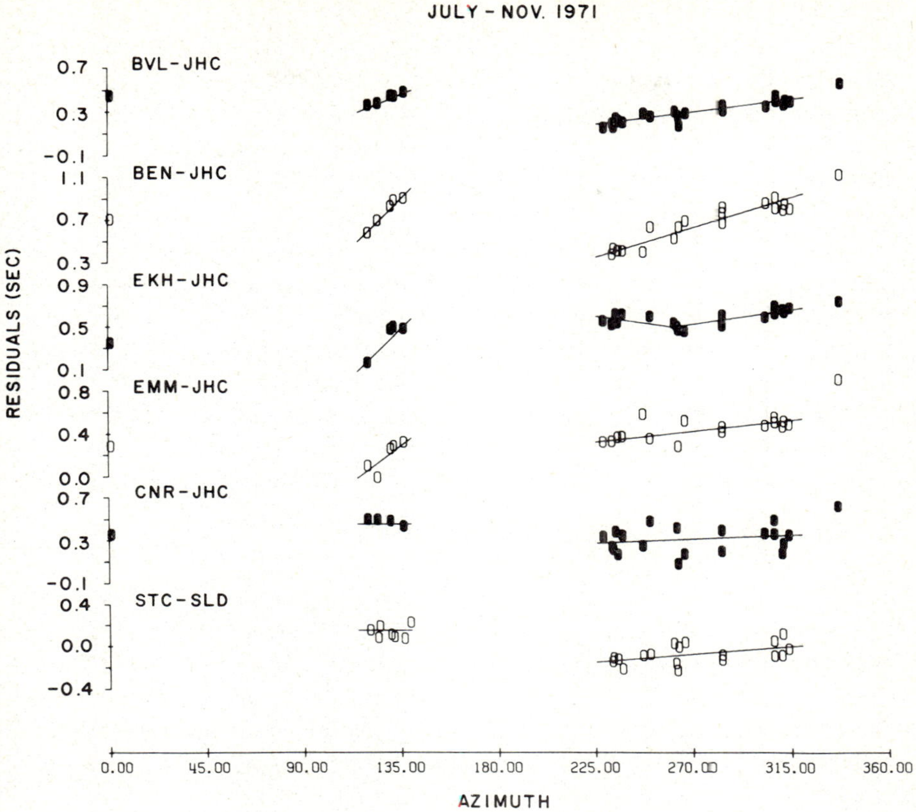

Figure 2
The residual vs. azimuth plots for July through November, 1971, from which the azimuthal corrections were determined for each station pair. The circles are 0.1 second high and each station's azimuthal correction has been plotted as a line passing through the data

magnitude 5.7–6.0 and larger events with a broad range of azimuthal coverage but a relatively narrow range of epicentral distances along any given azimuth.

Due to the presence of cultural noise and the frequency response of the seismometers in the network, events below the magnitude 6.0–5.7 level are not detected. Because of emergent arrivals, first arrivals were picked using visual waveform matching from station to station for a given event and first peak times plus some later peak times were also recorded. An analysis of the first arrival and peak arrivals for each event of this study shows that the waveform is generally coherent to within the reading error for the first second or so of the record. Thus, although the first break arrivals tend to scatter more due to the emergent character of the signals, the first peak arrivals proved to be an effective measure of the teleseismic *P*-arrival time and were therefore used in the subsequent analysis.

Individual station residuals were calculated using the HERRIN and TAGGART [16] tables and differenced with respect to an off-fault reference station so as to eliminate source effects. The reference stations used are station JHC for the NCER stations and station SLD for the EML station. Station pair separations are 50 km or less so that the error expected from the two-station residual technique for any limited range of azimuths is believed to be less than 0.05 sec. Further, reading errors in the arrival times for these stations are believed to be less than 0.05 sec also.

To provide enough data for an effective search for *P*-velocity variations, a wide range in source azimuth coverage had to be used. Since there is known to be a large variation in residuals with azimuth of approach in central California (OTSUKA [17–18] and NUTTLI and BOLT [19]), an azimuthal correction has been determined by fitting linear segments to the observed residual vs. azimuthal plot for a normal period prior to the 1972 Bear Valley earthquake. Figure 2 shows the residual vs. azimuth plot for July through November, 1971, plus the azimuthal correction fit to this data. After determining an azimuthal correction for each station pair the correction has been subtracted from the observed residuals effectively reducing the data along each azimuth to a common datum.

3. Results

The azimuthally corrected two-station residuals as a function of time are shown in Fig. 4. The events that have been used to compile these time plots are from azimuths between 115° to 140° and 225° to 320°. Only for these azimuth ranges are there sufficient data to establish an azimuthal trend. The horizontal reference lines are the means of each station-pair's corrected residuals. The number at the end of each plot is the standard deviation (in seconds) for that plot's residuals.

The relative amplitudes of arrivals vary from station to station depending on the magnification. The seismograph system response curve is given in WESSON *et al.* [26]

Table 1
Table of station amplifier attenuations which result in different trace amplitudes for different stations during the study period

Station	Attenuation
BEN	42 dB
BVL	30 dB
CNR	42 dB
EKH	42 dB
EMM	36 dB
JHC	36 dB
SLD	42 dB†
STC	48 dB†

† Equivalent attentuation for EML stations.

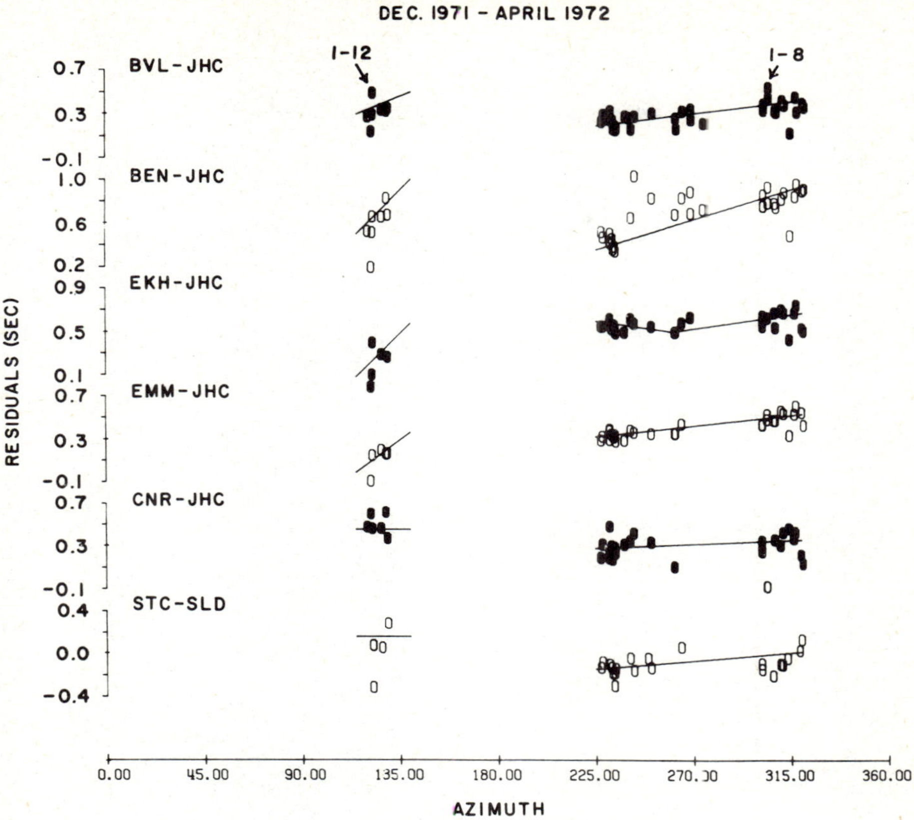

Figure 3

The residual vs. azimuth plots for December, 1971, through April, 1972. The circles are 0.1 second high and the azimuthal correction from Fig. 2 has been plotted as a line for reference

for a standard attenuation setting of 42 dB. Periodic calibration at each station shows that the station magnification is governed mostly by the attenuation setting. Table 1 shows each station's attenuation during the study period. Notice that BVL has the lowest attenuation and therefore the highest magnification. A higher magnification than normal results in more impulsive first arrivals, sharper peaks, and more reliable picking of first arrivals. Hence the arrival information at station BVL is much better in quality than the other stations. The attenuation settings at the reference station JHC and at station EMM are also lower than normal and first arrivals are more reliably picked at these stations also.

The uppermost plot of Fig. 4 is the residual vs. time plot for station BVL located 2 km from the Bear Valley epicenter. This plot is composed of the best arrival information due to the higher magnification at JHC and particularly at BVL. Generally the residuals do not vary much in time. However, the events on January 8 and 12 show a 0.16 and 0.15 second delay, which is a deviation slightly larger than two standard

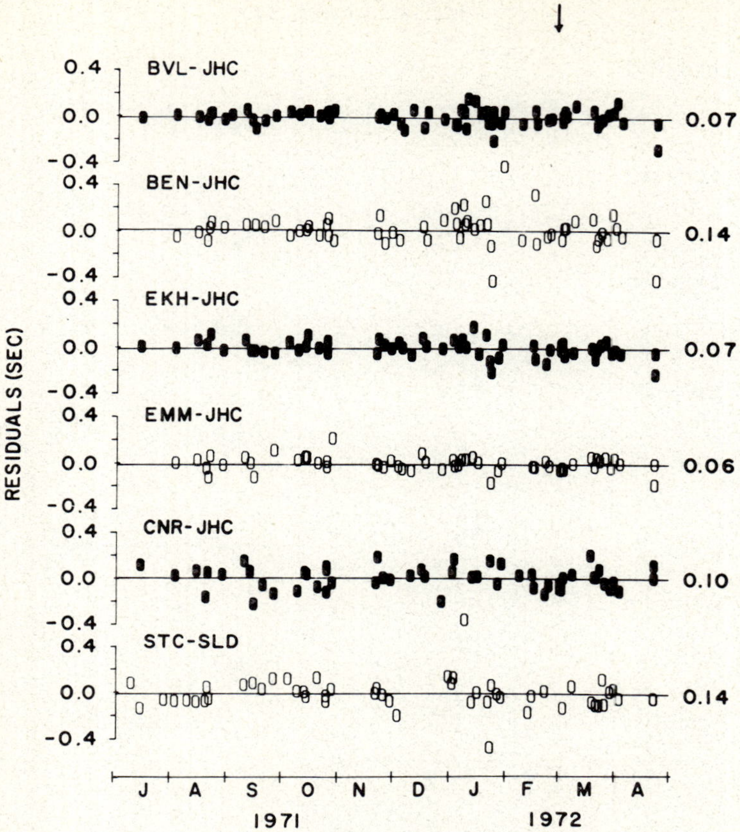

Figure 4
The azimuthally corrected residual vs. time plots for each station pair used in this study. The horizontal reference lines represent the mean value for the corrected residuals and the numbers at the end of each plot are the standard deviations for each plot in seconds. The circles are 0.1 second high

deviations from normal and suggests a *P*-velocity anomaly beneath station BVL during mid-January. The duration of this anomalous behavior is notably quite short (about one week) and the apparent anomaly has disappeared fully one month before the Bear Valley earthquake in February. To show the control along the azimuths from which these two arrivals came, Fig. 3 presents the residual vs. azimuth plot for events occurring in December, 1971, through April, 1972. The delayed events of January 8 and 12 have been indicated on the BVL data of Fig. 3. One can see that there are other events along these same azimuths in Figs. 2 and 3 and that the events of January 8 and 12 are truly delayed with respect to these other events. Returning to Fig. 4, there are two other separate events on January 22 and April 21 whose arrivals are early by two standard deviations or more. These events also appear to be anomalously early on the other JHC referenced time plots which could imply a variation in velocity of JHC instead of the other stations or at worst the slight mispicking of the arrival at JHC.

The residual vs. time plot for station BEN does exhibit an increased scatter during January and February, 1972, which may suggest a delay in arrivals at about the same time as the delay at station BVL. However, the larger standard deviation for the BEN data set casts doubt on the significance of this trend. Furthermore, over one-half of the data points are within 0.1 sec of the mean and there is no suggestion of a continuous delay in teleseismic arrivals. Lastly an examination of Fig. 3 reveals that the delayed arrivals are from the range 230° to 270° from the north during this time period, which suggests that the cause of the anomalous behavior is not due to a shallow anomalous zone associated with the impending Bear Valley earthquake to the northwest. Such a shallow zone would delay all arrivals at BEN regardless of azimuth.

The residual plot for station EKH is generally quite flat with only one point in January having a delay outside two standard deviations from the mean, in addition to the two JHC anomalous points discussed above. Although one may have the feeling that there might be a small decrease in velocity beneath EKH in January, 1972, the evidence is indeed weak since the delay would be on the order of the error in the data or only based on one data point.

Station EMM has a flat residual vs. time plot with a small standard deviation. No evidence for any anomalous behavior during January and February of 1972 is evident. There is one delayed arrival in October, but it is an isolated observation and definitely not associated with the Bear Valley event.

Stations CNR and STC have a generally increased data scatter in their residual vs. time plots. This is due in part to the higher attenuation at these stations and also in part to these two stations being located right in the fault zone which further attenuates incoming seismic signals. The greater attenuation effects the signal amplitude so that the signal peaks have a more rounded appearance increasing the first arrival picking error. Within the data scatter there appears to be no change in *P*-velocities at these stations.

On the BEN and STC time plots there are isolated data points with deviations of 0.4 sec or better. These data are not believed to represent real velocity variations in the crust beneath those stations.

4. *Discussion*

The reliability of the residual data for station BVL is supported by the quality of the arrivals at stations BVL and JHC and by the low standard deviation of the residuals over time. The existence of a *P*-velocity delay beneath stations BVL has support from two consecutive events with residual variations corresponding to two standard deviations from the mean and from the fact that the source regions for these two events are different and well controlled. Trends at other stations are within the data scatter or are composed of single delayed datum and can not be substantiated. The exception are the delayed arrivals at BEN that originate in source regions along one limited range of azimuths. This limited delay must be due to a source to the southwest of BEN and deeper in the crust in order to have such a limited effect. BUFE *et al.* [27] indicate a magnitude 4.6 event occurred in the vicinity of BEN 3 days after the main Bear Valley event. A deep event ($h = 11.5$ km) for central California, this smaller event may be associated with the observed anomalous behavior at BEN, but a more sophisticated ray path study is required to show this than has been undertaken here.

The teleseismic raypath information limits how far the anomalous zone extended out from the San Andreas fault trace. Station BVL is 1.5 km southwest of the fault trace and the teleseismic raypaths lie to the southwest of a line parallel to the fault and passing through station BVL. Since the teleseismic raypaths are essentially vertical (5° or less from the vertical in the upper crust according to OTSUKA [17]), there is little difference in raypath with azimuth. Thus an anomalous zone could not have extended beyond three kilometers southwestward from the fault trace prior to January 8 and after January 12 based on a comparison of direct and laterally refracted raypaths. If the two data points on Jaunary 8 and 12 are truly anomalous, then the boundary must have been at least 3.5 km southwest of the fault trace during this period but probably did not extend much beyond that distance at its maximum. This agrees with the limits imposed by the local blast data of MCEVILLY and JOHNSON [8–9].

The results at the other neighboring stations provide a limit to the extent of an anomalous zone along the San Andreas from Bear Valley. Since no significant shallow velocity anomaly has been detected in the teleseismic data for the stations STC and BEN, any anomalous *P*-velocity zone that might have existed could have extended along the fault only as far as the vicinity of these stations. These stations being 7 to 13 km from the epicenter, respectively, we would then have a maximum along the strike anomalous zone extent of about 8 to 15 km. This result corresponds well with ROBINSON *et al.*'s [10] data for near horizontal raypaths and would imply from a 0.30 sec delay an average *P*-velocity change of 15% over a 10 km path.

The small delay of 0.15 sec indicated in the teleseismic data for station BVL implies a small vertical extent to an existing anomalous zone. This size delay leads us to believe that the vertical extent of the anomalous zone is at least 5 km for a 15% change in *P*-velocity and probably not much thicker than 10 km assuming that the anomalous zone is confined to the seismically active portion of the upper crust in central California.

The correspondence in time of the suggested anomaly in the nearly vertical rayed teleseismic data of this study with the more horizontal rayed data of ROBINSON *et al.* [10] is remarkable. Robinson *et al.* find an anomalous zone exists from January 1 to about January 25 while this teleseismic data suggests anomalous behavior over a shorter period of January 8 to about January 15. Considering that station BVL is 1.5 km southwest of the fault trace, this discrepancy in the length of the anomalous period is not surprising in the light of a small anomalous zone.

The size of the *P*-velocity anomaly associated with the Bear Valley earthquake, a strike-slip event, is quite small when compared to the anomalous zone sizes observed for thrust-type events. Also the precursory and anomalous time periods of the anomalous *P*-velocity zone appear to be much shorter than that predicted by SCHOLZ *et al.* [20]

for a magnitude 5.0 event. NUR *et al.*'s [21] suggestion that strike-slip events should exhibit smaller anomalous zone sizes than trust-type events due to the confining effects of the surrounding crust seems to be supported by all the data available for the 1972 Bear Valley earthquake. The small size of the zone may partially explain the abbreviated precursory and anomalous time periods.

Another contributing factor, also presented by NUR *et al.* [21], is the larger amount of dilatancy required to overcome a higher initial pore pressure or larger fluid flow in a more porous rock before undersaturation can occur and be maintained. There is some evidence for the existence of above normal pore pressures in the coast range east of the San Andreas (BERRY [22]). Berry has proposed that these high pore pressures, ranging up to lithostatic pressure, may be a main contributing factor to the observed behavior of the San Andreas fault in the vicinity of Bear Valley. If higher than normal pore pressures exist in the vicinity of Bear Valley, then the size of any observed anomalous zone could be limited by the presence of high pore pressure as shown by NUR *et al.* [21]. The presence of high pore pressure could also delay the apparent onset of dilatancy and cause the anomalous zone to disappear more rapidly, thus abbreviating the length of the anomalous decrease in time and contributing to the shortened time from anomalous *P*-velocity onset to the occurrence of the earthquake.

The last point to consider is the limits that the teleseismic data imposes on the mechanism causing the observed *P*-velocity anomaly. There have been two general classes of mechanisms proposed as the cause of anomalous velocity zones. The first involves the simple shear loading of a dry crust as the cause of dilatant behavior in the crust. The preferential opening and closing of cracks in certain directions is the cause of the observed velocity changes. NUR [23] has shown in theory and in laboratory observation that this kind of mechanism produces an anisotropic velocity exhibiting decreased *P*-velocity along the axis of least stress, *increased* *P*-velocity along the axis of greatest stress, and *no significant* *P*-velocity change along the intermediate stress axis. GUPTA [28] performed a modified uniaxial stress test on limestone in which one axis perpendicular to the major stress axis was confined. His results indicate a large decrease in *P*-velocity along the minor stress axis, a very small increase in velocity along the major stress axis, and a very small decrease in velocity along the intermediate stress axis becoming significant only at uniaxial loading greater than 90% of the failure strength of virgin limestone. These results agree fairly well with those predicted by NUR [23], except for the accelerated decrease in velocity along the intermediate stress axis which is probably caused by uniaxial loading induced dilatancy overcoming the confining stress along the intermediate axis just before failure, an effect not expected under pure shear loading.

The second general class of mechanisms is based on a change in the bulk modulus of a portion of the crust such as in the well-known dilatancy–fluid flow theory. This theory postulates a wet crust in which pore water interacts with dilatancy. Here the observed *P*-velocity anomalies are due to the change in bulk modulus of a rock-water system when dilatancy causes the opening of dry pore space in a previously saturated portion of the crust. This effect has been discussed extensively by Nur [24], Whitcomb *et al.* [3], and Scholz *et al.* [20] and a general lowering of *P*-velocity would be expected. There have been few laboratory or theoretical studies of possible anisotropy within this zone of lower *P*-velocity. Recently, however, Anderson *et al.* [25] numerically examined the properties of highly oriented, penny-shaped, ellipsoidal cracks on *P*-velocity and found the greatest velocity decrease occurring along directions parallel to the minor stress axis and that the effect tapered off to no velocity change in any direction perpendicular to the minor stress axis.

The results of this study tend to support a change in bulk properties of the crust as the mechanism causing the observed change in *P*-velocity beneath station BVL prior to the February, 1972, Bear Valley earthquake. The observation of a lower *P*-velocity along the intermediate stress axis (vertical for strike-slip faulting) would preclude the simple shear stress in a dry crust mechanism since no significant *P*-velocity change along the intermediate axis is predicted by the shear loading model. The best argument for the bulk modulus mechanism comes from Robinson *et al.*'s [10] observation of a velocity decrease along the *major* stress axis, which is contrary to the behavior predicted by the shear loading model. At present, the only viable bulk modulus mechanism is the dilatancy–fluid flow theory.

5. Conclusions

This study shows that an azimuthal correction can be determined for two-station teleseismic residuals which will allow the incorporation of teleseismic data over a wider azimuth to source range in order to look for time variations in *P*-velocity. As a result of applying this approach, teleseismic data from several stations of the NCER network in the vicinity of Bear Valley, California suggest the existence of a *P*-velocity anomaly beneath Bear Valley prior to the February 24, 1972 event and shows that the vertical and horizontal extent of any anomalous zone for this strike-slip event is quite small. The maximum possible dimensions vary between 5 and 10 km relative to the epicenter, depending on direction. Also the anomalous time period of strike-slip events in central California may be shorter than expected due to the small size of an anomalous zone and possibly due to the effects of high pore pressure in some portions of the central California crust. For this Bear Valley anomaly all the seismic data supports dilatancy–fluid flow as the mechanism causing the observed *P*-velocity variations. Further observations both in central California and in other seismic regions will help to determine whether a bulk modulus mechanism or a simple shear in a dry crust mechanism is more generally applicable as the cause of *P*-velocity variations prior to moderate or large earthquakes.

Acknowledgements

We wish to thank the National Center for Earthquake Research for allowing us access to their seismograph records. This research was supported by the Earth Science Section, National Science Foundation, NSF Grant GA 36135X.

REFERENCES

[1] A. M. SEMENOV, *Variations in the travel-time of transverse and longitudinal waves before violent earthquakes*, Bull. Acad. Sci. USSR, Phys. Solid Earth *3* (1969), 245–248

[2] I. L. NERSESOV, A. N. SEMENOV and I. G. SIMBIREVA, *Space-time distribution of the travel-time ratios of transverse and longitudinal waves in the Garm area*, in *The Physical Basis of Foreshocks* (Nauka Publ. 1969).

[3] J. H. WHITCOMB, J. D. GARMANY and D. L. ANDERSON, *Earthquake prediction: variation of seismic velocities before the San Fernando earthquake*, Science *180* (1973), 632–635.

[4] Y. P. AGGARWAL, L. R. SYKES, J. ARMBRUSTER and M. L. SBAR, *Premonitory changes in seismic velocities and prediction of earthquakes*, Nature *241* (1973), 101–104.

[5] G. S. STEWART, *Prediction of the Pt. Mugu earthquake by two methods*, Stanford Univ. Publ. Geol. Sci. *13* (1973), 473–478.

[6] W. H. BAKUN, R. M. STEWART and D. TOCHER, *Variations* in V_p/V_s *in Bear Valley in 1972*, Stanford Univ. Publ. Geol. Sci. *13* (1973), 453–462.

[7] R. BROWN, *Precursory changes in* V_p/V_s *before strike-slip events*, Stanford Univ. Publ. Geol. Sci. *13* (1974), 463–472.

[8] T. V. MCEVILLY and L. R. JOHNSON, *Earthquakes of strike-slip type in central California: evidence on the question of dilatancy*, Science *182* (1973), 581–584.

[9] T. V. MCEVILLY and L. R. JOHNSON, *Stability of P and S velocities from Central California quarry blasts*, Bull. Seis. Soc. Am. *64* (1974), 343–353.

[10] R. ROBINSON, R. L. WESSON and W. L. ELLSWORTH, *Variation of P-wave velocity before the Bear Valley, California, earthquake of 24 January 1972*, Science *184* (1974), 1281–1283.

[11] R. ROBINSON, R. L. WESSON, W. L. ELLSWORTH and J. A. STEPPE, *Variation in P-wave velocity before earthquakes along the San Andreas fault*, Abstract, EOS, Transactions of the Am. Geophys. Union *55* (1974), 695.

[12] M. WYSS and D. J. HOLCOMB, *Earthquake prediction based on station residuals*, Nature *249* (1973), 139–140.

[13] T. UTSU, *Temporal variations in travel time residuals of P waves from Nevada sources*, J. Phys. Earth *21* (1973), 475–480.

[14] C. H. CRAMER and R. L. KOVACH, *A search for teleseismic travel-time anomalies along the San Andreas fault zone*, Geophys. Res. Letters *1* (1974), 90–92.

[15] M. WYSS and A. C. JOHNSTON, *A search for teleseismic P residual changes before large earthquakes in New Zealand*, J. Geophys. Res. *79* (1974), 3283–3290.

[16] E. HERRIN and J. N. TAGGART, *1968 seismological tables for P-phases*, Bull. Seis. Soc. Am. *58* (1968), 1193–1241.

[17] M. OTSUKA, *Azimuth and slowness anomalies of seismic waves measured on the central California seismographic array. Part I. Observations*, Bull. Seis. Soc. Am *56* (1966), 223–239.

[18] M. OTSUKA, *Azimuth and slowness anomalies of seismic waves measured on the central California seismographic array. Part II. Interpretation*, Bull. Seis. Soc. Am. *56* (1966), 655–675.

[19] O. NUTTLI and B. BOLT, *P wave residuals as a function of azimuth. 2. Undulations of the mantle low-velocity layer as an explanation*, J. Geophys. Res. *74* (1969), 6594–6602.

[20] C. H. SCHOLZ, L. R. SYKES and Y. P. AGGARWAL, *Earthquake prediction: A physical basis*, Science *181* (1973), 803–810.

[21] A. NUR, M. L. BELL and P. TALWANI, *Fluid flow and faulting 1: A detailed study of the dilatancy mechanism and premonitory velocity changes*, Stanford Univ. Publ. Geol. Sci. *13* (1973), 391–404.

[22] F. A. F. BERRY, *High fluid potentials in California coast ranges and their tectonic significance*, Bull. Am. Assoc. Petrol. Geol. *57* (1973), 1219–1249.

[23] A. NUR, *Effects of stress on velocity anisotropy in rocks with cracks*, J. Geophys. Res. *76* (1971), 2022–2034.

[24] A. NUR, *Dilatancy, pore fluids, and premonitory variations of* T_s/T_p *travel times*, Bull. Seis. Soc. Am. *62* (1972), 1217–1222.

[25] D. L. ANDERSON, B. MINSTER and D. COLE, *The effect of oriented cracks on seismic velocities*, J. Geophys. Res. *79* (1974), 4011–4015.

[26] R. L. Wesson, F. W. Lester and K. L. Meagher, *Catalog of earthquakes along the San Andreas fault system in central California, January–March 1973*, US Dept. Interior, Geol. Survey, Open-File Report *74–186* (1974), 6.

[27] C. G. Bufe, J. H. Pfluke and R. L. Wesson, *Premonitary vertical migration of microearthquakes in central California – Evidence of dilatancy biasing?*, Geophys. Res. Letters *1* (1974), 221–224.

[28] I. N. Gupta, *Seismic velocities in rock subjected to axial loading up to shear fracture*, J. Geophys. Res. *78* (1973), 6936–6942.

(Received 7th November 1974)

Pageoph,
Vol. 113 (1975)

Birkhäuser Verlag, Basel

Quarry Blast Sources and Earthquake Prediction: The Parkfield, California, Earthquake of June 28, 1966

By DAVID M. BOORE[1]), THOMAS V. MCEVILLY[2]) and ALLAN LINDH[1])

Summary – Quarry blasts provide an excellent source of waves with which to study temporal variations of seismic travel times. Using sources from several active quarries we found no time changes at Gold Hill (GHC), California from 7 months before the 1966 Parkfield mainshock to at least 13 months following the event. GHC is within 0.5 km of the zone of ground breakage associated with the earthquake. This event was one of the largest to occur in central California since 1906. The negative result of our experiment does not rule out the existence of an anomaly prior to the event, rather it emphasizes the difficulty of detecting such anomalies without a fortuitous combination of sources and stations. The routine monitoring of temporal changes in seismic velocity along the San Andreas fault may require not only dense networks of high quality seismic stations, but also sources whose spatial location and origin time can be controlled.

1. *Introduction*

Quarries which detonate large explosions at regular intervals provide a reliable source of waves with which to probe for systematic changes in seismic velocities; uncertainties due to focal depth and epicentral location are not important and sources can be outside the zone of seismic activity. In central California a number of quarries are available. In a previous study (MCEVILLY and JOHNSON [1]), blasts from the Natividad quarry (Fig. 1) were used. Although the scatter in the data was quite small (less than ± 0.1 sec), no anomalous velocity behavior was found, leading the authors to suggest a maximum dimension of 10 km for the volume experiencing premonitory velocity changes. During the same time period, however, an anomaly of 0.3–0.4 sec was observed for an earthquake of magnitude 5.1 (BRK) in the Bear Valley region of the San Andreas fault, using local earthquakes, 20–60 km distant, recorded at a station in Bear Valley (ROBINSON *et al.* [2]). The apparent inconsistency between the study using regional earthquakes and that using quarry blasts is easily explained by lack of blast–receiver travel paths in the small anomalous region and by the absence of a blast during the anomalous time period. The two studies, along with that of CRAMER and KOVACH ([3]), emphasize the difficulty in observing travel time variations along the San Andreas fault; a fortuitous combination of sources and receivers is required.

[1]) Geophysics Department, Stanford University, Stanford, California 94305.

[2]) Seismographic Station, Department of Geology and Geophysics, University of California, Berkeley, California 94720.

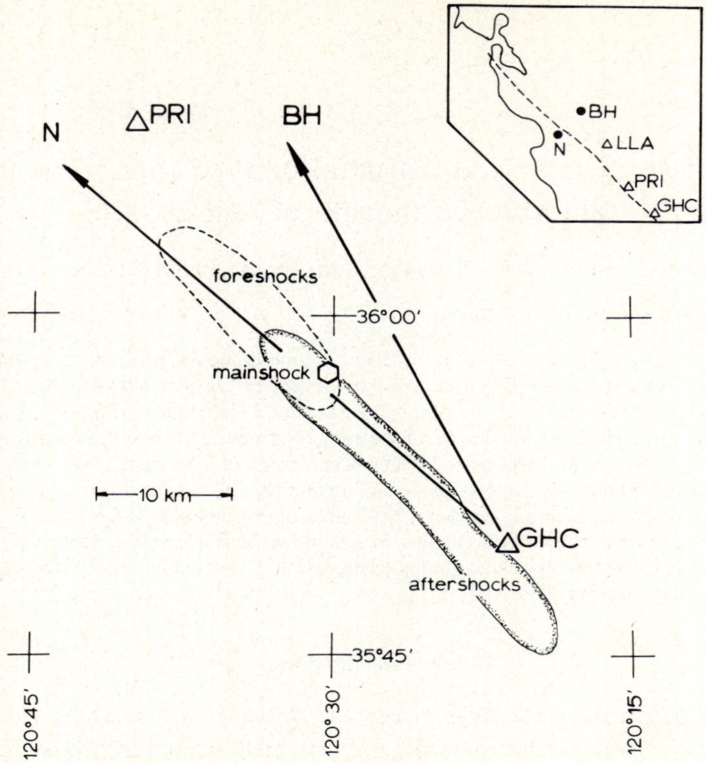

Figure 1
Location map showing the stations Gold Hill (GHC) and Priest (PRI), with arrows pointing to the Basalt Hill (BH) and Natividad (N) quarries. Regions of fore- and aftershock activity are shown schematically. Inset: location of stations and sources

2. Data

In this study blasts at the Basalt Hill (BH) and Natividad (N) quarries (Fig. 1) were used to search for variations in seismic velocity in the vicinity of the June 28, 1966, earthquake near Parkfield. The Parkfield earthquake ($M_L = 5.5$, $m_b = 5.8$, $M_s = 6.4$) is one of the most significant central California earthquakes since the great 1906 San Francisco earthquake. We expected that the source–receiver geometry and the frequency of blasts would provide an almost ideal experiment. Basalt Hill quarry blasts occurred almost weekly in early 1966, and a high quality, short period seismic station was installed at Gold Hill (GHC) 8 months prior to the earthquake. As seen in Fig. 1, GHC is near the southern end of the aftershock area. The arrival times at GHC were compared with those from PRI and LLA. Discussions of the stations can be found in EATON *et al.* ([4]) and MCEVILLY and JOHNSON ([1]). The timed arrival corresponded to a uniquely identifiable phase within 1 second of the initial *P*-wave

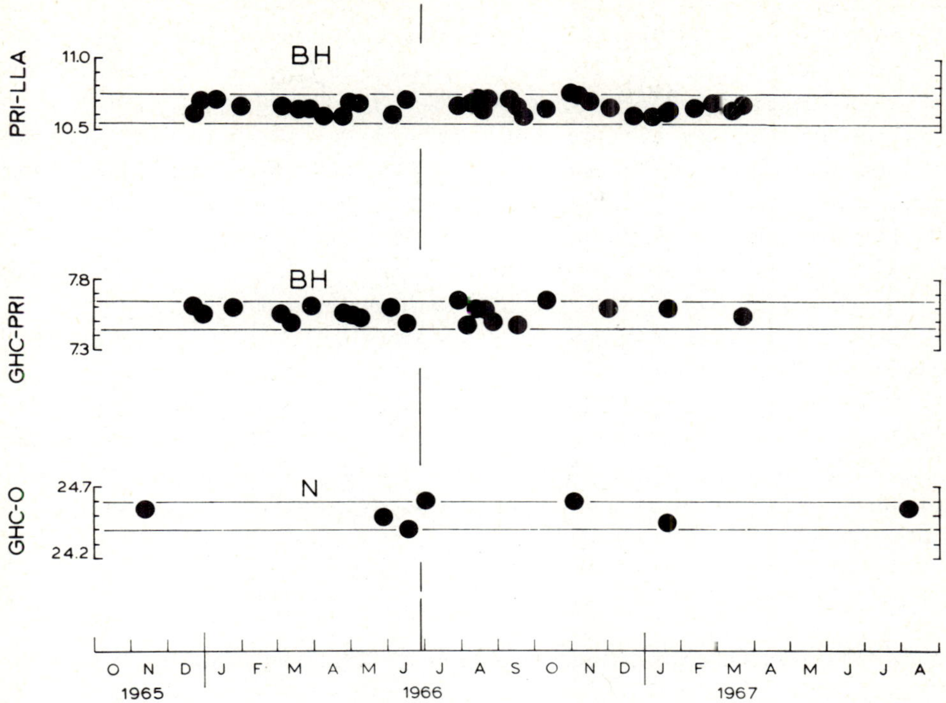

Figure 2
The residuals from BH and N as recorded at LLA, PRI, and GHC. Units of ordinate are seconds; vertical line shows the date of the Parkfield mainshock. The light lines through the data are ± 0.1 sec from the estimated mean and correspond to the approximate uncertainty in our readings

arrival. Transparent overlays were used to aid in the identification of the phase. Uncertainties in the arrival time are about ± 0.1 sec for the GHC data and about ± 0.05 sec for the PRI and LLA data.

The data are shown in Fig. 2. The time differences GHC–PRI and PRI–LLA are shown in the top bands for the Basalt Hill (BH) quarry blasts (148 km from GHC). The data from the Natividad (N) quarry (149 km from GHC), in the third band, are referenced to the origin time of the blast (as given by McEvilly and Johnson [1]). No temporal variations of travel time are apparent within the accuracy of our readings. Data from other, somewhat less accurate, sources are in accord with the stable results shown by the quarry blast observations (Boore *et al.* [5]).

3. Discussion

The lack of any variations in travel time can be explained in many ways. If, by analogy with the results of Robinson *et al.* ([2]) for a magnitude 5 earthquake along the fault to the north, we assume that a decrease in seismic velocity was associated with the Parkfield earthquake, then there is an outside chance that the anomaly had returned to zero before the installation of GHC 8 months prior to the earthquake and thus was missed by our measurements. This seems unlikely to us (we estimate a precursor time of 140–500 days based on the results of ROBINSON *et al.* [2] and SCHOLZ *et al.* [6]).

There are several more likely explanations involving the spatial extent of the region of anomalous velocity. If the region were only 5 km in depth and several km in width, velocity changes of 10–15% would be undetected both because of the small effect on the rays traveling through the zone and because of diffraction around the zone. On the other hand, a larger zone of low velocity material would be undetected if it existed in the region of the foreshocks to the north of the main shock epicenter.

Overall, it is disconcerting that, with a sensitivity of ± 0.1 sec at a station essentially within the aftershock zone, one of the largest central California earthquakes since 1906 showed no velocity precursor. The implication for earthquake prediction along the San Andreas fault is that dense networks of high quality seismic stations and, ideally, controlled sources are required. Furthermore, it is possible that the zone of anomalous velocity change may not coincide with the full zone of subsequent earthquake rupture.

Acknowledgements

We thank the Office of Earthquake Research, US Geological Survey, for providing access to data and facilities. Discussions with ROBERT A. PAGE stimulated the research in this paper.

REFERENCES

[1] T. V. MCEVILLY and L. R. JOHNSON, *Stability of P and S velocities from central California quarry blasts*, Bull. Seism. Soc. Am. *64* (1974), 343–353.

[2] R. ROBINSON, R. L. WESSON and W. L. ELLSWORTH, *Variations of P-wave velocity before the Bear Valley, California, earthquake of 24 February 1972*, Science *184* (1974), 1281–1283.

[3] C. H. CRAMER and R. L. KOVACH, *A search for teleseismic travel-time anomalies along the San Andreas fault zone*, Geophys. Res. Lett. *1* (1974), 90–92.

[4] J. P. EATON, M. E. O'NEILL and J. N. MURDOCK, *Aftershocks of the 1966 Parkfield-Cholame, California, earthquake: a detailed study*, Bull. Seism. Soc. Am. *60* (1970), 1151–1197.

[5] D. M. BOORE, A. LINDH, T. V. MCEVILLY and W. TOLMACHOFF, *A search for travel time changes associated with the Parkfield, California, earthquake of 1966*, Bull. Seism. Soc. Am. *65* (1975), in press.

[6] C. H. SCHOLZ, L. R. SYKES and Y. P. AGGARWAL, *The physical basis for earthquake prediction*, Science *181* (1973), 803–810.

(Received 12th December 1974)

Pageoph,
Vol. 113 (1975) Birkhäuser Verlag, Basel

A Search for Precursors to the Sitka, 1972, Earthquake: Sea Level, Magnetic Field, and *P*-Residuals

By MAX WYSS[1])

Summary – The Sitka earthquake of 30 July, 1972, was a shallow strike-slip fracture of 170 km length and it closed a gap defined by SYKES (1971). A tide gauge station, a magnetic observatory and a seismological station are located 40 km to the east of the center of the rupture. A comparison of the sea level at Sitka with that at three tide gauges up to 430 km from Sitka indicates that the sea level at Sitka remained constant for 32 years before the earthquake (standard deviation ± 1.5 cm) with the exception of the years 1966 and 1967 when the land may have been elevated by 3 cm. Preliminary analysis of mean teleseismic-*P*-residuals suggests that a *P*-wave delay may have existed under the station Sitka during 1966 and 1967. The quiet annual means of the horizontal component of the magnetic field at Sitka were compared to that at College, Alaska, and Victoria, Canada, for all years since the establishment of these stations in 1941 and 1957 respectively. Starting 7.5 years before the earthquake the field differences appear to have decreased by approximately 20γ, gradually returning to normal the year after the earthquake.

1. *Introduction*

The Sitka earthquake occurred on 30 July, 1972, 40 km offshore from Sitka Island, Alaska, along the Fairweather fault. The aftershock area was approximately 170 km long (PAGE and GAWTHROP, 1974), with hypocenters ranging from 0 to 10 km (PAGE, personal communication, 1974). The focal mechanism and the aftershocks show that right lateral strike-slip motion occurred across an almost vertical plane, striking N24°W. The rupture filled a seismic gap defined by SYKES (1971) (Fig. 1). The tectonic setting is practically identical to that of the Northern San Andreas Fault accommodating slip between the Pacific plate and the American plate. The magnitude determinations were M_b (NOAA) = 6.5, M_s (NOAA) = 7.6, M(PAS) = 7.2, M(BKS) = 7.1. We shall consider an average magnitude of $M = 7.1$ for the discussion.

The purpose of this paper is to determine whether any of the geophysical data collected on Sitka Island (57°03.5′N, 135°19.51′W) indicated the approach of this earthquake. Three kinds of data are available: (1) Sea level has been recorded since 1938. (2) The earth's magnetic field has been monitored continuously since 1902. (3) A seismograph station has been operating since before 1961.

The dimensions of the volumes in which precursory anomalies occur are often

[1]) Cooperative Institute for Research in Environmental Sciences, University of Colorado/NOAA, Boulder, Colorado 80302.

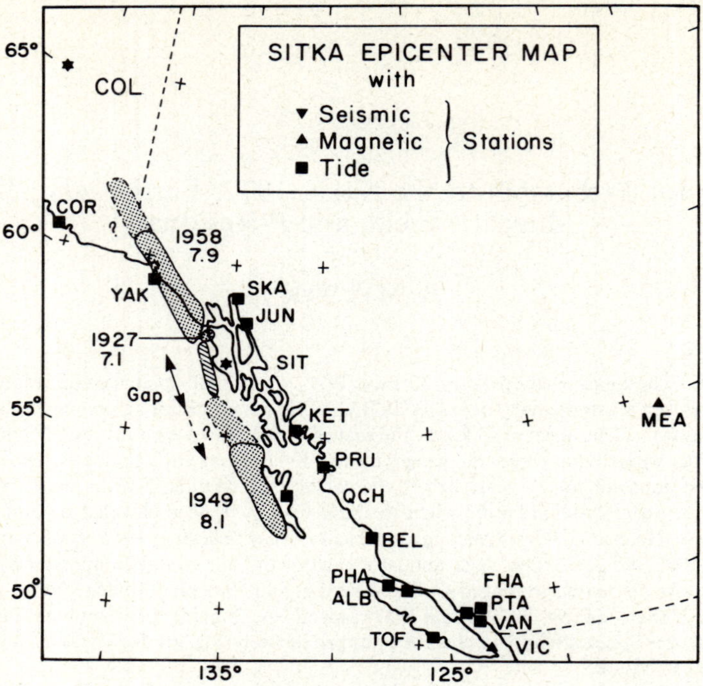

Figure 1
Map showing the aftershock area of the Sitka earthquake (shaded), aftershock areas of recent large earthquakes north and south of Sitka (stippled after SYKES, 1971) and seismic, magnetic as well as tide gauge stations

several times larger than the dimensions of the respective earthquakes. The measurements on Sitka Island were taken at 40 km from the center of the rupture. Half the rupture length was 85 km. If precursory anomalies occurred before this earthquake we may expect that they could have been observed on Sitka Island, unless the anomalous volume was very narrow and long.

2. *Vertical displacements*

If dilatancy precedes shallow earthquakes the earth's crust will expand in the direction of the least principal stress. If the intermediate principal stress is not much larger than the least principal stress one could expect also expansion in the intermediate direction. One may therefore find dilatancy uplift to precede strike-slip earthquakes. The pattern of possible uplift in space and time has been discussed by NUR (1974) and WYSS (this issue). HANKS (1974) has demonstrated that it is important to find more constraints to the dilatancy hypothesis for the earthquake source mechanism, and that it is especially important to demonstrate how much dilatancy uplift, if any, may occur before earthquakes.

The annual mean sea levels at Sitka for the years 1938 through 1972 (HICKS and CROSBY, 1974) are plotted as the top curve in Fig. 2. In this figure sea level has been plotted as increasing downward so that corresponding land uplift is in the upward direction. It is evident that no changes in water level exceeding the scatter took place before the earthquake. The standard deviation of the 24-year sample before the earthquake is ±3 cm.

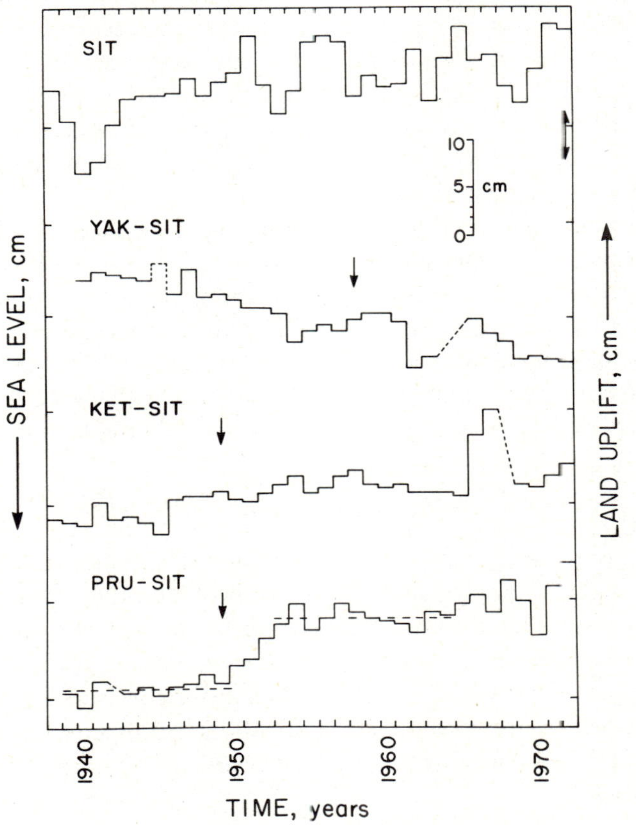

Figure 2
Annual mean sea level (increasing downwards) as a function of time at Sitka and the difference of annual mean sea level between the three closest stations and Sitka

A better means of identifying small amounts of uplift may be obtained by comparing the mean annual sea level at Sitka with that at nearby tide gauges where meteorological conditions are thought to be similar to those at Sitka. The locations of all continuously recording tide gauge stations are shown in Fig. 1. The land at Skagway (SKA) and Juneau (JUN) is rising rapidly (approximately 2 cm/year), evidently in response to decreasing ice load (HICKS and SHOFNOS, 1965). For this reason these two stations have not been used as control stations. The annual mean sea levels at SIT were subtracted from those at Yakutat (YAK), Ketchikan (KET) and Prince Rupert (PRU) and plotted in Fig. 2. The data from the station YAK are included even though this station is estimated to rise 0.7 cm/y (HICKS and SHOFNOS, 1965), and we must estimate possible uplift at SIT with respect to the long-term trend of YAK-SIT.

Two large earthquakes which occurred north and south of Sitka may have affected the mean sea levels at the reference stations, and make the comparison difficult. These were the Queen Charlotte (1949) and the Fairweather (1958) earthquakes whose aftershock areas are shown in Fig. 1, and whose times of occurrence are indicated by arrows in Fig. 2. The sea level changes that may have been associated with these earthquakes (see PRU–SIT Fig. 2) are discussed in detail by WYSS (this issue). Further, the possibility of errors comes to mind when one observes the unusually large differences between KET and SIT in 1966 and 1967, which are due to exceptionally high sea level reports from KET during those years.

Allowing for these undesired properties of the reference stations, we may state that in all three comparisons it appears that the land at Sitka stood a little higher than the average during 1966–1968. The standard deviations of the differences in sea level are approximately 1.5 during 13 years before 1966. It appears that during 1966 to 1968 Sitka may have been uplifted by approximately 3 cm $\pm$ 3 cm. The best available estimate of vertical movements at Sitka is obtained from 1/3 (YAK + KET + PRU) – SIT. This is shown in Fig. 7, and it also suggests a 3 cm uplift during 1966 and 1967.

3. The magnetic field at Sitka

The possibility of a tectonomagnetic precursor to the Sitka earthquake can be estimated by comparing the magnetic field at Sitka with that at the observatories nearest to Sitka (Fig. 1). Although the horizontal component, H, at Sitka varied by hundreds of γ during this century (Fig. 3) it is the quietest of the three magnetic field components and will be used in all computations. H at Sitka may be considered to be composed of H_c, the dipolar field of the earth; H_A, the non-dipolar component originating inside the earth; H_s, the field due to solar and ionospheric activity; and possibly of H_T, a tectonic component.

$$H(\text{SIT}) = H_c(\text{SIT}) + H_A(\text{SIT}) + H_s(\text{SIT}) + H_T(\text{SIT}).$$

If neighboring stations are free of magnetic changes due to tectonic stresses, the field may be written as

$$H(\text{VIC}) = H_c(\text{VIC}) + H_A(\text{VIC}) + H_s(\text{VIC}).$$

Since Sitka is located approximately halfway on a great circle path between Victoria and College, we may estimate the secular changes in the sum of $H_c + H_A + H_s$ at Sitka by taking the difference $H(\text{VIC}) - H(\text{COL})$. The horizontal quiet annual means of Victoria and College are compared to those at Sitka in Fig. 3. The difference $H(\text{VIC}) - H(\text{COL})$ is shown in Fig. 4 and it is evident that the horizontal field difference between

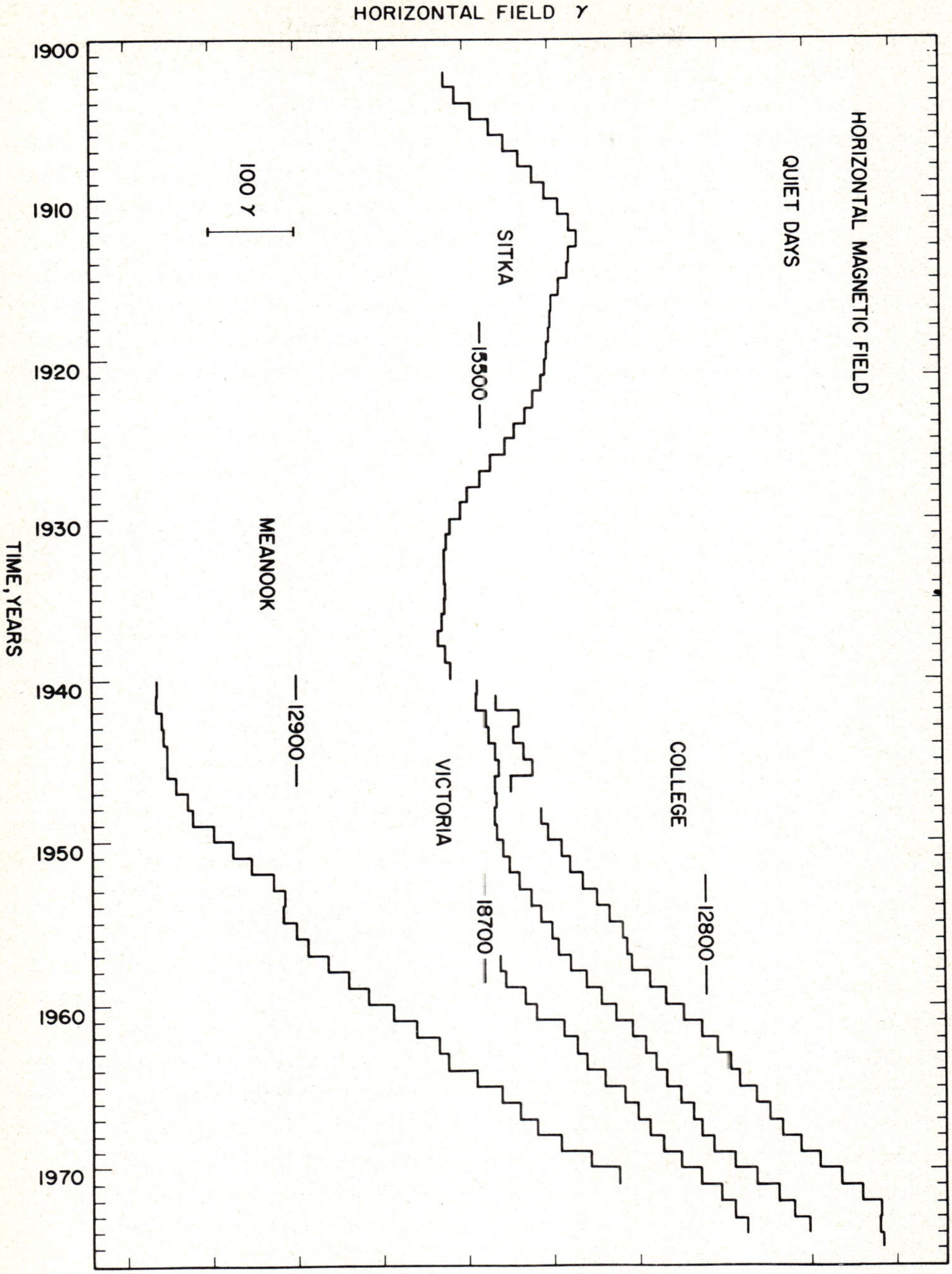

Figure 3
Annual mean horizontal magnetic field (quiet days) as a function of time at Sitka and the three closest observatories to it

VIC and COL was constant, $H(\mathrm{VIC}) - H(\mathrm{COL}) = 6000 \pm 8\gamma$, during the last 17 years. The largest changes between yearly mean differences (VIC – COL) occurred at the beginning of the record and between 1972, 1973. The variations at the beginning are apparently due to less reliable data from VIC where a permanent building was established in 1961 (see arrow in Fig. 4). The jump at the end of the record is due to a change in trend at COL. However, on the whole the constancy of the difference (VIC – COL) is remarkable and from this we may conclude that changes in large-scale features of the non-dipolar field H_A, and of H_c and H_s at Sitka can be eliminated by taking the differences between the control stations and Sitka. We obtain $H(\mathrm{control}) - H(\mathrm{SIT}) = \mathrm{constant} + H_T(\mathrm{SIT})$, where constant ≈ 3000 (see Fig. 4). The station Meanook, located to the southeast of Sitka, cannot be included in the comparison because its field (Fig. 3) changes rapidly with respect to the other control stations VIC and COL.

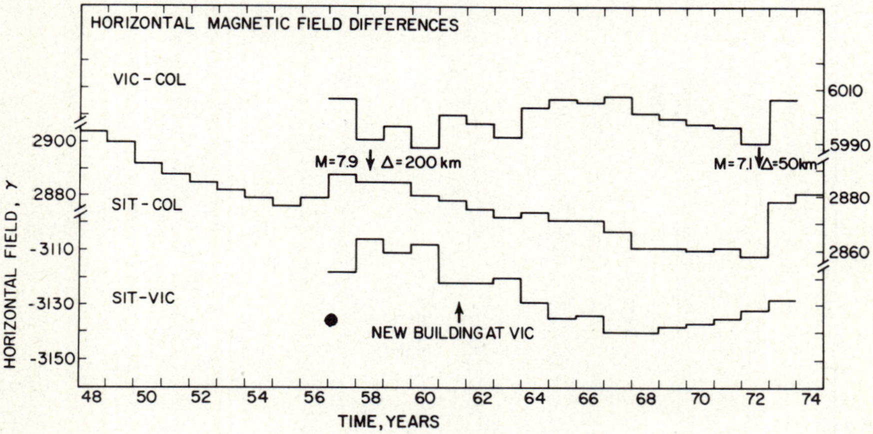

Figure 4
Differences of the annual mean horizontal magnetic field between the stations Sitka, College and Victoria as a function of time

The differences (SIT – COL) and (SIT – VIC) are less constant in time than (VIC – COL) (Fig. 4). Both differences between SIT and the reference stations show a bay before the 1972 Sitka earthquake. This bay is interpreted as a 20 γ reduction of the H field at Sitka starting about 7.5 years before the earthquake and returning slowly back to normal over recent years. It is likely that this magnetic field decrease was related to the Sitka earthquake. It must be emphasized that the 20γ increase of (SIT – COL) in 1973 is not considered to be related to the earthquake because (SIT – VIC) shows no such step and because the record at COL clearly changes trend in 1973 (Fig. 3).

The early years of (SIT – COL) data suggest that there may exist a trend of the difference to decrease by 1γ/year. Superimposed on this trend there appear to exist the anomaly before the Sitka and another anomaly before the 1958 Yakutat earthquake. This shock had a 400 km long aftershock zone of which the closest point was located 100 km from the Sitka observatory. The epicenter was at 200 km distance (Fig. 1).

The sum of $1/2[(\mathrm{SIT}-\mathrm{COL})+(\mathrm{SIT}-\mathrm{VIC})]$ is the best estimate available of magnetic field variations at Sitka (Fig. 7). In this figure the Victoria values which pre-dated the establishment of the present building at that observatory were adjusted by -14γ. The amplitude of the magnetic anomaly is estimated to be -20γ, and its development began 7.5 years before the earthquake.

It is remarkable that this possible precursor time happens to fit exactly the value expected for a magnitude 7.1 earthquake (SCHOLZ *et al.*, 1973). Could this be just a coincidence? Yes, it could, and the data to rule out this possibility are not available because a much denser network of magnetic stations would be needed to define non-dipolar anomalies of a few hundred km dimension. An anomaly of -20γ amplitude and 200 km dimension could have drifted over SIT without affecting COL or VIC. Solar activity, on the other hand, is ruled out as a source of the anomaly because all stations are affected in phase and at approximately the same amplitude of 10γ.

4. *P-wave velocity*

If the teleseismic *P*-arrival times at a seismic station are delayed by more than 0.2 sec, the effect can be detected by considering mean residuals at that station. Such a delay was detected before the Gisborne, 1966, earthquake using the residuals published by the International Seismological Center (ISC) (SUTTON, 1974; WYSS and JOHNSTON, 1974).

For Sitka, the *P*-residual data set consisted of 1418 values between ± 5 sec and from earthquakes at a distance $\Delta \geqslant 15°$ from Sitka. Only those residuals were used which were obtained from earthquakes located by means of more than 25 readings. Of these residuals 683 were reported by the ISC, 450 by the Preliminary Determination of Epicenters (PDE) and 233 by the International Seismological Summary (ISS). Since only large signals are recorded at Sitka, 90% of the used events were located by more than 100 stations except for the PDE data (1971–1973) where 32% were located with more than 100 stations.

First 150 consecutive residuals of the ISC and PDE set starting in January 1964 were averaged (dotted line in Fig. 5). Since the ISS residuals are given to the nearest second only, all the 233 events for the years 1961–1963 were included as one average in Fig. 5. The resulting first plot of the average residuals (dotted line in Fig. 5) suggested that the average residuals from the three data sources agreed with each other within the standard deviation of the averages and the residuals from the years 1966 and 1967 deviated from the long-term mean by 0.5 sec.

The possibility that an aftershock sequence or a peculiar source distribution might have caused a bias during the anomalous period was ruled out by the following tests. The average of residuals of earthquakes from depths greater than 80 km (dashed-dotted line in Fig. 5) also showed the same anomalous period. The residuals of Gutenberg–Richter regions with similar average residuals were merged into six composite

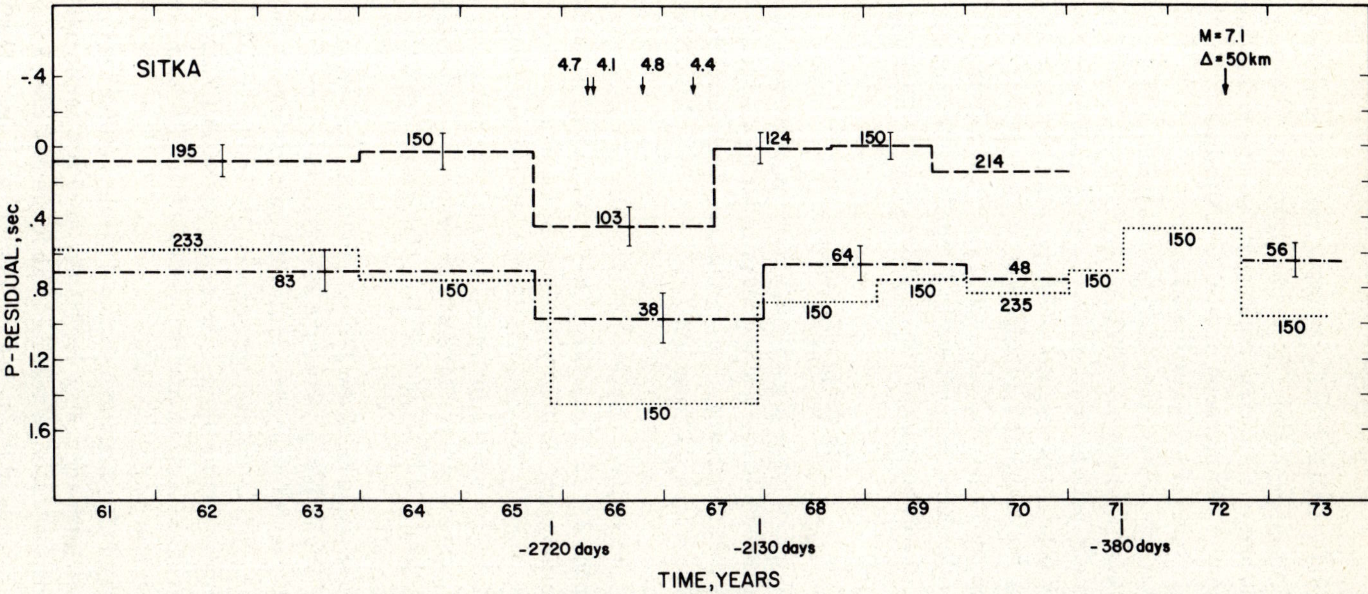

Figure 5

Mean P-wave residuals as a function of time at Sitka. Arrows indicate occurrence of earthquakes within 100 km from Sitka with magnitude shown. Dotted line includes all residuals selected, dash-dotted includes only residuals from events originating deeper than 80 km, and dashed shows path corrected residuals for all events. Number of residuals per average are indicated along these curves

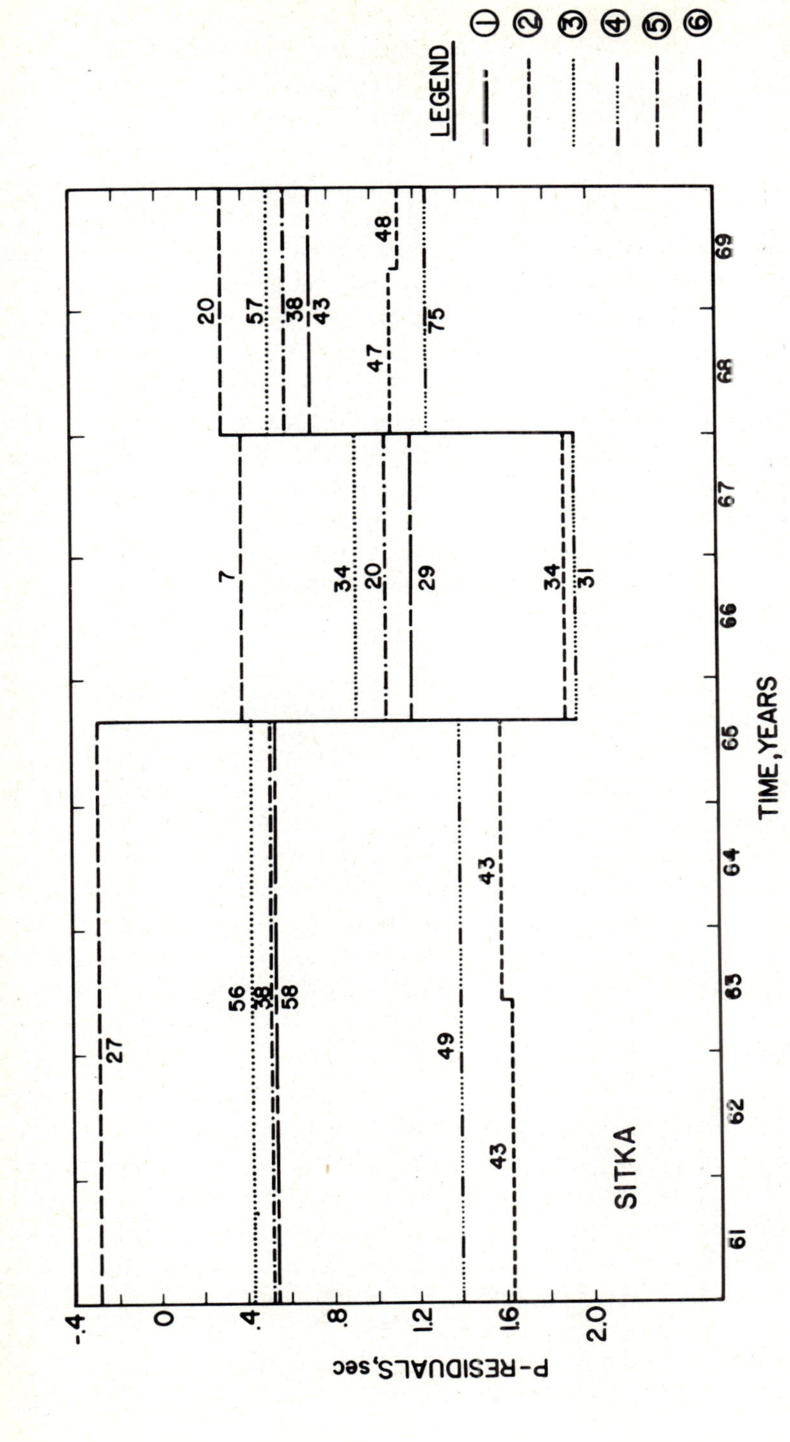

Figure 6

Mean P-wave residuals at Sitka grouped by region of origin. These groups include the following Gutenberg-Richter regions: 1–1; 2–19, 41, 46; 3–15, 16, 17, 18, 20, 21; 4-Eurasia; 5–2, 3, 4, 5, 6, 8, 12, 34; 6–7, 13, 14, 24 regions. The average residuals of each of the composite regions increases during the anomalous time period (Fig. 6). The difference between residuals for the same earthquake recorded at College and at Sitka (not shown) exhibit the same pattern as the deep earthquakes; a marginal increase during 1966–1967.

There are several problems with this data set. The deep earthquakes, which are the most reliable sources, do not show a significant difference between the pre-anomalous and the anomalous period. The error bars in Fig. 5 are the standard deviations of the mean. In the case of deep sources only, the anomaly is only two standard deviations different from the earliest sample period. This is not sufficient to make a convincing case for the anomaly. This uncertainty may be clarified by recomputing the residuals for 1961 through 1963 to the nearest tenth of a second and by re-reading all the records. To improve the data all the records should be re-read because in 1968 a change in recording procedure was made. Starting in 1968 the Sitka records were telemetered to Palmer, Alaska. From that time on, most readings were obtained from the develicorder records at Palmer with Palmer time corrections. Before that time the readings were obtained from the Sitka paper records using the Sitka time corrections. Because of the above problems, the *P*-residual results can be considered preliminary only.

5. Discussion and conclusions

During the decade preceding the Sitka earthquake, only 4 events larger than approximately magnitude 3.5 occurred within 100 km of Sitka (Fig. 5). Such a lack of local earthquakes would have made it difficult to study changes in velocities with an array of local stations. Teleseismic arrivals are also not abundant because the proximity of the ocean makes Sitka a noisy low-gain seismic station. The residual data that are available suggest that during 1966 and 1967 there may have been a decrease in *P*-velocity beneath Sitka. It is interesting though that these same two years are also marked by comparatively low sea level which can be interpreted as a temporary uplift of 3 cm (Fig. 7). This rather small upper limit for uplift was obtained at a distance of one-quarter of the rupture length. If dilatancy did occur, one might have expected a larger uplift. However, the aftershocks indicated the rupture extended only to approximately 10 km depth (PAGE and GAWTHROP, 1974). An anomaly associated with such a shallow feature may decrease very rapidly away from the fault. If we assume an average shear strain of 10^{-5} between Sitka and the inaccessible fault trace, the uplift at the fault would have been an order of magnitude larger than at Sitka. For a 10 km dilatant crust, an uplift of 3 cm implies approximately $\Delta V/V = 10^{-5}$ (where ΔV is the increase in volume V). Depending on the incremental crack growth and the initial porosity (HANKS, 1974) as well as depending on the aspect ratio of cracks (NUR *et al.*, 1973), dilatancy in the vertical direction could have been sufficient to cause a 10% drop in v_p in the top 10 km of the crust as the *P*-residual data suggest.

The magnetic data in Figs. 4 and 7 clearly show that the horizontal field decreased temporarily near Sitka before the Sitka as well as the Yakutat earthquake. That these

20γ anomalies were causally related to the respective earthquakes is suggested by the fact that the field difference with respect to the control stations seemed to return to pre-anomaly values after the event. However, the data of future years will be needed to firmly establish what the 'normal' differences should be. Also suggestive is the fact that the magnetic anomaly before the Sitka earthquake started 7.5 years before the event which is the expected precursor time for earthquakes of magnitude about 7 (SCHOLZ *et al.*, 1973).

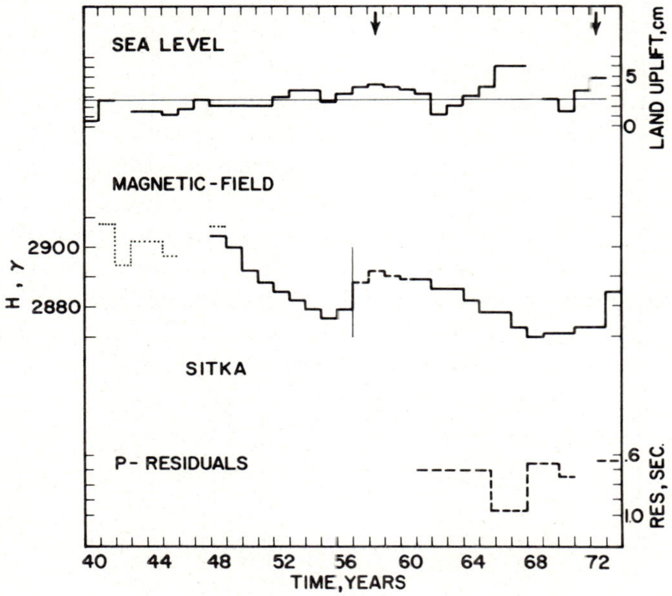

Figure 7

Summary estimate of uplift, magnetic field change and residual change before earthquakes near Sitka. Sea level (increasing downward) is estimated from (YAK + KET + PRU)/3 − SIT. The magnetic field is obtained from SIT $-\frac{\text{COL}+\text{VIC}}{2}$ after 1956; before that the curve shows SIT − COL where the dotted quiet means were estimated by subtracting 18γ from the all day means. The residual means are the same as in Fig. 5 for sources with depth larger than 80 km

RIKITAKE (1974) considers only two magnetic precursors as reliably demonstrated; the one before the Niigata earthquake (FUJITA, 1965) and the one at Tanabe (TAZIMA, 1966). In these cases the amplitudes reached 20γ and 35γ, respectively, approximately the same amount as in the present case. On theoretical grounds SHAMSI and STACEY (1969) and NAGATA (1970) expect tectonomagnetic affects associated with earthquakes to be less than 10γ. Their calculations are based on piezomagnetic laboratory experiments below 1.5 kbar where rocks respond elastically to stress. The calculations, furthermore, are based on the elastic response of a half-space to stresses of less than 100 bar. The case of 100 bar stress drops at a few kbar ambient stress level has not been considered. More importantly a first study of the piezomagnetic behavior of dilating rocks (MARTIN and WYSS, this issue) shows that dilatancy and creep are associated with non-linear and large changes of the remanent magnetization in laboratory samples. The theoretical studies, therefore, may not be accurate in estimating the magnitude of tectonomagnetic precursors. In addition, measurements by HASBROOK and ALLEN (1972) have shown that on Amchitka a 60 bar stress change caused by Cannikin effected a 9γ change in the magnetic field at a station 3 km from the explosion. It appears, therefore, that the 20γ anomaly suggested at Sitka is not unreasonably large. To demonstrate conclusively that the magnetic anomaly at Sitka was not due to a westward drifting non-dipolar field anomaly, one would need a much tighter array of magnetic observations than was available.

Obvious places to search for precursors are seismic gaps, that is, portions of active faults which have not broken recently. Off the coast of Sitka a seismic gap of 150 km minimum length was defined by SYKES (1971). On shore, 40 km east of the center of this gap there exist observatories routinely recording sea level, the magnetic field and seismic events. The data from these observatories were not enough to predict the Sitka earthquake. However, they were suggestive enough that a special study with many observation points might have been initiated. In summary one may only say that at 40 km from the Sitka fault dilatancy uplift has not exceeded 3 cm and was confined to the years 1966 and 1967. A 10% *P*-velocity reduction may have existed in the crust under Sitka during 1966 and 1967, but the data are marginal. The horizontal magnetic field at Sitka decreased by 20γ starting 7.5 years before the earthquake and returned to normal the year after the event.

Acknowledgements

I thank T. RIKITAKE for suggesting the study of the magnetic data at Sitka. I am grateful to K. L. SVENDSEN, Solid Earth and Marine Geophysical Data Service NOAA, and G. J. VAN BECK, Energy, Mines and Resources Canada, for making available to me the magnetic data of the US and Canadian observatories respectively. I thank E. LARSEN and C. KISSLINGER for their numerous critical remarks on the manuscript. This study was sponsored by the US Geological Survey Contrast No. 14-08-0001-14571.

REFERENCES

N. FUJITA, *The magnetic disturbances accompanying the Niigata earthquake*, J. Geodetic Soc. Japan *11* (1965), 8.

T. C. HANKS, *Constraints on the dilatancy–diffusion model of the earthquake mechanism*, J. Geophys. Res. *79* (1974), 3023.

W. P. HASBROUCK and J. H. ALLEN, *Quasi-static magnetic field changes associated with the Cannikin nuclear explosion*, Bull. Seism. Soc. Am. *62* (1972), 1479.

S. D. HICKS and W. SHOFNOS, *The determination of land emergence from sea level observations in south east Alaska*, J. Geophys. Res. *70* (1965), 3315

S. D. HICKS and J. E. CROSBY, *Trends and variability of yearly mean sea level 1893–1972*, NOAA Technical Memorandum NOS13 National Ocean Survey (1974).

R. J. MARTIN, III, and M. WYSS, *Rock magnetism and volumetric strain in uniaxial failure tests*, Pure appl. Geophys., this issue, 1975.

T. NAGATA, Tectonomagnetism, IAGA Bull. *27* (1970), 12.

A. NUR, M. L. BELL and P. TALWANI, *Fluid flow and faulting, 1: A detailed study of the dilatancy mechanism and premonitory velocity changes*, in Proc. Conference on Tectonic Problems *13* (1973), 391.

A. NUR, *Matsushiro, Japan, earthquake swarm: confirmation of the dilatancy–fluid diffusion model*, Geology *2* (1974), 217.

R. PAGE and W. GAWTHROP, *The Sitka, Alaska, earthquake of 30 July 72 and its aftershocks*, Abstract, Seism. Soc. Amer. meeting May 1973.

T. RIKITAKE, *Earthquake Precursors, in preparation.*

C. H. SCHOLZ, L. R. SYKES and Y. P. AGGARWAL, *The physical basis for earthquake prediction*, Science *181* (1973), 803.

S. SHAMSI and F. D. STACEY, *Dislocation models and seismomagnetic calculations for California 1906 and Alaska 1964 earthquakes*, Bull. Seism. Soc. Am. *59* (1969), 4, 1435.

D. J. SUTTON, *A fall in P-wave velocity before the Gisborne, New Zealand, earthquake of 1966*, Bull. Seism. Soc. Am. *64* (1974), 1501.

L. R. SYKES, *Aftershock zone of great earthquakes, seismicity gaps, and earthquake prediction for Alaska and the Aleutians*, J. Geophys. Res. *76* (1971), 8021.

M. TAZIMA, *Accuracy of recent magnetic survey and locally anomalous behaviour of the geomagnetic secular variation in Japan*, Ph.D. Thesis, University of Tokyo (1966).

M. WYSS and A. C. JOHNSTON, *A search for teleseismic P-residual changes before large earthquakes in New Zealand*, J. Geophys. Res. *79* (1974), 3283.

M. WYSS, *Mean sea level before and after some great strike-slip earthquakes*, in preparation.

Pageoph,
Vol. 113 (1975) Birkhäuser Verlag, Basel

Deep Borehole Observation of the Earth's Crust Activities Around Tokyo – Introduction of the Iwatsuki Observatory

By HIROSHI TAKAHASHI[1,2]) and KAZUO HAMADA[1,3])

Summary – As a part of the Japanese earthquake prediction programme, deep borehole observation of the earth's crustal activities has been planned in and around Tokyo by the National Research Center for Disaster Prevention. The first place selected was in the city of Iwatsuki, about 30 km from the centre of Tokyo. Starting in 1970, the main construction was completed at the end of 1973. The tentative observation of micro-earthquakes and crustal movement by tiltmeters has been carried out since May, 1973. As expected, noise at the bottom of the hole (3500 m depth) is extremely small, about 1/1000–1/300 of that at the surface. We could, possibly, detect earthquakes with magnitudes greater than 2.0 within 50 km radius from the Iwatsuki observatory.

1. Observation of the earth's crust activities around Tokyo, the capital of Japan

From the viewpoint of disaster prevention and earthquake prediction, Tokyo is the most important place in Japan and is assigned as 'a special region for observation' by the Coordinating Committee for Earthquake Prediction of Japan, because of the city's function as a centre of political, economic, and cultural activities of the Japanese people. Unfortunately, Tokyo, which has a large population (more than 11 million, exceeding 10% of the whole of Japan), has repeatedly suffered considerable damage from great earthquakes such as the great Kwanto earthquake of 1923. Figure 1 shows the location of historical earthquakes that caused damage in Tokyo (1600–1969). High-sensitivity observations of earthquakes and the movement of the earth's crust, which supply important data for earthquake prediction, are difficult around Tokyo because of high-level background noise generated by active industry and concentrated transportation. In addition, the Tokyo region is covered by thick, soft layers of the Quaternary and Tertiary, furthermore, there is subsidence of the ground, caused by the pumping of ground water and natural gas. The depth of the basement, which is thought to be pre-Tertiary, is illustrated in Fig. 2. Such disturbances make high-sensitivity observation of crustal activities very difficult at the surface in this region. That is why deep borehole observation has been planned in and around Tokyo, although it is more expensive than the ordinary observation on the ground surface.

Our National Research Center has carried out four borehole observations since the

[1]) National Research Center for Disaster Prevention, Ginza 6-15-1, Chuo-ku, Tokyo, Japan.

[2]) Head of the Second Research Division.

[3]) Chief of the Earthquake Prediction Laboratory.

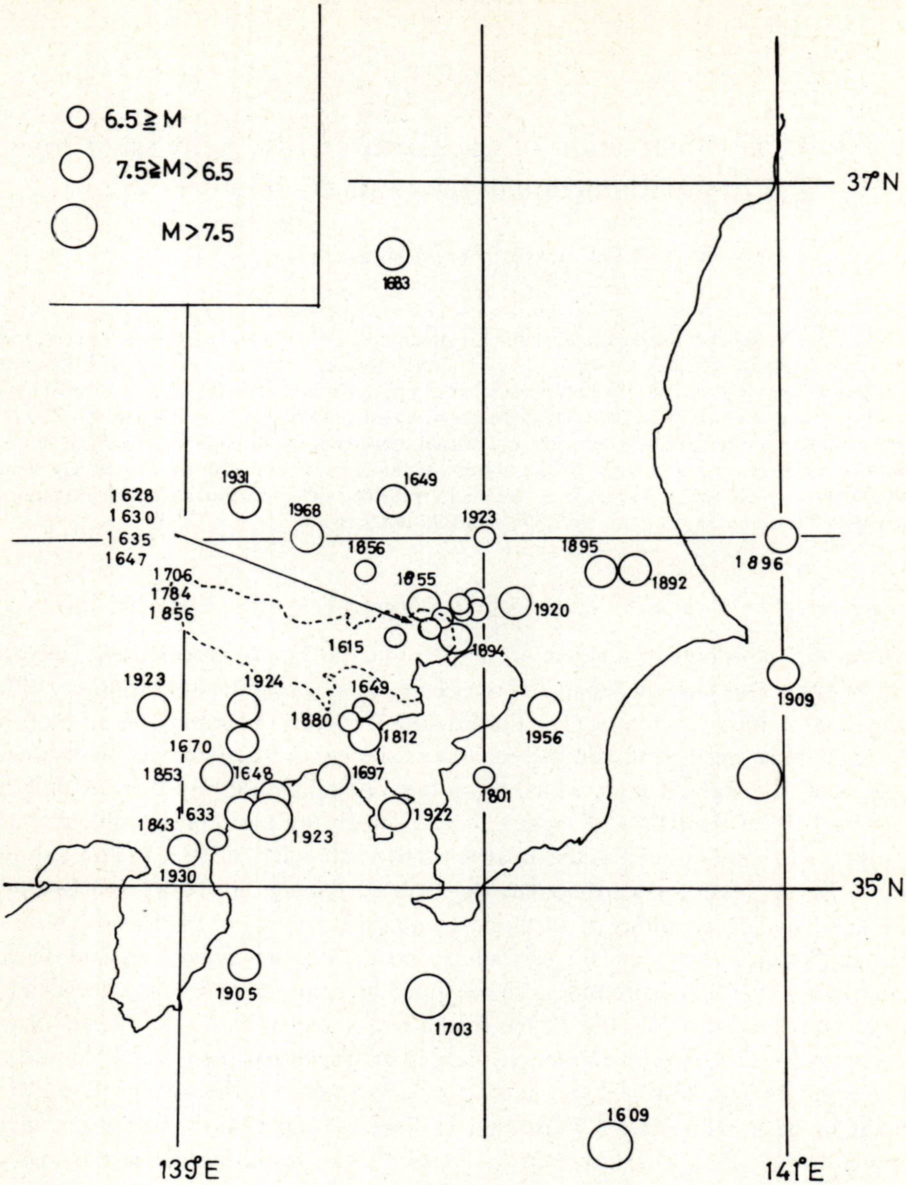

Figure 1
Earthquakes that caused damage in Tokyo (1600–1969) modified from KAMINUMA *et al.* (1973). Tokyo is surrounded by a broken line

1966 Matsushiro earthquake swarm (e.g. TAKAHASHI *et al.*, 1967). Also, the Meteorological Research Institute of Japan Meteorological Agency has carried out six borehole observations since 1967 (e.g. HIRONO *et al.*, 1969). As a part of the Japanese earthquake

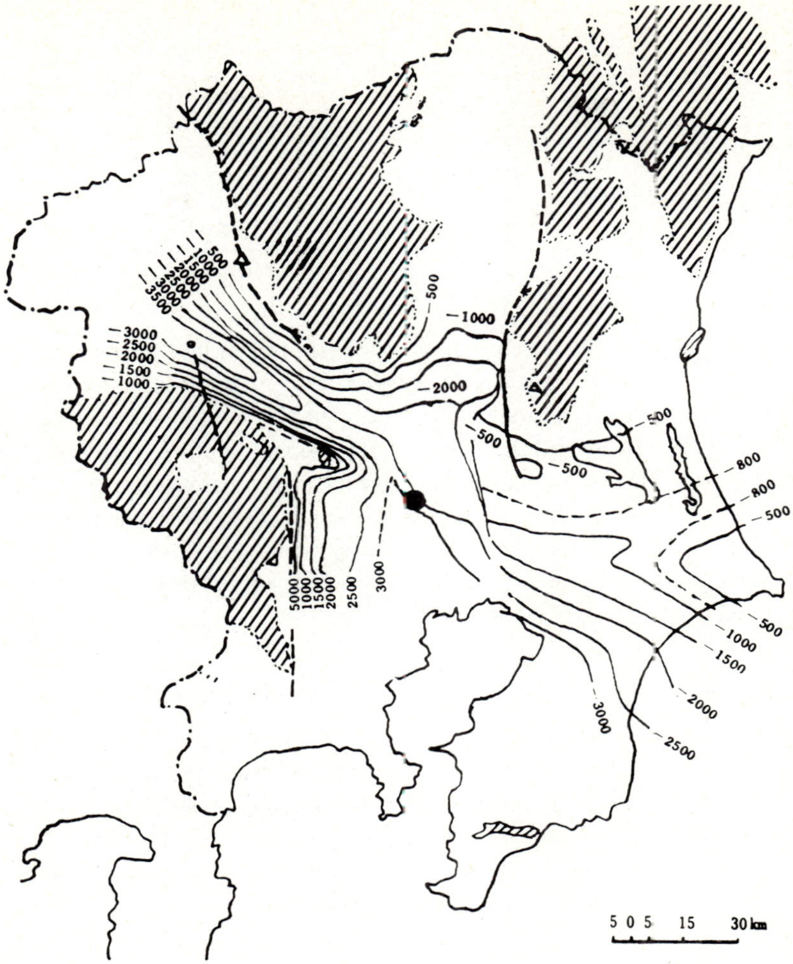

Figure 2
Contour lines of the basement depth in metres, in the southern part of the Kwanto region after FUKUDA (1962). ▨, exposed basement; ●, the Iwatsuki observatory

prediction programme, at least three deep borehole observatories of high sensitivity were planned in the capital area by the National Research Center for Disaster Prevention. This network is expected to supply important data for earthquake prediction. The third deep borehole observatory will be completed in a few years, according to this project. The first place selected was in the city of Iwatsuki, about 30 km from the centre of Tokyo, where geological structures were relatively well known compared with other places. Starting in 1970, the construction was completed at the end of 1973, costing about 600 million yen. Tentative observations of micro-earthquakes and crustal movement by tiltmeters have been carried out since May, 1973. Locations and sensors

Table 1
Locations and sensors at the Iwatsuki observatory

Sensors	Overall Sensitivity
A three-components set of velocity seismometers (1 Hz)	magnification of 1,000,000 at 10 Hz
A three-components set of forced balance accelerometers	5 mgal–30 gal
A two-components set of tiltmeters	max. 0.02″
Two sets of thermometers	max. 0.1°

Location: 35°55′33″N, 139°44′17″E, 3500 m below sea level (● mark in Fig. 2)

are given in Fig. 2 and Table 1. Sensors, attached equipment and control units have to work in difficult conditions, such as a 300–400 atm and at temperatures of about 86°C. We have to set the sensors at the bottom of the hole by remote control. All equipment in the hole can be lifted up for maintenance and/or repair if necessary. Very advanced techniques were required to solve the difficulties encountered in carrying out the deep borehole observations under these conditions. Even the construction of the hole itself was quite different from the ordinary boring and casing.

The following is a brief report of the observation system and low-level background noises obtained at the Iwatsuki observatory.

2. Borehole and its geological and geophysical surrounding

The completed borehole and casing are shown in Fig. 3. The hole is filled with water containing rust-proofing chemicals. A rotary type oil well drilling rig was used for construction of the hole. The construction was commenced in February, 1971, and was completed at the end of September of the same year. The depth is 3510 m, the inclination from the vertical direction is 1° 15′ at the bottom. Non-magnetic casing was used for the lowest 10 m. The whole borehole was cemented to fix the casing against the rock wall. A cement mix for high pressure and high temperature with expansion cement was used below 2000 m depth to ensure rigid contact between the casing and the surrounding rocks. Sampling cores and cuttings were taken at every 250 m depth and every 20 m depth, respectively. For every test piece, geology, elastic constants and obliqueness of the layer were examined besides velocity and density logs.

The geology of the site is outlined as follows. Ordinary deposits of the Quarternary and Tertiary are above 2864 m depth, the late-Cretaceous or later quartz porphyry are between 2864 m and 3346 m depths, and metamorphic rocks are below that. Main results are shown in Fig. 4.

3. Observation system

The sensors, control units, transmitter including the modulator, and the vessel that is to be fixed at the bottom of the hole are illustrated in Fig. 5. The vessel is heat-

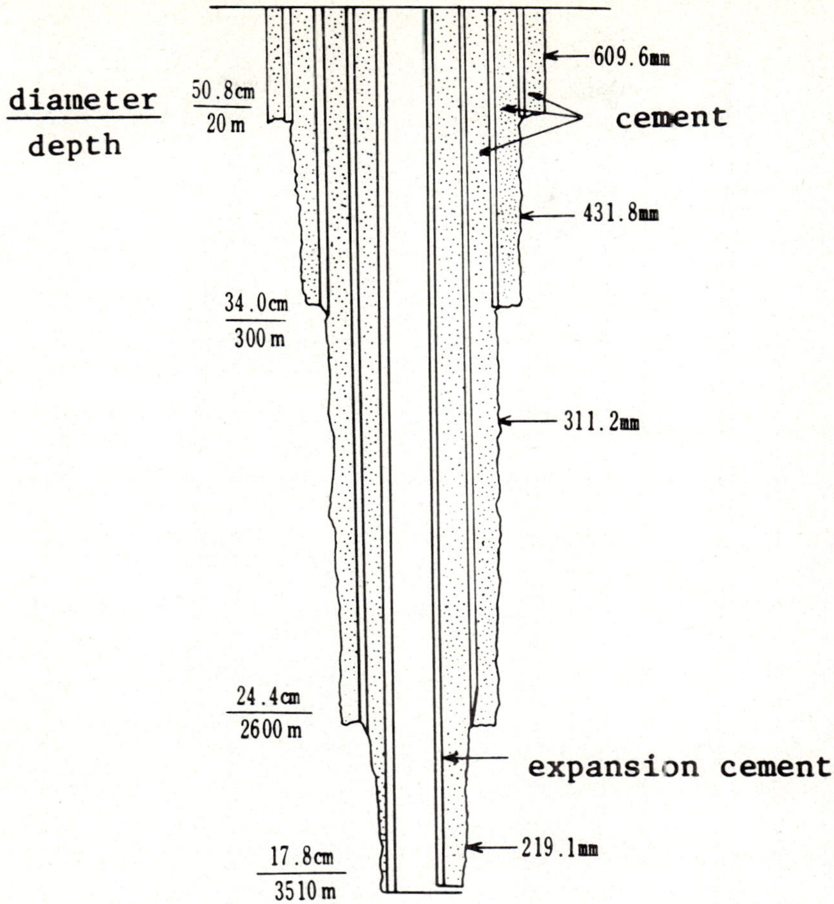

Figure 3
Borehole and casing at the Iwatsuki observatory

and pressure-proof stainless steel, 9 m long and 14 cm in external diameter. It weighs 870 kg in air and 670 kg in water. The signal cable connects the sensors at the bottom of the hole with recording equipments at the surface through the modulator and demodulator. It consists of 19 lines insulated with plastic which is covered with duplicated stainless wires as seen in Fig. 6. It is 26 mm in diameter and of 2.29 t/km weight in water. To move up and down such a heavy cable connected with the vessel of the sensors, a special lifting winch was made which was driven by an oil-pressure engine.

A schematic diagram of the whole observation system is shown in Fig. 7, where the equipment to the left of the cable is set in the vessel and that to the right is installed in the observatory at the surface. Explanations of symbols and codes are given in the figure caption in detail. Overall frequency characteristics are illustrated in Fig. 8.

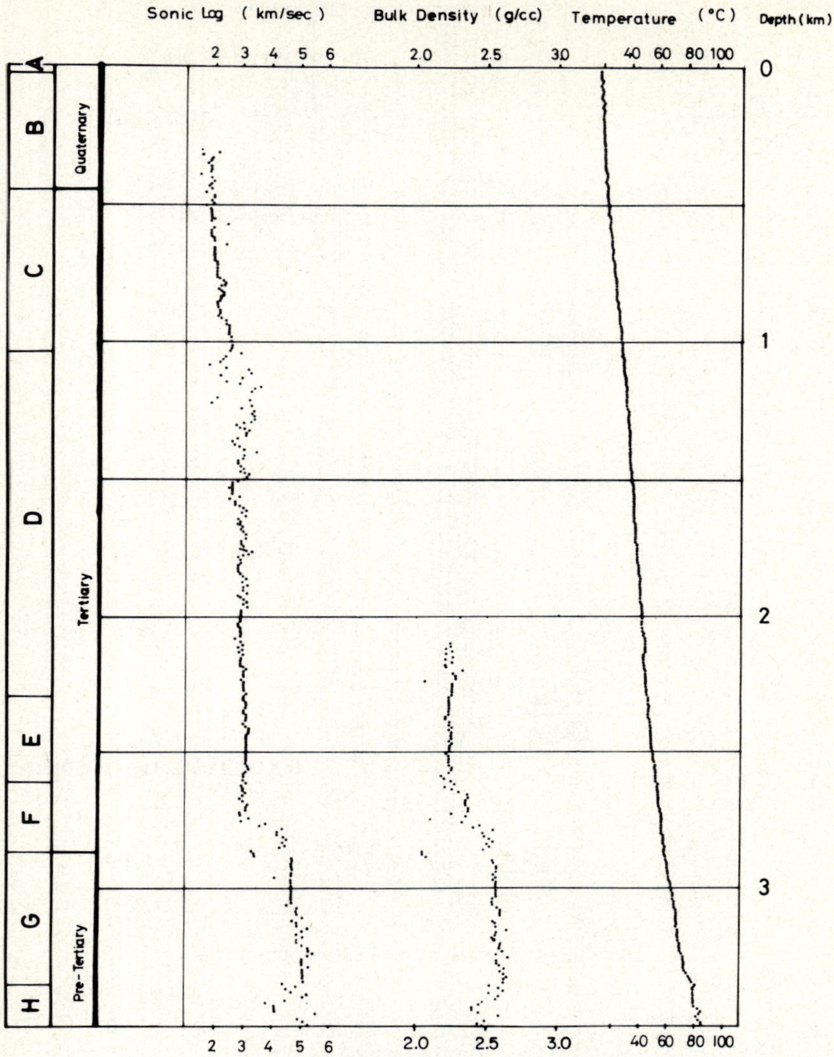

Figure 4
Geology, sonic velocity, density, and temperature at the Iwatsuki observatory site. A (0–20 m depth) sand and mud, Holocene. B (20–447 m) gravel, silt, sand, and mud, Pleistocene. C (447–1039 m) sand, silt, and gravel, Plio-Pleistocene. D (1039–2291 m) tuff, sandstone, and conglomerate, Miocene. E (2291–2610 m) sandstone, Miocene. F (2610–2864 m) mudstone, and conglomerate (quartz porphyry), Miocene. G (2864–3346 m) quartz porphyry rarely intervened by tuff, pre-Miocene. H (3346–3510 m) metamorphic rocks, pre-Tertiary

To transmit the signal of 24 channels two groups of FM method by multi-carriers were designed whose carrier frequencies were 40 to 128 kHz with the interval of 8 kHz for 12 channels.

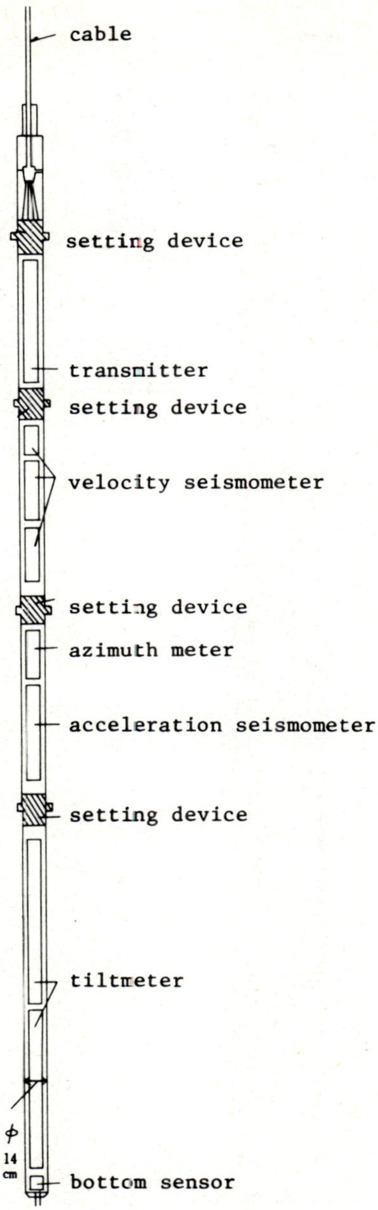

Figure 5
Vessel and its contents

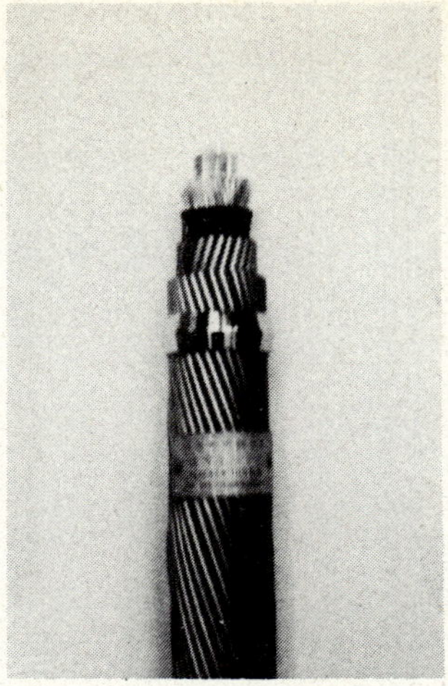

Figure 6
Signal cable

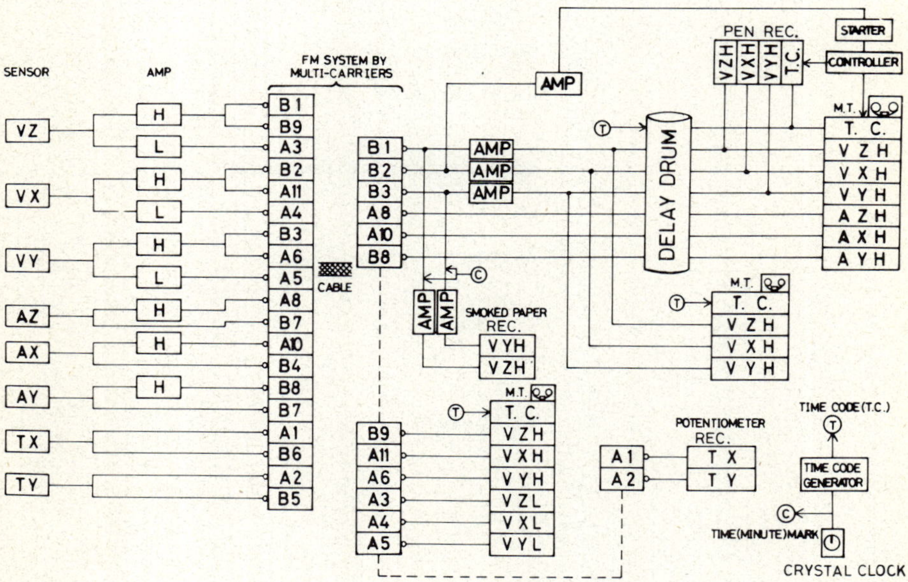

Figure 7
Schematic diagram of the whole observation systems. VZ, VX, VY; three components of velocity seismometers (Z; vertical). AZ, AX, AY; three components of acceleration seismometers. VZH, VXH, VYH; three components of high-gain velocity seismometers. VZL, VXL, VYL; three components of low-gain velocity seismometers. AZH, AXH, AYH; three components of high-gain acceleration seismometers. H L ; high-gain and low-gain amp. ◁□ □▷; channel of FM method by multi-carriers. TX, TY; two components of tiltmeters

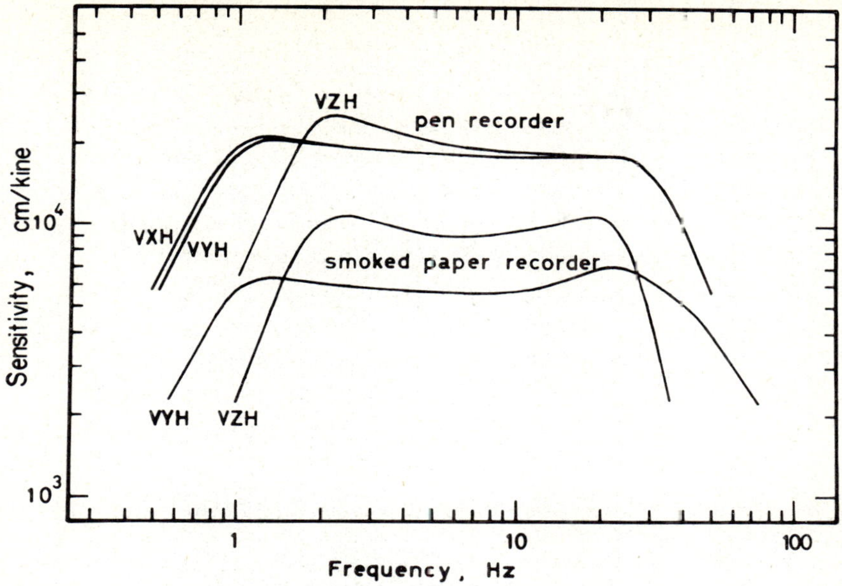

Figure 8
Overall frequency characteristics

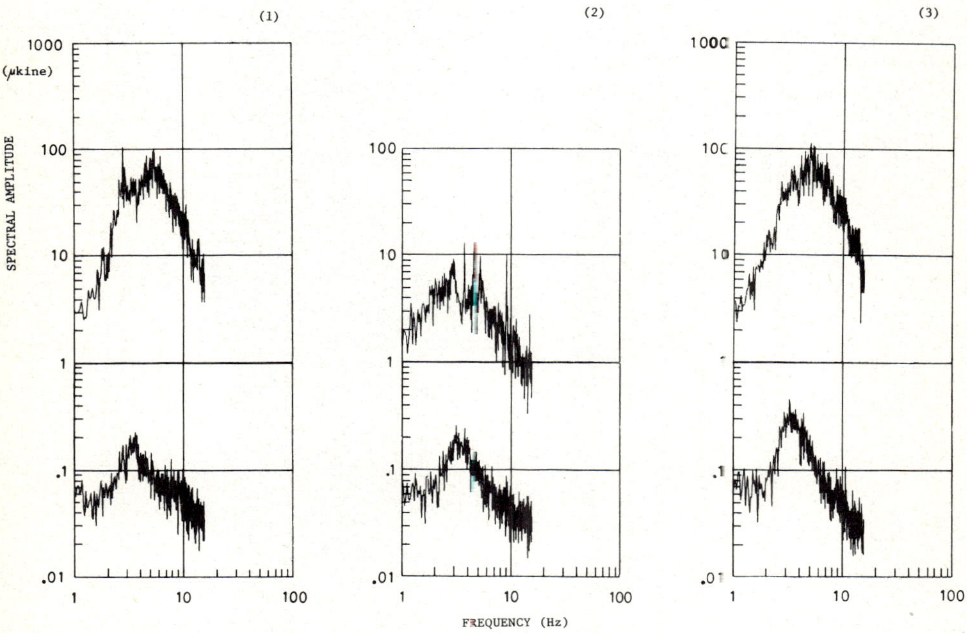

Figure 9
Background noises in spectral amplitude at the bottom of the hole and at the surface. Below: at the bottom of the hole. Above: at the surface. 1) Vertical, night-time, the case of heavy cars passing by the observatory; 2) vertical, night-time, the case of no heavy cars passing by the observatory; 3) vertical, daytime, the case of heavy cars passing by the observatory

4. Background noise and expected detection capability of micro-earthquakes

Figure 9 shows background noise measured by the velocity seismometer in the hole and on the ground for both day and night. As seen in the figure, spectral amplitude of background noise in the hole are about 1/1000–1/300 of those at the surface and fairly stable regardless of the local disturbance such as trucks passing near the observatory. As expected, noise at a depth of 3500 m is extremely small compared with those at the surface, and are even small when compared with that of other Japanese seismic stations, which lie on a rigid basement of a quiet location in the country. As a result of the recent observation at the Iwatsuki observatory, we could, possibly, detect earthquakes with magnitudes greater than 2.0 within 50 km radius from the observatory.

References

O. Fukuda, *Test boring at the Kasukabe region in the southern part of Kwanto district* (Preannouncement), Geology Monthly *100* (1962), 1–16 (in Japanese).

T. Hirono, S. Suyehiro, M. Furuta and K. Sato, *Noise attenuation in shallow holes (II) – Improvement of signal to noise ratio*, Papers in Meteorology and Geophysics *20* (1969), 189–206.

K. Kaminuma, T. Iwata, I. Kayano and M. Ohtake, *Illustrations of earthquakes in Japan*, Quick research report, Earthq. Res. Inst. *9* (1973), 135 (in Japanese).

H. Takahashi, M. Takahashi and H. Suzuki, *Studies on the underground structures of the Matsushiro earthquake area by test boring*, Notes of cooperative research for disaster prevention, National Research Center for Disaster Prevention *5* (1967), 57–69.

H. Takahashi, O. Fukuda, N. Oyagi, Y. Fujita and K. Ogusa, *Geology of the Tokyo deep-bore-hole observation site*, Abstract of the Joint Meeting of the Five Japanese Societies of Geology and its Relations, *424* (1971) (in Japanese).

(Received 5th November 1974)

Pageoph,
Vol. 113 (1975) Birkhäuser Verlag, Basel

Hydraulic Fracturing Stress Measurements near the Hohenzollern-Graben-Structure, SW Germany

By F. RUMMEL and R. JUNG[1])

Summary – Hydraulic fracturing stress measurements have been performed in a limestone quarry near the Hohenzollern-Graben, a fault structure in SW Germany. The values of the two horizontal principal stresses were 24 and 15 bars at a depth of 25 m. The magnitude and the direction of the stresses agree with the results obtained by door-stopper measurements at the same location and the direction of the maximum horizontal principal stress derived from fault plane solutions of shallow earthquakes of this tectonic active area.

Zusammenfassung – In der Nähe des Hohenzollern-Grabens/Schwäbische Alb/SW-Deutschland wurden absolute Spannungsmessungen nach der Methode des Hydraulic Fracturing durchgeführt. Die Bestimmung der beiden horizontalen Hauptspannungen lieferte Werte von 24 und 15 bar für eine Tiefe von 25 m. Als Richtung für die größte Hauptspannung ergab sich N30°W. Diese Werte stimmen sowohl mit den Ergebnissen von 'Doorstopper'-Messungen, die an der gleichen Stelle vorgenommen wurden, als auch mit den Ergebnissen überein, die sich aud Erdbeben dieses tektonisch aktiven Gebietes ableiten lassen

1. Introduction

Most of the current tectonic activities, such as earthquakes, are concentrated along the boundaries of relatively large stiff and brittle crustal plates, which, on the basis of plate-tectonics, are in a more or less continuous motion relative to each other. For the interpretation of the motion of these plates as well as for an understanding of the intra-plate tectonic activity (SYKES and SBAR, 1973) it would be most desirable to know the distribution of the magnitudes of the stresses within these crustal plates and the directions of the principal stresses relative to the plate boundaries (RALEIGH, 1974).

For central Europe, in particular for southwest Germany, data on the directions of the principal stresses are available from fault plane solutions of shallow earthquakes in the Upper Rhine Graben and its boundary region (SCHICK, 1970; SCHNEIDER, 1971; AHORNER and SCHNEIDER, 1974; AHORNER, 1974). According to these data the maximum principal horizontal stress is orientated NW–SE at an average angle of N30°W ± 10°. A similar orientation has been found from fault plane solutions of earthquakes in the Lower Rhine area (AHORNER, 1967) and the western Alps (PAVONI and PETERSCHMITT, 1974).

This general direction of the principal horizontal stress axis in south-western Germany was confirmed by *in situ* surface measurements of absolute stresses using

[1]) Address: Institut für Geophysik, Ruhr-Universität Bochum, 463 Bochum.

a stress–relieve method (door-stopper method after LEEMAN, 1969). The results of these measurements were presented by GREINER (1974a, 1974b), indicating a maximum horizontal principal stress of the order of 20 bars, acting in the direction N28°W.

In order to obtain further absolute stress data for central Europe, in particular at greater depths, a stress measurement programme is being developed using the technique of hydraulic fracturing in deep boreholes. Although this new stress measuring technique is currently used by several investigators the interpretation of the data is often difficult, dubious and based on various assumptions such as impermeability of the rock to the pressurizing fluid (RALEIGH *et al.*, 1972; ROGIERS, 1974; RUMMEL, 1974a, 1974b). Therefore, as the first step towards this programme it was intended to compare the results of hydraulic fracturing measurements with stress data obtained by a conventional stress relief technique at the same *in-situ* location.

2. Location

The experiments were carried out in a quarry of Upper Jurassic limestone in the western part of the Swabian Jura, south of Tübingen, south-west Germany (coordinates: 9°05′E/48°10′27″N). The area is known for its high seismic activity with focal depths mainly between 2 and 12 km and magnitudes up to $M = 5.6$ (SCHICK and SCHNEIDER, 1973). The only significant tectonic structure of this area is the Hohenzollern-Graben, a fault structure which extends ESE to WNW for a distance of about 30 km. The test site was located about 10 km south of this structure (Fig. 1). The

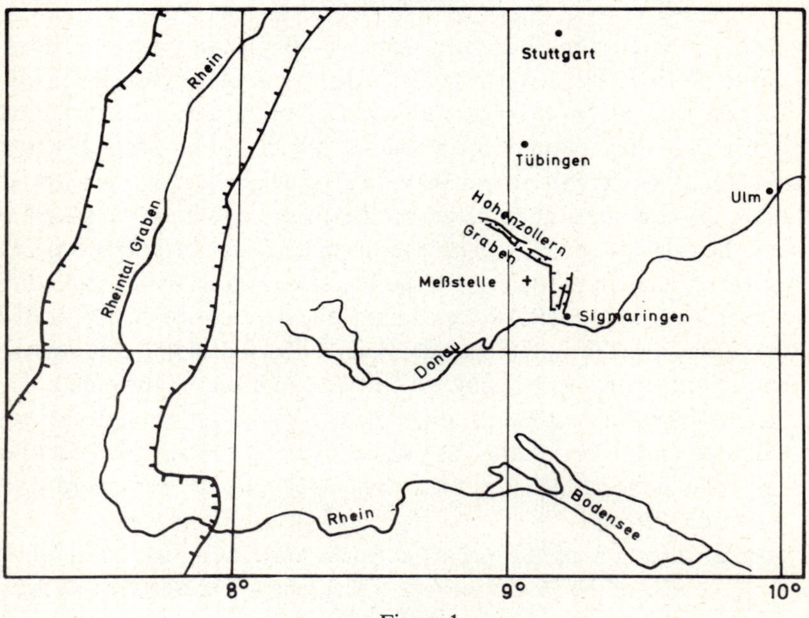

Figure 1
Location of the test site (+)

location has already been selected for the stress measurements mentioned above using the door-stopper method prior to our measurements (GREINER, 1974a, 1974b). The horizontal boreholes were core-drilled to a depth of about 20 m into the quarry wall and were orientated parallel, perpendicular and under 45° to the direction of the maximum principal stress indicated by the fault plane solutions of earthquakes in this area. For the hydro-frac experiments an additional vertical hole was drilled at the centre of the quarry floor to a depth of 30 m. To obtain a sufficiently smooth surface of the borehole wall the final 10 m section of the borehole was also core-drilled. The diameter of all the boreholes was 85 mm.

3. *Measurement technique*

The method of hydraulic fracturing is commonly known in the oil industry to increase the productivity of oil wells by inducing cracks in the oil bearing rock formation. SCHEIDEGGER (1960, 1962), KEHLE (1964) and FAIRHURST (1964) suggested using the technique for *in situ* stress measurements. Since then the method was applied in various tests by SCHOENFELDT (1970a, 1970b), HAIMSON (1972) and ROGIERS (1973) in the US.

The method is based on the assumption that a fracture will initiate at the borehole wall, when the fluid pressure applied in the wells is sufficient to induce a tangential stress equal to the tensile strength of the rock, assuming further impermeability of the rock. It is also expected that the fracture initiates and propagates in a plane perpendicular to the direction of the least principal tectonic stress. The condition for the origination of a radial (vertical) fracture in a vertical borehole is then given by the simple relation

$$p_F = 3\sigma_{H,\min} - \sigma_{H,\max} - T \tag{2}$$

where p_F is the critical fluid pressure in the borehole, T is the tensile strength and $\sigma_{H,\max}$ and $\sigma_{H,\min}$ are the horizontal principal stresses. Since p_F is measured during the test and T may be determined in laboratory tests on rock cores, $\sigma_{H,\max}$ can be estimated if one assumes that $\sigma_{H,\min}$ is equal to the fluid pressure in the fracture, which is required to just keep the fracture open after it has propagated sufficiently far from the well.

For pressurizing a section of the borehole in our experiments a $2\frac{3}{4}$ inch diameter straddle packer assembly (Lynes, Houston, Texas) was used which had 3 ft packer elements and allowed to apply the pressurizing fluid to a 3 ft long section of the borehole. The fluid pressure was applied by an electro-hydraulic pump which gave a constant pumping rate of 3.2 l per min (Weber Hydraulik, type 7). As pressurizing fluid a water–oil emulsion was used. The fluid pressure was measured down-hole by an electric pressure cell attached directly to the straddle packer assembly and recorded at the surface.

Prior to each test the section of the borehole wall to be pressurized was inspected by an impression packer (Lynes), giving a direct picture of pre-existing cracks and joints. The impression packer was also used after each test to give an impression of the induced crack and its orientation. The pressure applied to this packer was 50 bars and kept constant over 30 min. A pressure of 50 bars was also applied to the hydro-frac tool prior to fluid injection to the borehole section exposed to the fluid.

4. *Experimental results*

Pressure vs. time curves for the *in situ* tests in the vertical borehole and at different locations in the horizontal holes are presented in Figs. 2 and 3 (a, b, c). In most cases,

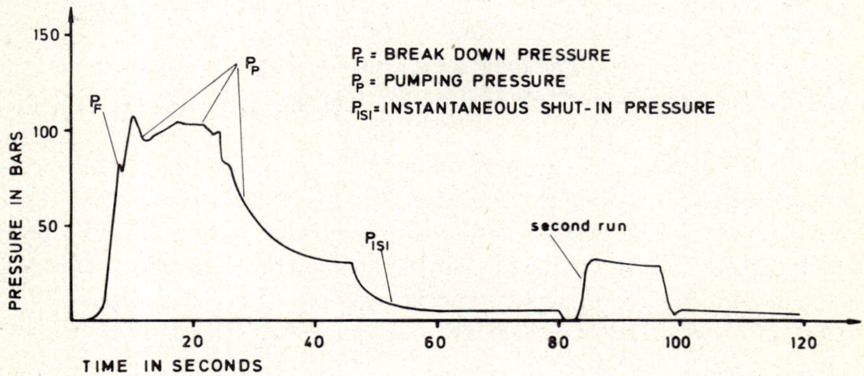

Figure 2
Pressure vs. time curve recorded in the vertical borehole at a depth of 25 m

especially for the horizontal holes, the fluid pressure increased continuously to a more or less constant value depending on the size and the number of pre-existing joints within the pressurized section. Distinct peaks in the curves (break-down pressures) were only observed for tests in borehole sections without pre-existing joints. As the impressions taken from these locations after the tests showed, radial fractures were initiated in this case with fracture planes parallel to the borehole axis. In the vertical hole the orientation of the initiated vertical fracture was N30°W. The fractures in horizontal holes were mainly horizontal or vertical, depending on the orientation of the borehole axis in respect to the direction of the maximum and minimum principal stresses acting, which were given by the far-field stress and the boundary condition of the free quarry face.

For the vertical hole a break-down pressure of 85 bars was observed at a depth of 25 m assuming that the first peak in the curve (Fig. 2) corresponds to the moment of fracture initiation. The second peak in the curve, the pressure plateau, and the subsequent pressure drop are interpreted as different stages of fracture propagation. After 45 sec of pumping at constant rate (3.2 l per min) the pressure generator was shut down and the well was shut-in. During this period the pressure rapidly dropped to a constant value of 6.5 bars. In subsequent injections the pressure increased to a more or less constant value, decreasing with each injection nearly to the value of the shut-in pressure. In the case of horizontal holes the pressure after shut-down, in general, rapidly dropped to zero, which may be due to the heavily jointed rock near the working face of the quarry.

The tensile strength of the rock was determined in laboratory hydraulic fracturing tests on rock cores from the boreholes. The cores were 70 mm in diameter and had an

3(a)

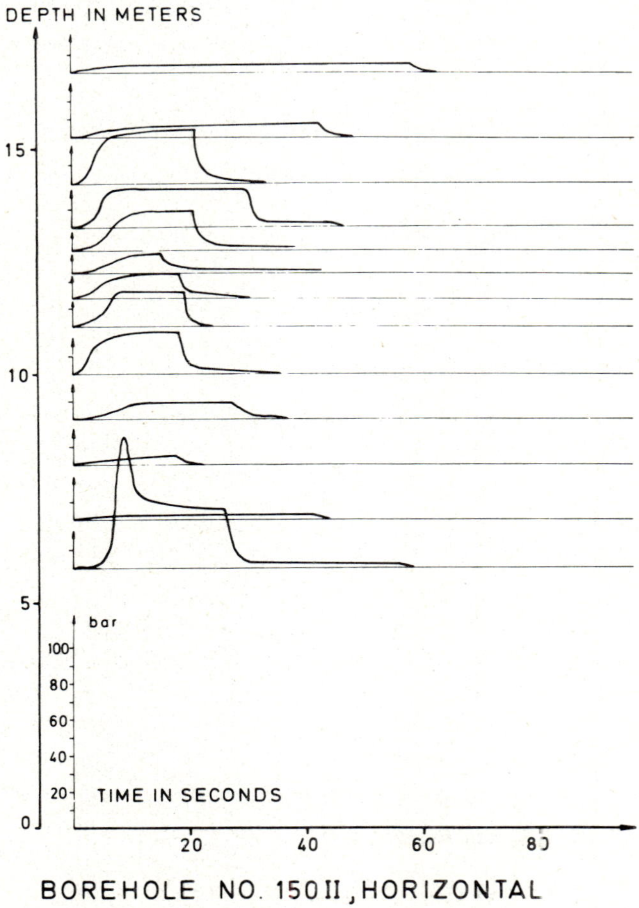

Figure 3(a)

3(b)

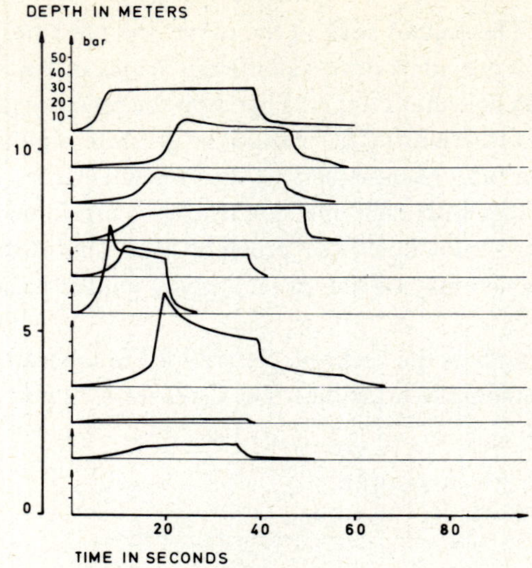

BOREHOLE No. 150 I, HORIZONTAL

Figure 3(b)

3(c)

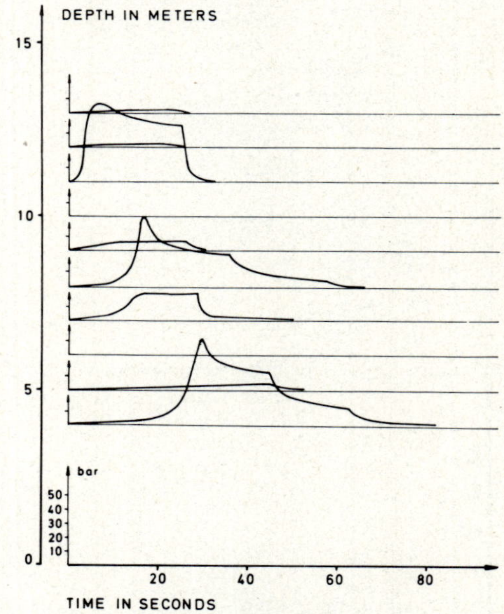

BOREHOLE NO. 60, HORIZONTAL

Figure 3(c)

Figure 3

Pressure vs. time curves recorded in different horizontal boreholes at various locations. (a, b) borehole axis parallel to $\sigma_{H,\max}$ (c) borehole axis perpendicular to $\sigma_{H,\max}$

axial hole of 7 mm in diameter. The internal fluid pressure was increased at a constant rate of pressurization similar to that used in the field tests. Fracture occurred in the form of radial cracks when the fluid pressure reached the value of the tensile strength of the specimens. Although limestone usually is considered as extremely homogeneous the tensile strength of the 10 specimens investigated varied between 100 and 210

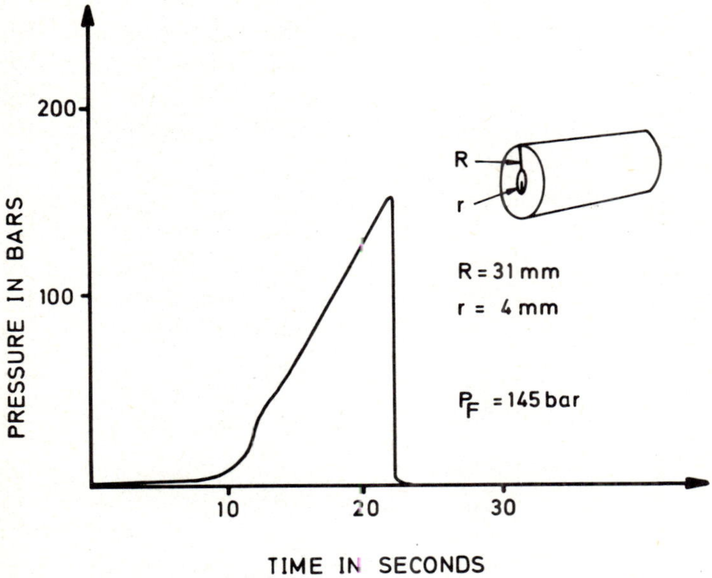

Figure 4
Pressure vs. time curve for a laboratory hydraulic fracturing test on a limestone core

bars, with an average value of 140 bars. The length of the pressurized section varied between 8 cm and 20 cm. An example of a typical pressure vs. time curve obtained in a laboratory hydro-frac test is given in Fig. 4.

5. *Determination of the* in situ *principal stresses*

According to equation 1 the maximum principal horizontal stress, $\sigma_{H,\max}$, can be derived from the pressure vs. time curve for a vertical borehole if a vertical fracture is initiated and propagates in an axial plane. In this case the minor principal horizontal stress, $\sigma_{H,\min}$, is equal to the shut-in pressure, p_{ISI}.

The impression taken from the vertical borehole after the hydro-frac test showed that a vertical fracture was induced. The shut-in pressure was 6.5 bars, a value which is equal to the value of the overburden pressure at the depth of 25 m. Since the measurement was performed at shallow depth, where commonly the initiation of horizontal fractures may be expected in tectonic active regions, it is concluded that the fracture plane became horizontal during propagation. Therefore the shut-in pressure p_{ISI} obtained from the pressure vs. time curve cannot be used in equation 1 valid for initiation of a vertical fracture.

According to the variation of the laboratory tensile strength data observed, which may be explained by the spatial distribution and the different magnitudes of pre-existing defects in the specimens, the assumption of a unique tensile strength value seems to be unreasonable. Further, considering a size effect the value of *in situ* tensile strength may be considerably smaller than the average value derived from laboratory experiments on small, more or less intact, rock samples. Therefore the laboratory tensile strength was not used for the interpretation of the field data by equation 1.

On the basis of these considerations the determination of the principal horizontal stresses, and the *in situ* tensile strength, was performed by only using the experimental data of the field measurements in the vertical and horizontal boreholes, under the assumption that the axis of the boreholes were parallel to the principal tectonic stresses and neglecting a stress concentration due to the quarry. Then the relations for the initiation of new fractures for the three holes perpendicular to each other are given by:

Vertical hole: $$p_{F1} = 3\sigma_{H,\min} - \sigma_{H,\max} - T \tag{2}$$

Horizontal hole parallel to $\sigma_{H,\max}$: $$p_{F2} = 3\sigma_{02} - \sigma_{H,\min} - T \tag{3}$$

Horizontal hole perpendicular to $\sigma_{H,\max}$: $$p_{F3} = 3\sigma_{03} - \sigma_{H,\max} - T \tag{4}$$

In the equations p_{F1}, p_{F2} and p_{F3} are the break-down pressures observed, and σ_0 are the values of the overburden pressures calculated for the test locations assuming a density of 2.6 g cm^{-3} for the limestone. For the calculation of the three unknowns

Table 1
Experimental data of field tests

borehole	depth m	P_F bar	P_o bar	orientation of fracture plane
vertical	25	85	6.5	N 30° W, vertical
horizontal parallel to $\sigma_{H,max}$	—	65	10	horizontal
horizontal perpendicular to $\sigma_{H,max}$	—	57	10	horizontal

$\sigma_{H,\max}$, $\sigma_{H,\min}$ and T only pressure vs. time curves with a distinct peak pressure were considered (Figs. 2 and 3).

Using the mean values given in Table 1 the following values for the horizontal principal stresses and the *in situ* tensile strength are obtained:

$$T = -63 \text{ bar}$$
$$\sigma_{H,\max} = 24 \text{ bar}$$
$$\sigma_{H,\min} = 15 \text{ bar}$$

(The negative value of T is obtained because of the assumption that fracture occurs if the tangential stress σ is equal to T and tensile stresses are taken as being negative.)

Considering these values the difference between the two principal horizontal stresses is not sufficient to satisfy the condition

$$(3\sigma_{H,\min} - \sigma_{H,\max}) < \sigma_0 + \nu(3\sigma_{H,\min} - \sigma_{H,\max}) \tag{5}$$

for the possible initiation of a vertical fracture at a shallow depth of 25 m (JAEGER and COOK, 1969). The fact, that a vertical fracture was initiated may be due to the straddle packer assembly used.

The orientation of the maximum horizontal stress, was N30°W according to the orientation of the induced vertical fracture at the wall of the vertical borehole. This orientation is equal to the general direction of the maximum principal stress, found from fault plane solutions of shallow earthquakes in this region as well as from earthquakes in the Rhine-Graben and its boundary regions. The direction of $\sigma_{H,\max}$ also agrees surprisingly well with the direction of the maximum horizontal stress derived from the door-stopper measurements in this quarry (GREINER, 1974a, 1974b).

The values of the horizontal stresses, are in good agreement with the values determined by the door-stopper method (GREINER, 1974a, 1974b) at the same location. The magnitudes of the horizontal stresses are, however, small compared to the high tectonic stresses reported for various regions in northern Europe, North America and South Africa, but considerably exceed the vertical stress given by the overburden. Extrapolating the value to greater depth and considering the effect of pore pressure as well as stress concentrations occurring in this area, assuming the Hohenzollern-Graben to be a fault structure, the occurrence of shallow earthquakes in this region may be explained.

Acknowledgements

The work was carried out within the 'Sonderforschungsbereich 77, Felsmechanik', Universität Karlsruhe. We are grateful to H. J. ALHEID, G. GREINER and Prof. J. H. ILLIES for help during the experimental stage and for many discussions.

REFERENCES

L. AHORNER, *Herdmechanismen rheinischer Erdbeben und der seismotektonische Beanspruchungsplan im nordwestlichen Mitteleuropa*, Veröff. Geol. Inst. Köln, Heft *13* (1967), 109–130.

L. AHORNER and G. SCHNEIDER, *Herdmechanismen von Erdbeben im Oberrhein-Graben und in seinen Randgebirgen. Approaches to Taphrogenesis*, edited by ILLIES, H. and K. FUCHS, S. (Stuttgart 1974), pp. 104–107.

L. AHORNER, *Seismicity and Neotectonics of the Rhine Graben System*. Proc. 'Afar-Symposium', Bad Bergzabern (1974).

C. FAIRHURST, *Measurement of* in-situ *rock stresses with particular reference to hydraulic fracturing*. Rock Mechanics and Engineering Geol. *2* (1964), 129-147.

G. GREINER, In-Situ *Stress Measurements in South West Germany, Approaches to Taphrogenesis*, edited by H. ILLIES and K. FUCHS (Stuttgart 1974).

G. GREINER, In-Situ *Spannungsmessungen, Seismotektonik und Gefüge der Südalb*. Jahresbericht 1973 (Sonderforschungsbereich – Felsmechanik – Universität Karlsruhe 1974).

B. C. HAIMSON, *Earthquake Related Stresses at Rangely, Colorado*. 14th Symposium Rock Mechanics, 1972.

J. C. JAEGER and N. G. W. COOK, *Fundamentals of Rock Mechanics* (Methuen, London 1969).

R. O. KEHLE, *Determination of tectonic stresses through analysis of hydraulic well fracturing*, J.G.R., *69* (1964), 259–273.

E. R. LEEMAN, *The doorstopper and triaxial rock stress measuring instruments developed by the C.S.I.R.* J. S. Afr. Inst. Min. Metall., *69* (1969), 305–339.

N. PAVONI and E. PETERSCHMITT, *Das Erdbeben von Jeure vom 21. Juni 1971 und seine Beziehungen zur Tektonik des Faltenjura. Approaches to Taphrogenesis*, edited by H. ILLIES and K. FUCHS (Stuttgart 1974), pp. 322–329.

C. B. RALEIGH, J. H. HEALY and J. D. BREDEHAFT, *Faulting and crustal stress at Rangely, Colorado. Flow and fracture of rocks*. AGU Monograph, *16* (1972), pp. 275–284.

C. B. RALEIGH, *Crustal stress and global tectonics*. Proc. 3rd Int. Congr. on Rock Mechanics, Denver 1974.

J. C. ROGIERS, *The development and evaluation of a field method for* in-situ *stress determination using hydraulic-fracturing*. Ph.D. Thesis, University of Minnesota (1974).

F. RUMMEL, *Experimentelle Untersuchungen zum Bruchvorgang in Gesteinen*. Habilitationsschrift Univ. Bochum (1974).

F. RUMMEL, *Experimental studies of tensile strength of weber sandstone in laboratory hydraulic fracturing tests*. USGS Report, Menlo Park (1974).

A. E. SCHEIDEGGER, *Stresses in the earth crust as determined from hydraulic fracturing data*, Geologie und Bauwesen *27* (1962), 45–53.

A. E. SCHEIDEGGER, *On the connection between tectonic stresses on well fracturing data*. Geophysica Pura e Applicata *46* (1960), 66–76.

R. SCHICK, *A method for Determining source parameters of small magnitude earthquakes*, Z. Geophysik. *36* (1970), 205–224.

G. SCHNEIDER, *Seismizität und Seismotektonik der Schwäbischen Alb* (Enke Verlag, Stuttgart 1971).

H. v. SCHOENFELDT, *An experimental Study of Open-Hole Hydraulic Fracturing as a Stress Measurement Method with Particular Emphasis on Field Tests*. Ph.D. Thesis, University of Minnesota (1970).

H. v. SCHOENFELDT, *Field Experiments on Hydraulic Fracturing*. 45th Annual Fall Meeting of the Society of Petroleum Engineers of Aime (1970).

L. SYKES and M. SBAR, *Intraplate earthquakes, lithospheric stresses and the driving mechanism of plate tectonics*. Nature *245* (1973), 298–302.

(Received 15th December 1974)